FOURIER SERIES AND BOUNDARY VALUE PROBLEMS

Brown and Churchill Series

Complex Variables and Applications, 7th Edition
Fourier Series and Boundary Value Problems, 7th Edition

The Walter Rudin Student Series in Advanced Mathematics

Bóna, Miklós: *Introduction to Enumerative Combinatorics*
Chartrand, Gary and Ping Zhang: *Introduction to Graph Theory*
Davis, Sheldon: *Topology*
Dumas, Bob and John E. McCarthy: *Transition to Higher Mathematics: Structure and Proof*
Rudin, Walter: *Functional Analysis,* 2nd Edition
Rudin, Walter: *Principles of Mathematical Analysis,* 3rd Edition
Rudin, Walter: *Real and Complex Analysis,* 3rd Edition
Simmons, George F. and Steven G. Krantz: *Differential Equations: Theory, Technique, and Practice*

Other McGraw-Hill Titles in Higher Mathematics

Ahlfors, Lars: *Complex Analysis,* 3rd Edition
Burton, David M.: *Elementary Number Theory,* 6th Edition
Burton, David M.: *The History of Mathematics: An Introduction,* 6th Edition
Hvidsten, Michael: *Geometry with Geometry Explorer*
Ledder, Glenn: *Differential Equations: A Modeling Approach*
Simmons, George F.: *Differential Equations with Applications and Historical Notes,* 2nd Edition

FOURIER SERIES AND BOUNDARY VALUE PROBLEMS

Seventh Edition

James Ward Brown

Professor of Mathematics
The University of Michigan—Dearborn

Ruel V. Churchill

Late Professor of Mathematics
The University of Michigan

Boston Burr Ridge, IL Dubuque, IA New York San Francisco St. Louis
Bangkok Bogotá Caracas Kuala Lumpur Lisbon London Madrid Mexico City
Milan Montreal New Delhi Santiago Seoul Singapore Sydney Taipei Toronto

The McGraw·Hill Companies

FOURIER SERIES AND BOUNDARY VALUE PROBLEMS, SEVENTH EDITION

Published by McGraw-Hill, a business unit of The McGraw-Hill Companies, Inc., 1221 Avenue of the Americas, New York, NY 10020. Copyright © 2008 by The McGraw-Hill Companies, Inc. All rights reserved. Copyright renewed 1959 by Ruel V. Churchill. All rights reserved. No part of this publication may be reproduced or distributed in any form or by any means, or stored in a database or retrieval system, without the prior written consent of The McGraw-Hill Companies, Inc., including, but not limited to, in any network or other electronic storage or transmission, or broadcast for distance learning.

Some ancillaries, including electronic and print components, may not be available to customers outside the United States.

This book is printed on acid-free paper.

1 2 3 4 5 6 7 8 9 0 BKW/BKW 0 9 8 7 6

ISBN 978-0-07-305193-2
MHID 0-07-305193-4

Publisher: *Elizabeth J. Haefele*
Senior Sponsoring Editor: *Elizabeth Covello*
Developmental Editor: *Dan Seibert*
Senior Marketing Manager: *Dawn R. Bercier*
Project Manager: *April R. Southwood*
Senior Production Supervisor: *Kara Kudronowicz*
Associate Design Coordinator: *Brenda A. Rolwes*
Cover Designer: *Studio Montage, St. Louis, Missouri*
(USE) Cover Image: *Brand X Pictures/PunchStock*
Lead Photo Research Coordinator: *Carrie K. Burger*
Supplement Producer: *Melissa M. Leick*
Compositor: *Interactive Composition Corporation*
Typeface: *10/12 Times-Ten-Roman*
Printer: *Bookmart Press*

Library of Congress Cataloging-in-Publication Data

Brown, James Ward.
Fourier series and boundary value problems/James Ward Brown, Ruel V. Churchill.—7th ed.
p. cm.—(Brown and Churchill series)
Includes index.
ISBN 978-0-07-305193-2—ISBN 0-07-305193-4 (hard copy : alk. paper)
1. Fourier series. 2. Functions, Orthogonal. 3. Boundary value problems. I. Churchill, Ruel Vance, 1899– II. Title.

QA404.B76 2008 2006020773
515′.2433—dc22 CIP

www.mhhe.com

ABOUT THE AUTHORS

JAMES WARD BROWN is Professor of Mathematics at The University of Michigan—Dearborn. He earned his A.B. in physics from Harvard University and his A.M. and Ph.D. in mathematics from The University of Michigan in Ann Arbor, where he was an Institute of Science and Technology Predoctoral Fellow. He is coauthor with Dr. Churchill of *Complex Variables and Applications,* now in its seventh edition. He has received a research grant from the National Science Foundation as well as a Distinguished Faculty Award from the Michigan Association of Governing Boards of Colleges and Universities. Dr. Brown is listed in *Who's Who in the World.*

RUEL V. CHURCHILL was, at the time of his death in 1987, Professor Emeritus of Mathematics at The University of Michigan, where he began teaching in 1922. He received his B.S. in physics from the University of Chicago and his M.S. in physics and Ph.D. in mathematics from The University of Michigan. He was coauthor with Dr. Brown of *Complex Variables and Applications,* a classic text that he first wrote almost 60 years ago. He was also the author of *Operational Mathematics*. Dr. Churchill held various offices in the Mathematical Association of America and in other mathematical societies and councils.

To the Memory of My Father,
George H. Brown,

and of My Long-Time Friend and Coauthor,
Ruel V. Churchill.

These Distinguished Men of Science for Years Influenced
The Careers of Many People, Including Myself.
J.W.B.

Joseph Fourier

JOSEPH FOURIER

JEAN BAPTISTE JOSEPH FOURIER was born in Auxerre, about 100 miles south of Paris, on March 21, 1768. His fame is based on his mathematical theory of heat conduction, a theory involving expansions of arbitrary functions in certain types of trigonometric series. Although such expansions had been investigated earlier, they bear his name because of his major contributions. Fourier series are now fundamental tools in science, and this book is an introduction to their theory and applications.

Fourier's life was varied and difficult at times. Orphaned by the age of 9, he became interested in mathematics at a military school run by the Benedictines in Auxerre. He was an active supporter of the Revolution and narrowly escaped imprisonment and execution on more than one occasion. After the Revolution, Fourier accompanied Napoleon to Egypt in order to set up an educational institution in the newly conquered territory. Shortly after the French withdrew in 1801, Napoleon appointed Fourier prefect of a department in southern France with headquarters in Grenoble.

It was in Grenoble that Fourier did his most important scientific work. Since his professional life was almost equally divided between politics and science and since it was intimately geared to the Revolution and Napoleon, his advancement of the frontiers of mathematical science is quite remarkable.

The final years of Fourier's life were spent in Paris, where he was Secretary of the Académie des Sciences and succeeded Laplace as President of the Council of the Ecole Polytechnique. He died at the age of 62 on May 16, 1830.

CONTENTS

PREFACE

This is an introductory treatment of Fourier series and their applications to boundary value problems in partial differential equations of engineering and physics. It is designed for students who have completed a first course in ordinary differential equations. In order that the book be accessible to as great a variety of readers as possible, there are footnotes to texts which give proofs of the more delicate results in advanced calculus that are occasionally needed. The physical applications, explained in some detail, are kept on a fairly elementary level.

The *first objective* of the book is to introduce the concept of orthonormal sets of functions and representations of arbitrary functions by series of functions from such sets. Representations of functions by Fourier series, involving sine and cosine functions, are given special attention. Fourier integral representations and expansions in series of Bessel functions and Legendre polynomials are also treated.

The *second objective* is a clear presentation of the classical method of separation of variables used in solving boundary value problems with the aid of those representations. In the final chapter, some attention is given to the verification of solutions and to their uniqueness; for the method cannot be presented properly without such considerations.

This seventh edition is a revision of the 2001 edition. The first two editions were written by Professor Churchill alone. While improvements appearing in earlier revisions have been retained with this one, there are a number of major changes here that should be mentioned.

Problem sets in this edition appear more frequently and focus more directly on the sections they follow. There has also been some rearrangement of chapters. The first two chapters now concentrate on Fourier series and their convergence. Hence those chapters can be used by someone who is interested only in an introduction to Fourier series, without partial differential equations. This rearrangement also allows ordinary Fourier series to be used as concrete examples when orthonormal sets are introduced. The chapter on orthonormal sets (Chap. 7) is now more self-contained and placed just before the ones in which it is needed, namely before the chapters on Sturm-Liouville problems (Chap. 8), Bessel functions (Chap. 9), and Legendre polynomials (Chap. 10). Also, the chapter on Fourier integrals and applications (Chap. 6) now appears immediately after the chapters on applications of Fourier series (Chaps. 4 and 5).

The entire book has been thoroughly rewritten, with special attention to suggestions from users. Such people have asked for clearer identification of material that can be skipped by someone who wants to reach applications more quickly. Following that suggestion, we now have a separate section in Chap. 3 which merely states the three-dimensional laplacian in cylindrical and spherical coordinates. That section is then followed by one, which can be skipped, deriving those results. Several users have pointed out how, in earlier editions, it could be difficult to use some of the theorems leading to series expansions involving Bessel functions of orders $n = 0, 1, 2, \ldots$. The difficulty is now rectified by stating the theorems separately when $n = 0$ and then when $n = 1, 2, \ldots$. The proofs, however, remain combined.

Various other changes serve to make this edition more attractive. The closed forms of solutions of boundary value problems for vibrating strings with prescribed initial conditions are, for example, no longer scattered but are grouped together in one section. Bessel's inequality and the related Parseval's equation now appear in the same section, rather than in different chapters. Also, sections on the zeros of Bessel functions and Rodrigues' formula for Legendre polynomials have been completely rewritten for clarity. Finally, an *Instructor's Solutions Manual* (ISBN: 978-0-07-329326-4; MHID: 0-07-329326-1), containing worked solutions to selected exercises from the problem sets, is available upon request to instructors who adopt the book.

This and earlier editions have benefited from the continued interest of a number of people, many of whom are current or former students. The late Ralph P. Boas, Jr., furnished the reference to Kronecker's extension of the method of integration by parts in Chap. 1, and the derivation of the laplacian in spherical coordinates in Chap. 3 was suggested by a note of R. P. Agnew's in the *American Mathematical Monthly,* vol. 60, 1953. The most important source of day-to-day support and encouragement was, in addition to the staff at McGraw-Hill, my wife Jacqueline Read Brown.

James Ward Brown

CHAPTER 1

FOURIER SERIES

This book is concerned with two general topics:

(*i*) One is the representation of a given function by an infinite series involving a prescribed set of functions.

(*ii*) The other is a method of solving boundary value problems in partial differential equations, with emphasis on equations that are prominent in physics and engineering.

Representations by series are encountered in solving such boundary value problems. The theories of those representations can be presented independently. They have such attractive features as the extension of concepts of geometry, vector analysis, and algebra into the field of mathematical analysis. Their mathematical precision is also pleasing. But they gain in unity and interest when presented in connection with boundary value problems.

The set of functions that make up the terms in the series representation is determined by the boundary value problem. Representations by Fourier series, which are certain types of series of sine and cosine functions, are associated with a large and important class of boundary value problems. We shall give special attention to the theory and application of Fourier series and their generalizations. But we shall also consider various related representations, concentrating on those involving so-called Fourier integrals and what are known as Fourier-Bessel and Legendre series.

In this chapter, we begin our discussion of Fourier series. Once the convergence of such series has been established (Chap. 2) and a variety of partial differential equations have been derived (Chap. 3), we shall see (Chap. 4) how such series are used in what is known as the Fourier method for solving boundary value problems.

The first section here is devoted to a description of a class of functions that is central to the theory of Fourier series.

1. PIECEWISE CONTINUOUS FUNCTIONS

Let a function f be continuous at all points of a bounded open interval $a < x < b$ except possibly for a finite set of points $x_1, x_2, \ldots, x_{n-1}$, where

$$a < x_1 < x_2 < \cdots < x_{n-1} < b.$$

If we write $x_0 = a$ and $x_n = b$, then f is continuous on each of the n open subintervals

$$x_0 < x < x_1, \quad x_1 < x < x_2, \quad \ldots, \quad x_{n-1} < x < x_n.$$

It is not necessarily continuous, or even defined, at their endpoints. But if in each of those subintervals, f has finite limits as x approaches the endpoints from the interior, f is said to be *piecewise continuous* on the interval $a < x < b$. More precisely, the one-sided limits

$$(1) \quad f(x_{k-1}+) = \lim_{\substack{x \to x_{k-1} \\ x > x_{k-1}}} f(x) \qquad \text{and} \qquad f(x_k-) = \lim_{\substack{x \to x_k \\ x < x_k}} f(x) \qquad (k = 1, 2, \ldots, n)$$

are required to exist.

Note that if the limiting values from the interior of a subinterval are assigned to f at the endpoints, then f is continuous on the *closed* subinterval. Since any function that is continuous on a closed bounded interval is bounded, it follows that f is bounded on the entire interval $a \le x \le b$. That is, there exists a nonnegative number M such that $|f(x)| \le M$ for all points x $(a \le x \le b)$ at which f is defined.

EXAMPLE 1. Consider the function f that has the values

$$f(x) = \begin{cases} x & \text{when } 0 < x < 1, \\ -1 & \text{when } 1 \le x < 2, \\ 1 & \text{when } 2 < x < 3, \end{cases}$$

and $f(3) = 0$. (See Fig. 1.) Although f is discontinuous at the points $x = 1$ and $x = 2$ in the interval $0 < x < 3$, it is nevertheless piecewise continuous on

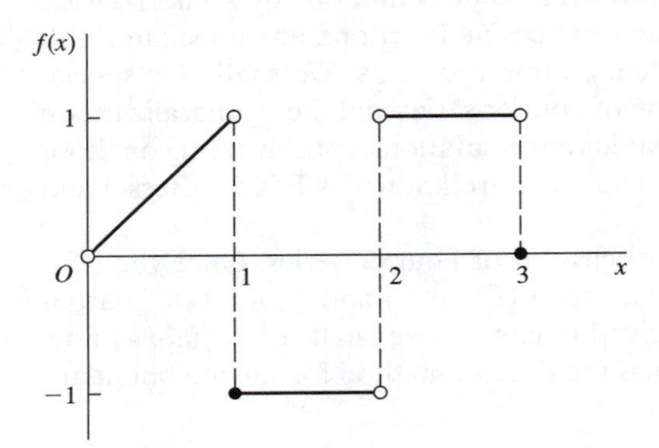

FIGURE 1

that interval. This is because the one-sided limits from the interior exist at the endpoints of each of the three open subintervals on which f is continuous. Note, for instance, that the right-hand limit at $x = 0$ is $f(0+) = 0$ and that the left-hand limit at $x = 1$ is $f(1-) = 1$.

A function is piecewise continuous on an interval $a < x < b$ if it is continuous on the *closed* interval $a \le x \le b$. Continuity on the *open* interval $a < x < b$ does not, however, imply piecewise continuity there, as Example 2 illustrates.

EXAMPLE 2. The function $f(x) = 1/x$ is continuous on the interval $0 < x < 1$, but it is not piecewise continuous there since $f(0+)$ fails to exist.

When a function f is piecewise continuous on an interval $a < x < b$, the integral of $f(x)$ from $x = a$ to $x = b$ always exists. It is the sum of the integrals of $f(x)$ over the open subintervals on which f is continuous:

$$\int_a^b f(x)\,dx = \int_a^{x_1} f(x)\,dx + \int_{x_1}^{x_2} f(x)\,dx + \cdots + \int_{x_{n-1}}^{b} f(x)\,dx. \tag{2}$$

The first integral on the right exists since it is defined as the integral over the interval $a \le x \le x_1$ of the continuous function whose values are $f(x)$ when $a < x < x_1$ and whose values at the endpoints $x = a$ and $x = x_1$ are $f(a+)$ and $f(x_1-)$, respectively. The remaining integrals on the right in equation (2) are similarly defined and therefore exist.

EXAMPLE 3. If f is the function in Example 1 and Fig. 1, then

$$\int_0^3 f(x)\,dx = \int_0^1 x\,dx + \int_1^2 (-1)\,dx + \int_2^3 1\,dx = \frac{1}{2} - 1 + 1 = \frac{1}{2}.$$

Observe that the value of the integral of $f(x)$ over each subinterval is unaffected by the values of f at the endpoints. The function is, in fact, not even defined at $x = 0$ and $x = 2$.

If two functions f_1 and f_2 are each piecewise continuous on an interval $a < x < b$, then there is a finite subdivision of the interval such that both functions are continuous on each closed subinterval when the functions are given their limiting values from the interior at the endpoints. Hence linear combinations $c_1 f_1 + c_2 f_2$ and products $f_1 f_2$ have that continuity on each subinterval and are themselves piecewise continuous on the interval $a < x < b$. The integrals

$$\int_a^b [c_1 f_1(x) + c_2 f_2(x)]\,dx \qquad \text{and} \qquad \int_a^b f_1(x) f_2(x)\,dx$$

must then exist.

We refer to the class of all piecewise continuous functions defined on an interval $a < x < b$ as a *function space* and denote it by $C_p(a, b)$. It is analogous to three-dimensional space, where linear combinations of vectors are well-defined vectors in that space. The analogy will be developed further in Chap. 7.

In this book we shall restrict our attention to functions that are piecewise continuous on bounded intervals; and the notion of piecewise continuity clearly applies regardless of whether the interval is open or closed.

2. FOURIER COSINE SERIES

Let f be any function in $C_p(0,\pi)$ and assume for the moment that $f(x)$ has a *Fourier cosine series* representation

$$f(x) = \frac{a_0}{2} + \sum_{n=1}^{\infty} a_n \cos nx \qquad (0 < x < \pi), \tag{1}$$

where a_0 and a_n $(n = 1, 2, \ldots)$ are constants. To find these constants, we also assume that series (1) and any related series that arises can be integrated term by term.

The constant a_0 is easily found by integrating each side of equation (1) from 0 to π and writing

$$\int_0^\pi f(x)\,dx = \frac{a_0}{2}\int_0^\pi dx + \sum_{n=1}^{\infty} a_n \int_0^\pi \cos nx\,dx,$$

or

$$\int_0^\pi f(x)\,dx = \frac{a_0}{2}[x]_0^\pi + \sum_{n=1}^{\infty} a_n \left[\frac{\sin nx}{n}\right]_0^\pi.$$

Inasmuch as $\sin n\pi = 0$ when n is an integer, this shows that

$$a_0 = \frac{2}{\pi}\int_0^\pi f(x)\,dx. \tag{2}$$

To find a_n $(n = 1, 2, \ldots)$, we write equation (1) as

$$f(x) = \frac{a_0}{2} + \sum_{m=1}^{\infty} a_m \cos mx \qquad (0 < x < \pi),$$

with a new index of summation, and then multiply each side by $\cos nx$, where n is any fixed positive integer. Integration of the resulting equation from 0 to π yields

$$\int_0^\pi f(x)\cos nx\,dx = \frac{a_0}{2}\int_0^\pi \cos nx\,dx + \sum_{m=1}^{\infty} a_m \int_0^\pi \cos mx \cos nx\,dx.$$

But

$$\int_0^\pi \cos nx\,dx = 0$$

and (see Problem 8, Sec. 5)

$$\int_0^\pi \cos mx \cos nx\,dx = \begin{cases} 0 & \text{when } m \neq n, \\ \pi/2 & \text{when } m = n. \end{cases}$$

Hence

$$\int_0^\pi f(x)\cos nx\,dx = a_n\frac{\pi}{2},$$

or

$$a_n = \frac{2}{\pi}\int_0^\pi f(x)\cos nx\,dx \qquad (n = 1, 2, \ldots). \tag{3}$$

Note that expression (2) for a_0 can be included with expression (3) when the integer n is allowed to run from $n = 0$, rather than from $n = 1$. This is the reason that $a_0/2$ was used instead of a_0 in series (1). Note, too, that $a_0/2$ is the mean, or average, value of $f(x)$ over the interval $0 < x < \pi$.

Because we cannot be certain at this time that representation (1) is actually valid for a specific f, we write

$$f(x) \sim \frac{a_0}{2} + \sum_{n=1}^{\infty} a_n \cos nx \qquad (0 < x < \pi), \tag{4}$$

where the tilde symbol $\sim$ merely denotes correspondence. Observe that correspondence (4), with coefficients (2) and (3), can be written more compactly as

$$f(x) \sim \frac{1}{\pi}\int_0^\pi f(s)\,ds + \frac{2}{\pi}\sum_{n=1}^{\infty}\cos nx \int_0^\pi f(s)\cos ns\,ds, \tag{5}$$

where s is used for the variable of integration in order to distinguish it from the free variable x.

The fact that f is piecewise continuous on the interval $0 < x < \pi$ ensures the existence of the integrals in expressions (2) and (3) for the coefficients in a cosine series. We shall, in Chap. 2, establish further conditions on f under which series (4) converges to $f(x)$ when $0 < x < \pi$, in which case correspondence (4) becomes an equality.

If f is defined on the interval $0 \le x \le \pi$ and series (4) converges to $f(x)$ for all x in that interval, the series also converges to the *even periodic extension, with period* 2π, of f on the entire x axis. That is, it converges to a function $F(x)$ having the properties

$$F(x) = f(x) \qquad \text{when } 0 \le x \le \pi \tag{6}$$

and

$$F(-x) = F(x), \qquad F(x + 2\pi) = F(x) \qquad \text{for all } x. \tag{7}$$

The reason for this is that each term in series (4) is itself even and periodic with period 2π. The graph of the extension $y = F(x)$ is obtained by reflecting the graph of $y = f(x)$ in the y axis, to give a graph for the interval $-\pi \le x \le \pi$, and then repeating that graph on the intervals $\pi \le x \le 3\pi$, $3\pi \le x \le 5\pi$, etc., as well as on the intervals $-3\pi \le x \le -\pi$, $-5\pi \le x \le -3\pi$, etc. It follows from these observations that if one is given a function f that is both even and periodic with period 2π, then the cosine series corresponding to $f(x)$ on the interval $0 < x < \pi$ represents $f(x)$ for all x when that series converges to it on the interval $0 \le x \le \pi$. Clearly, a cosine series cannot represent a function $f(x)$ for all x if $f(x)$ is not both even and periodic with period 2π.

3. EXAMPLES

Examples 1 and 2 here illustrate the material in Sec. 2.

EXAMPLE 1. Let us find the coefficients in the Fourier cosine series correspondence

$$f(x) \sim \frac{a_0}{2} + \sum_{n=1}^{\infty} a_n \cos nx \qquad (0 < x < \pi) \tag{1}$$

when $f(x) = x$ $(0 < x < \pi)$. It is easy to see that

$$a_0 = \frac{2}{\pi}\int_0^{\pi} x\,dx = \pi;$$

and, using integration by parts, we find that

$$a_n = \frac{2}{\pi}\int_0^{\pi} x\cos nx\,dx = \frac{2}{\pi}\left\{\left[\frac{x\sin nx}{n}\right]_0^{\pi} - \frac{1}{n}\int_0^{\pi}\sin nx\,dx\right\} \qquad (n = 1, 2, \ldots).$$

Since

$$\sin n\pi = 0 \qquad \text{and} \qquad \cos n\pi = (-1)^n$$

when n is an integer, this reduces to

$$a_n = \frac{2}{\pi}\cdot\frac{(-1)^n - 1}{n^2} \qquad (n = 1, 2, \ldots).$$

Note that a_0 needed to be found separately in order to avoid division by zero.

For the function $f(x)$ here, correspondence (1) evidently becomes

$$x \sim \frac{\pi}{2} + \frac{2}{\pi}\sum_{n=1}^{\infty}\frac{(-1)^n - 1}{n^2}\cos nx \qquad (0 < x < \pi).$$

Since $(-1)^n - 1 = 0$ when n is even, the series can be written more efficiently by summing only the terms that occur when n is odd. This is accomplished by replacing n by $2n-1$ wherever it appears after the summation symbol and in order starting the summation from $n = 1$. The result is

$$x \sim \frac{\pi}{2} - \frac{4}{\pi}\sum_{n=1}^{\infty}\frac{\cos(2n-1)x}{(2n-1)^2} \qquad (0 < x < \pi). \tag{2}$$

Conditions in Sec. 13 (Chap. 2) will ensure that correspondence (2) is actually an equality when $0 \le x \le \pi$. The even periodic extension to which the series converges is shown in Fig. 2, which tells us that

$$|x| = \frac{\pi}{2} - \frac{4}{\pi}\sum_{n=1}^{\infty}\frac{\cos(2n-1)x}{(2n-1)^2} \qquad (-\pi \le x \le \pi).$$

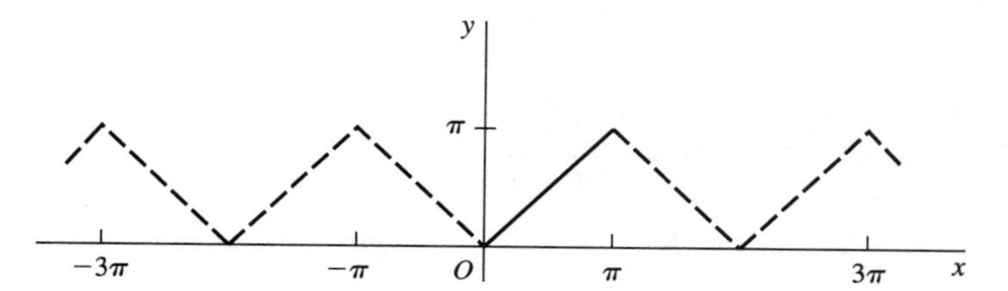

FIGURE 2

EXAMPLE 2. In this example, we shall find the Fourier cosine series for the function $f(x) = \sin x$ on the interval $0 < x < \pi$. The trigonometric identity

$$2 \sin A \cos B = \sin(A + B) + \sin(A - B)$$

enables us to write

$$\begin{aligned} a_n &= \frac{2}{\pi} \int_0^\pi \sin x \cos nx \, dx \\ &= \frac{1}{\pi} \int_0^\pi [\sin(1+n)x + \sin(1-n)x] \, dx \qquad (n = 0, 1, 2, \ldots). \end{aligned}$$

Hence, when $n \neq 1$,

$$a_n = \frac{1}{\pi} \left[-\frac{\cos(1+n)x}{1+n} - \frac{\cos(1-n)x}{1-n} \right]_0^\pi = \frac{2}{\pi} \cdot \frac{1 + (-1)^n}{1 - n^2};$$

and when $n = 1$, the coefficient is

$$a_1 = \frac{1}{\pi} \int_0^\pi \sin 2x \, dx = 0.$$

The desired cosine series correspondence is, then,

$$\sin x \sim \frac{2}{\pi} + \frac{2}{\pi} \sum_{n=2}^{\infty} \frac{1 + (-1)^n}{1 - n^2} \cos nx \qquad (0 < x < \pi).$$

Observe that $1 + (-1)^n = 0$ when n is *odd* (compare with Example 1). To sum the terms occurring when n is even, we replace n by $2n$ in this correspondence and write

$$\sin x \sim \frac{2}{\pi} - \frac{4}{\pi} \sum_{n=1}^{\infty} \frac{\cos 2nx}{4n^2 - 1} \qquad (0 < x < \pi). \tag{3}$$

The function $\sin x$ will, in fact, satisfy conditions in Sec. 13 ensuring that the correspondence here is an equality for each value of x in the interval $0 \le x \le \pi$. Thus, at each point on the x axis, the series converges to the even periodic extension, with period 2π, of $\sin x$ $(0 \le x \le \pi)$. That extension, shown in Fig. 3, is the function $y = |\sin x|$.

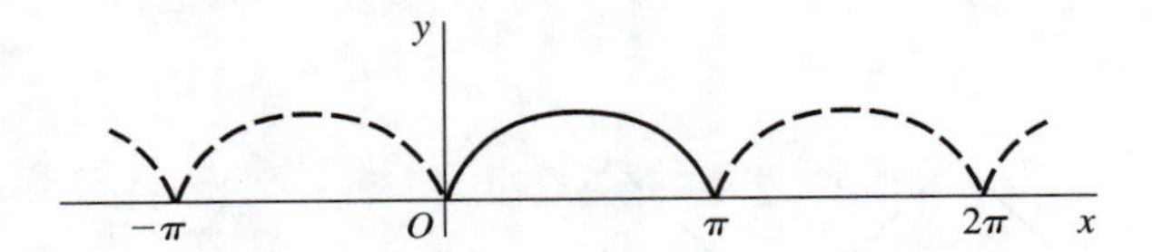

FIGURE 3

4. FOURIER SINE SERIES

We assume here that when f is in $C_p(0, \pi)$, there is a *Fourier sine series* representation

$$f(x) = \sum_{n=1}^{\infty} b_n \sin nx \qquad (0 < x < \pi), \tag{1}$$

where the coefficients b_n $(n = 1, 2, \ldots)$ are constants. The b_n can be found in a way similar to that used in Sec. 2 to find the coefficients in a cosine series. This time we write

$$f(x) = \sum_{m=1}^{\infty} b_m \sin mx \qquad (0 < x < \pi)$$

and multiply each side by $\sin nx$, where n is any fixed positive integer. Assuming that term-by-term integration is valid, we find that

$$\int_0^{\pi} f(x) \sin nx \, dx = \sum_{m=1}^{\infty} b_m \int_0^{\pi} \sin mx \sin nx \, dx.$$

Then, because (Problem 9, Sec. 5)

$$\int_0^{\pi} \sin mx \sin nx \, dx = \begin{cases} 0 & \text{when } m \neq n, \\ \pi/2 & \text{when } m = n, \end{cases}$$

this reduces to

$$\int_0^{\pi} f(x) \sin nx \, dx = b_n \frac{\pi}{2}.$$

That is,

$$b_n = \frac{2}{\pi} \int_0^{\pi} f(x) \sin nx \, dx \qquad (n = 1, 2, \ldots). \tag{2}$$

Inasmuch as we have only *assumed* the validity of representation (1), we use the tilde symbol $\sim$, as we did in Sec. 2, to denote correspondence:

$$f(x) \sim \sum_{n=1}^{\infty} b_n \sin nx \qquad (0 < x < \pi). \tag{3}$$

Expression (2) can, of course, be used to put this correspondence in the form

$$f(x) \sim \frac{2}{\pi} \sum_{n=1}^{\infty} \sin nx \int_0^{\pi} f(s) \sin ns \, ds. \tag{4}$$

Suppose that f is defined on the open interval $0 < x < \pi$ and that series (3) converges to $f(x)$ there. Since series (3) clearly converges to zero when $x = 0$ and when $x = \pi$, it converges to $f(x)$ for all x in the closed interval $0 \le x \le \pi$ if f is assigned the values $f(0) = 0$ and $f(\pi) = 0$. Remarks similar to ones in Sec. 2, regarding cosine series, show that series (3) then converges to the *odd periodic extension, with period* 2π, of f for all values of x. This time, the extension is the function $F(x)$ defined by means of the equations

$$F(x) = f(x) \qquad \text{when } 0 \le x \le \pi \tag{5}$$

and

$$F(-x) = -F(x), \qquad F(x + 2\pi) = F(x) \qquad \text{for all } x. \tag{6}$$

The extension F is odd and periodic with period 2π since each term $b_n \sin nx$ in series (3) has those properties. The graph of $y = F(x)$ is symmetric with respect to the origin and can be obtained by first reflecting the graph of $y = f(x)$ in the y axis, then reflecting the result in the x axis, and finally repeating the graph found for the interval $-\pi \le x \le \pi$ every 2π units along the entire x axis. Evidently, a Fourier sine series on the interval $0 < x < \pi$ can also be used to represent a given function that is defined for all x and is both odd and periodic with period 2π, provided that the representation is valid when $0 \le x \le \pi$.

5. EXAMPLES

We now illustrate some methods for finding Fourier sine series.

EXAMPLE 1. For the sine series corresponding to the function $f(x) = x$ on the interval $0 < x < \pi$ (Fig. 4), we refer to expression (2), Sec. 4, and use integration by parts to write

$$b_n = \frac{2}{\pi}\int_0^\pi x \sin nx\, dx = \frac{2}{\pi}\left\{\left[-\frac{x\cos nx}{n}\right]_0^\pi + \frac{1}{n}\int_0^\pi \cos nx\, dx\right\} = 2\,\frac{(-1)^{n+1}}{n}$$
$$(n = 1, 2, \ldots).$$

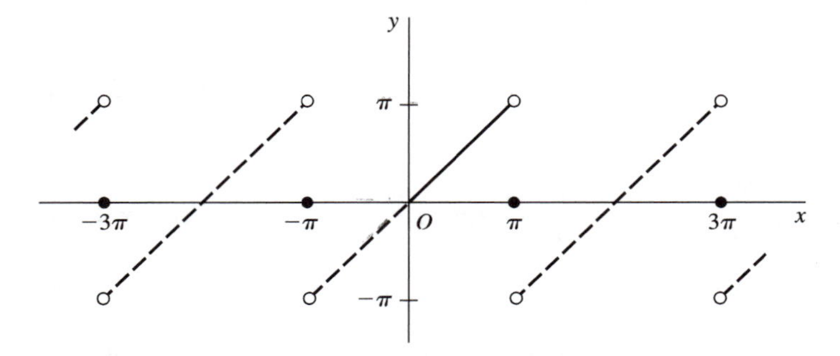

FIGURE 4

Thus

$$x \sim 2\sum_{n=1}^{\infty}\frac{(-1)^{n+1}}{n}\sin nx \qquad (0 < x < \pi). \tag{1}$$

Our theory will show that the series converges to $f(x)$ when $0 < x < \pi$. Hence it converges to the odd periodic function $y = F(x)$ that is graphed in Fig. 4. The fact that the series converges to zero when $x = 0, \pm\pi, \pm 3\pi, \pm 5\pi, \ldots$ is in agreement with our theory, which will tell us that it must converge to the mean value of the one-sided limits (Sec. 1) of $F(x)$ at each of those points.

In the evaluation of integrals representing Fourier coefficients, it is sometimes necessary to apply integration by parts more than once. We now give an example in which this can be accomplished by means of a single formula due to L. Kronecker (1823–1891). We preface the example with a statement of that formula.[†]

Let $p(x)$ be a polynomial of degree m, and suppose that $f(x)$ is continuous. Then, except for an arbitrary additive constant,

$$\int p(x)\,f(x)\,dx = pF_1 - p'F_2 + p''F_3 - \cdots + (-1)^m p^{(m)}F_{m+1} \tag{2}$$

where p is successively differentiated until it becomes zero, where F_1 denotes an indefinite integral of f, where F_2 is an indefinite integral of F_1, etc., and where alternating signs are affixed to the terms. Note that the differentiation of p begins with the *second* term, whereas the integration of f begins with the *first* term. The formula, which is readily verified by differentiating its right-hand side to obtain $p(x)\,f(x)$, could even have been used to evaluate the integral in Example 1, where only one integration by parts was needed.

EXAMPLE 2. To illustrate the advantage of formula (2) when successive integration by parts is required, let us find the Fourier sine series for the function $f(x) = x^3$ on the interval $0 < x < \pi$. With the aid of that formula, we may write

$$\begin{aligned} b_n &= \frac{2}{\pi}\int_0^{\pi} x^3 \sin nx\,dx \\ &= \frac{2}{\pi}\left[(x^3)\left(-\frac{\cos nx}{n}\right) - (3x^2)\left(-\frac{\sin nx}{n^2}\right) + (6x)\left(\frac{\cos nx}{n^3}\right) - (6)\left(\frac{\sin nx}{n^4}\right)\right]_0^{\pi} \\ &= 2\,(-1)^{n+1}\frac{(n\pi)^2 - 6}{n^3} \qquad (n = 1, 2, \ldots). \end{aligned}$$

Hence

$$x^3 \sim 2\sum_{n=1}^{\infty}(-1)^{n+1}\frac{(n\pi)^2 - 6}{n^3}\sin nx \qquad (0 < x < \pi). \tag{3}$$

[†]Kronecker actually treated the problem more extensively in papers that originally appeared in the *Berlin Sitzungsberichte* (1885, 1889).

As was the case in Example 1, the series converges to the given function on the interval $0 < x < \pi$. Since x^3 is an odd function whose value is zero when $x = 0$, this series represents x^3 on the larger interval $-\pi < x < \pi$ too.

We conclude this section by pointing out a computational aid that is useful in finding the coefficients b_n $(n = 1, 2, \ldots)$ in the Fourier sine series for a linear combination $c_1 f_1(x) + c_2 f_2(x)$ of two functions $f_1(x)$ and $f_2(x)$ whose sine series are already known. Namely, since the expression

$$b_n = \frac{2}{\pi}\int_0^\pi [c_1 f_1(x) + c_2 f_2(x)] \sin nx\, dx$$

can be written

$$b_n = c_1\left[\frac{2}{\pi}\int_0^\pi f_1(x)\sin nx\, dx\right] + c_2\left[\frac{2}{\pi}\int_0^\pi f_2(x)\sin nx\, dx\right],$$

it is clear that each b_n is simply the same linear combination of the nth coefficients in the sine series for the individual functions $f_1(x)$ and $f_2(x)$. Such an observation applies as well in finding coefficients in cosine and other types of series encountered in this and later chapters.

EXAMPLE 3. In view of the sine series for x and x^3 found in Examples 1 and 2, respectively, the coefficients b_n in the sine series corresponding to the function

$$f(x) = x(\pi^2 - x^2) = \pi^2 x - x^3 \qquad (0 < x < \pi)$$

are

$$b_n = \pi^2\left[2\,\frac{(-1)^{n+1}}{n}\right] - \left[2\,(-1)^{n+1}\frac{(n\pi)^2 - 6}{n^3}\right] = 12\,\frac{(-1)^{n+1}}{n^3} \qquad (n = 1, 2, \ldots).$$

Thus

$$x(\pi^2 - x^2) \sim 12\sum_{n=1}^{\infty}\frac{(-1)^{n+1}}{n^3}\sin nx \qquad (0 < x < \pi). \tag{4}$$

PROBLEMS

For the functions f in Problems 1 through 3, find (*a*) the Fourier cosine series and (*b*) the Fourier sine series on the interval $0 < x < \pi$.

1. $f(x) = 1 \qquad (0 < x < \pi)$.

Answers: (*a*) 1; (*b*) $\displaystyle\sum_{n=1}^{\infty}\frac{2[1-(-1)^n]}{n\pi}\sin nx = \frac{4}{\pi}\sum_{n=1}^{\infty}\frac{\sin(2n-1)x}{2n-1}$.

2. $f(x) = \pi - x \qquad (0 < x < \pi)$.

Answers: (*a*) $\displaystyle\frac{\pi}{2} + \frac{4}{\pi}\sum_{n=1}^{\infty}\frac{\cos(2n-1)x}{(2n-1)^2}$; (*b*) $\displaystyle 2\sum_{n=1}^{\infty}\frac{\sin nx}{n}$.

3. $f(x) = x^2 \qquad (0 < x < \pi).$

Answers: (*a*) $\dfrac{\pi^2}{3} + 4\sum_{n=1}^{\infty} \dfrac{(-1)^n}{n^2} \cos nx$;

(*b*) $2\pi^2 \sum_{n=1}^{\infty} \left[\dfrac{(-1)^{n+1}}{n\pi} - 2\,\dfrac{1-(-1)^n}{(n\pi)^3} \right] \sin nx.$

4. Find the Fourier cosine series on the interval $0 < x < \pi$ that corresponds to the function f defined by the equations

$$f(x) = \begin{cases} 1 & \text{when } 0 < x < \dfrac{\pi}{2}, \\ 0 & \text{when } \dfrac{\pi}{2} < x < \pi. \end{cases}$$

Suggestion: Note that

$$a_n = \frac{2}{\pi} \int_0^{\pi/2} \cos nx \, dx \qquad (n = 0, 1, 2, \ldots)$$

and that

$$\sin \frac{(2n-1)\pi}{2} = \sin n\pi \cos \frac{\pi}{2} - \cos n\pi \sin \frac{\pi}{2} = (-1)^{n+1} \qquad (n = 1, 2, \ldots).$$

Answer: $\dfrac{1}{2} + \dfrac{2}{\pi} \sum_{n=1}^{\infty} \dfrac{(-1)^{n+1}}{2n-1} \cos(2n-1)x.$

5. By referring to the sine series for x in Example 1, Sec. 5, and the one found for x^2 in Problem 3(*b*) above, show that

$$x(\pi - x) \sim \frac{8}{\pi} \sum_{n=1}^{\infty} \frac{\sin(2n-1)x}{(2n-1)^3} \qquad (0 < x < \pi).$$

6. Show that

$$x^4 \sim \frac{\pi^4}{5} + 8 \sum_{n=1}^{\infty} (-1)^n \frac{(n\pi)^2 - 6}{n^4} \cos nx \qquad (0 < x < \pi).$$

Given that this correspondence is actually an equality when $0 \le x \le \pi$, sketch the function represented by the series for all x.

7. Verify Kronecker's formula (2), Sec. 5.

8. Use the trigonometric identity

$$2 \cos A \cos B = \cos(A - B) + \cos(A + B)$$

to show that

$$\int_0^{\pi} \cos mx \cos nx \, dx = \begin{cases} 0 & \text{when } m \ne n, \\ \pi/2 & \text{when } m = n, \end{cases}$$

where m and n are positive integers.

9. Use the trigonometric identity

$$2 \sin A \sin B = \cos(A - B) - \cos(A + B)$$

to show that

$$\int_0^{\pi} \sin mx \sin nx \, dx = \begin{cases} 0 & \text{when } m \ne n, \\ \pi/2 & \text{when } m = n, \end{cases}$$

where m and n are positive integers.

10. With the aid of the integration formula obtained in Problem 9, find the Fourier sine series corresponding to the function $f(x) = \sin x$ on the interval $0 < x < \pi$.

Answer: $\sin x$.

6. FOURIER SERIES

Consider a function f in $C_p(-\pi, \pi)$ and write

$$f(x) = g(x) + h(x), \tag{1}$$

where

$$g(x) = \frac{f(x) + f(-x)}{2} \quad \text{and} \quad h(x) = \frac{f(x) - f(-x)}{2}. \tag{2}$$

The function $g(x)$ is evidently *even*, and $h(x)$ is *odd*. That is,

$$g(-x) = g(x) \quad \text{and} \quad h(-x) = -h(x)$$

for each point x in the interval $-\pi < x < \pi$ at which these functions are defined.

According to Secs. 2 and 4,

$$g(x) \sim \frac{a_0}{2} + \sum_{n=1}^{\infty} a_n \cos nx \qquad (0 < x < \pi), \tag{3}$$

where

$$a_n = \frac{2}{\pi} \int_0^{\pi} g(x) \cos nx \, dx \qquad (n = 0, 1, 2, \ldots), \tag{4}$$

and

$$h(x) \sim \sum_{n=1}^{\infty} b_n \sin nx \qquad (0 < x < \pi), \tag{5}$$

where

$$b_n = \frac{2}{\pi} \int_0^{\pi} h(x) \sin nx \, dx \qquad (n = 1, 2, \ldots). \tag{6}$$

When correspondence (3) is an equality that is valid for $0 < x < \pi$, the equation also holds on the interval $-\pi < x < 0$ since each side of the correspondence is an even function. A similar remark applies to correspondence (5) since each side there is an odd function. Because $f(x)$ is the sum of $g(x)$ and $h(x)$, this suggests that the correspondence

$$f(x) \sim \frac{a_0}{2} + \sum_{n=1}^{\infty} (a_n \cos nx + b_n \sin nx) \qquad (-\pi < x < \pi) \tag{7}$$

may be an equality under certain circumstances.

In view of the first of equations (2), expression (4) for the coefficients a_n can be written

$$a_n = \frac{1}{\pi}\left[\int_0^\pi f(x)\cos nx\,dx + \int_0^\pi f(-s)\cos ns\,ds\right].$$

By making the substitution $x = -s$ in the second of these two integrals, we find that

$$a_n = \frac{1}{\pi}\left[\int_0^\pi f(x)\cos nx\,dx + \int_{-\pi}^0 f(x)\cos nx\,dx\right],$$

or

$$a_n = \frac{1}{\pi}\int_{-\pi}^{\pi} f(x)\cos nx\,dx \qquad (n = 0, 1, 2, \ldots). \tag{8}$$

Likewise,

$$b_n = \frac{1}{\pi}\int_{-\pi}^{\pi} f(x)\sin nx\,dx \qquad (n = 1, 2, \ldots). \tag{9}$$

Correspondence (7), when combined with expressions (8) and (9) for the constants a_n and b_n, becomes

$$f(x) \sim \frac{1}{2\pi}\int_{-\pi}^{\pi} f(s)\,ds + \frac{1}{\pi}\sum_{n=1}^{\infty}\left[\cos nx\int_{-\pi}^{\pi} f(s)\cos ns\,ds + \sin nx\int_{-\pi}^{\pi} f(s)\sin ns\,ds\right].$$

The trigonometric identity

$$\cos(A - B) = \cos A\cos B + \sin A\sin B$$

then enables us to write the correspondence in the form

$$f(x) \sim \frac{1}{2\pi}\int_{-\pi}^{\pi} f(s)\,ds + \frac{1}{\pi}\sum_{n=1}^{\infty}\int_{-\pi}^{\pi} f(s)\cos n(s - x)\,ds. \tag{10}$$

Note that the term

$$\frac{1}{2\pi}\int_{-\pi}^{\pi} f(s)\,ds$$

here, which is the same as the term $a_0/2$ in series (7), is the mean, or average, value of $f(x)$ over the interval $-\pi < x < \pi$.

The form (10) of correspondence (7) will be the starting point of the proof in Sec. 12 of our theorem ensuring the convergence of the Fourier series to $f(x)$ on the interval $-\pi < x < \pi$.

Series (7), with coefficients (8) and (9), is the *Fourier series* corresponding to $f(x)$ on the interval $-\pi < x < \pi$. Suppose that the series converges to $f(x)$

when $-\pi < x < \pi$. Then, in view of the periodicity of its terms, it converges to a function $y = F(x)$ that coincides with $y = f(x)$ on $-\pi < x < \pi$ and whose graph there is repeated every 2π units along the x axis. The function F is, therefore, the *periodic extension, with period* 2π, of f. If, on the other hand, f is a given periodic function, with period 2π, series (7) represents $f(x)$ everywhere when it converges to $f(x)$ on the interval $-\pi \le x \le \pi$.

It may be that the given function f in $C_p(-\pi, \pi)$ is *even* on the interval $-\pi < x < \pi$. That is, $f(-x) = f(x)$ for all such values of x. Then

$$f(-x)\cos(-nx) = f(x)\cos nx \qquad (n = 0, 1, 2, \ldots)$$

and

$$f(-x)\sin(-nx) = -f(x)\sin nx \qquad (n = 1, 2, \ldots)$$

when $-\pi < x < \pi$; and we see that $f(x)\cos nx$ and $f(x)\sin nx$ are even and odd, respectively. Because the graph of $y = f(x)\cos nx$ is symmetric with respect to the y axis and the graph of $y = f(x)\sin nx$ is symmetric with respect to the origin, it follows that expressions (8) and (9) reduce to

$$a_n = \frac{2}{\pi}\int_0^\pi f(x)\cos nx\,dx \qquad (n = 0, 1, 2, \ldots)$$

and $b_n = 0$ $(n = 1, 2, \ldots)$. Series (7) thus becomes a Fourier cosine series (Sec. 2) for $f(x)$ on the interval $0 < x < \pi$.

Similarly, if f is *odd* on the interval $-\pi < x < \pi$, it follows from expressions (8) and (9) that $a_n = 0$ $(n = 0, 1, 2, \ldots)$ and

$$b_n = \frac{2}{\pi}\int_0^\pi f(x)\sin nx\,dx \qquad (n = 1, 2, \ldots).$$

In this case, series (7) becomes a Fourier sine series (Sec. 4) for the function $f(x)$ on $0 < x < \pi$.

7. EXAMPLES

We include here three examples of Fourier series on the interval $-\pi < x < \pi$ that illustrate points made in Sec. 6.

EXAMPLE 1. Let us find the Fourier series corresponding to the function $f(x)$ that is defined on the fundamental interval $-\pi < x < \pi$ as follows:

$$(1) \qquad f(x) = \begin{cases} 0 & \text{when } -\pi < x \le 0, \\ x & \text{when } \quad 0 < x < \pi. \end{cases}$$

The graph of $y = f(x)$ is indicated by the bold line segments in Fig. 5 that are solid.

According to expression (8), Sec. 6,

$$a_n = \frac{1}{\pi}\left(\int_{-\pi}^0 0\cos nx\,dx + \int_0^\pi x\cos nx\,dx\right) = \frac{1}{\pi}\int_0^\pi x\cos nx\,dx \qquad (n = 0, 1, 2, \ldots).$$

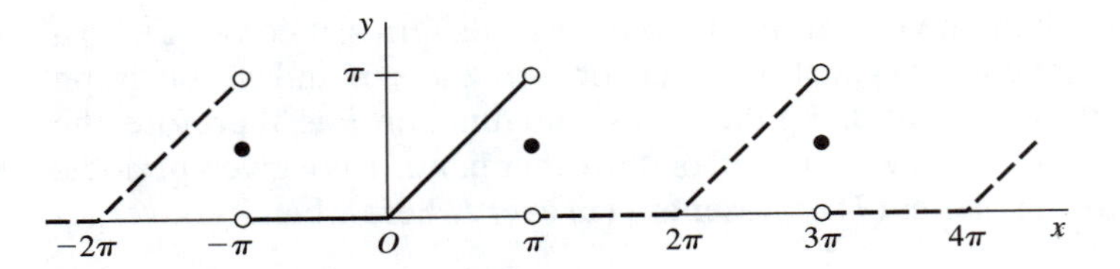

FIGURE 5

By applying integration by parts, or Kronecker's method (Sec. 5), one can show that

$$a_n = \frac{(-1)^n - 1}{\pi n^2}$$

when $n = 1, 2, \ldots$. To avoid division by zero, we must evaluate the integral for a_0 separately:

$$a_0 = \frac{1}{\pi}\int_0^\pi x\,dx = \frac{\pi}{2}.$$

Expression (9), Sec. 6, tells us that

$$\begin{aligned} b_n &= \frac{1}{\pi}\left(\int_{-\pi}^0 0 \sin nx\,dx + \int_0^\pi x \sin nx\,dx\right) \\ &= \frac{1}{\pi}\int_0^\pi x \sin nx\,dx = \frac{(-1)^{n+1}}{n} \end{aligned}$$

for all positive integers $n = 1, 2, \ldots$. Hence, on the interval $-\pi < x < \pi$,

$$f(x) \sim \frac{\pi}{4} + \sum_{n=1}^{\infty}\left[\frac{(-1)^n - 1}{\pi n^2}\cos nx + \frac{(-1)^{n+1}}{n}\sin nx\right]. \tag{2}$$

This series will be shown to converge to $f(x)$ on the fundamental interval, as well as to the periodic extension $F(x)$ that is indicated in Fig. 5, where the graph of $y = F(x)$ is sketched. As in Example 1, Sec. 5, the series must converge to the mean value of the one-sided limits of the periodic extension at each of the discontinuities $x = \pm\pi, \pm 3\pi, \pm 5\pi, \ldots$. Here the mean values are all $\pi/2$.

EXAMPLE 2. The function $f(x) = |\sin x|$ $(-\pi < x < \pi)$ is even. Hence the Fourier series corresponding to $f(x)$ on the interval $-\pi < x < \pi$ is actually the Fourier cosine series for the function

$$f(x) = |\sin x| = \sin x \qquad (0 < x < \pi).$$

That series has already been found in Example 2, Sec. 3; and, by referring to correspondence (3) there, we see that

$$|\sin x| \sim \frac{2}{\pi} - \frac{4}{\pi}\sum_{n=1}^{\infty}\frac{\cos 2nx}{4n^2 - 1} \qquad (-\pi < x < \pi). \tag{3}$$

EXAMPLE 3. Since the function $f(x) = x\ (-\pi < x < \pi)$ is odd, the Fourier series for f on $-\pi < x < \pi$ is simply the Fourier sine series for that function on $0 < x < \pi$. Hence correspondence (1) in Example 1, Sec. 5, is also a correspondence on the larger interval $-\pi < x < \pi$:

$$x \sim 2\sum_{n=1}^{\infty} \frac{(-1)^{n+1}}{n} \sin nx \qquad (-\pi < x < \pi). \tag{4}$$

Similarly, correspondence (3) in Example 2, Sec. 5, can be written

$$x^3 \sim 2\sum_{n=1}^{\infty} (-1)^{n+1} \frac{(n\pi)^2 - 6}{n^3} \sin nx \qquad (-\pi < x < \pi). \tag{5}$$

PROBLEMS

Find the Fourier series on the interval $-\pi < x < \pi$ that corresponds to each of the functions in Problems 1 through 6.

1. $f(x) = \begin{cases} -\pi/2 & \text{when } -\pi < x < 0, \\ \pi/2 & \text{when } \quad 0 < x < \pi. \end{cases}$

Answer: $2\sum_{n=1}^{\infty} \dfrac{\sin(2n-1)x}{2n-1}$.

2. $f(x)$ is the function such that the graph of $y = f(x)$ consists of the two bold line segments shown in Fig. 6.

Answer: $\dfrac{3}{2} + 2\sum_{n=1}^{\infty} \left[\dfrac{1-(-1)^n}{(n\pi)^2} \cos nx + \dfrac{(-1)^{n+1}}{n\pi} \sin nx \right]$.

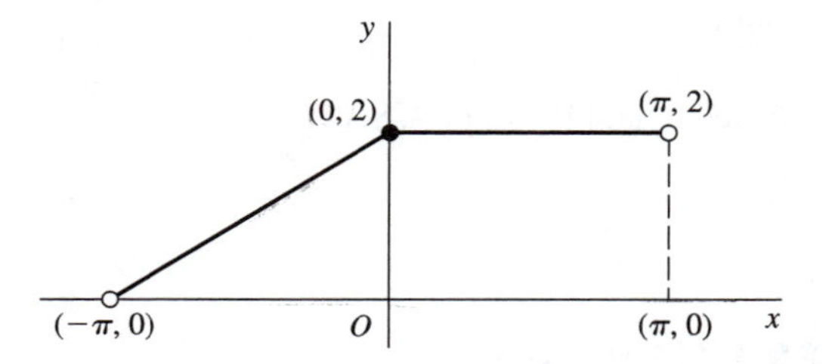

FIGURE 6

3. $f(x) = x + \dfrac{1}{4}x^2\ (-\pi < x < \pi)$.

Suggestion: Use the series (4) for x in Example 3, Sec. 7, and the one for x^2 in Problem 3(*a*), Sec. 5.

Answer: $\dfrac{\pi^2}{12} + \sum_{n=1}^{\infty} (-1)^n \left(\dfrac{\cos nx}{n^2} - \dfrac{2\sin nx}{n} \right)$.

4. $f(x) = e^{ax}$ $(-\pi < x < \pi)$, where $a \neq 0$.

Suggestion: Use Euler's formula $e^{i\theta} = \cos\theta + i\sin\theta$, where $i = \sqrt{-1}$, to write

$$a_n + ib_n = \frac{1}{\pi}\int_{-\pi}^{\pi} f(x)e^{inx}\,dx \qquad (n = 1, 2, \ldots).$$

Then, after evaluating this single integral, equate real parts and imaginary parts.[†]

Answer: $$\frac{\sinh a\pi}{a\pi} + \frac{2\sinh a\pi}{\pi}\sum_{n=1}^{\infty}\frac{(-1)^n}{a^2+n^2}(a\cos nx - n\sin nx).$$

5. $f(x) = \cosh ax$ $(-\pi < x < \pi)$, where $a \neq 0$.

Suggestion: Use the series found in Problem 4.

Answer: $$\frac{\sinh a\pi}{a\pi}\left[1 + 2a^2\sum_{n=1}^{\infty}\frac{(-1)^n}{a^2+n^2}\cos nx\right].$$

6. $f(x) = \cos ax$ $(-\pi < x < \pi)$, where $a \neq 0, \pm 1, \pm 2, \ldots$.

Suggestion: With the aid of Euler's formula, stated in the suggestion with Problem 4, write

$$\cos ax = \frac{e^{iax} + e^{-iax}}{2}.$$

Then use the series already obtained in that earlier problem.

Answer: $$\frac{2a\sin a\pi}{\pi}\left[\frac{1}{2a^2} + \sum_{n=1}^{\infty}\frac{(-1)^{n+1}}{n^2-a^2}\cos nx\right].$$

7. Find the Fourier series on the interval $-\pi < x < \pi$ for the function f defined by the equations

$$f(x) = \begin{cases} 0 & \text{when } -\pi \le x \le 0, \\ \sin x & \text{when } \quad 0 < x \le \pi. \end{cases}$$

Then, given that the series converges to $f(x)$ when $-\pi \le x \le \pi$, describe graphically the function that is represented by the series for all x $(-\infty < x < \infty)$.

Suggestion: To find the series, write the function in the form

$$f(x) = \frac{\sin x + |\sin x|}{2} \qquad (-\pi \le x \le \pi)$$

and then use the results in Problem 10, Sec. 5, and Example 2, Sec. 7.

Answer: $$\frac{1}{\pi} + \frac{1}{2}\sin x - \frac{2}{\pi}\sum_{n=1}^{\infty}\frac{\cos 2nx}{4n^2-1}.$$

8. ADAPTATIONS TO OTHER INTERVALS

Let f denote a piecewise continuous function of x on an interval $-c < x < c$ of the x axis, and define the related function

$$g(s) = f\left(\frac{cs}{\pi}\right) \qquad (-\pi < s < \pi) \tag{1}$$

[†]For a justification of Euler's formula and background on complex-variable methods, see the authors' book (2004), listed in the Bibliography.

of s. The Fourier series corresponding to this new function on $-\pi < s < \pi$ is, according to Sec. 6,

$$f\left(\frac{cs}{\pi}\right) \sim \frac{a_0}{2} + \sum_{n=1}^{\infty} (a_n \cos ns + b_n \sin ns) \qquad (-\pi < s < \pi) \tag{2}$$

where

$$a_n = \frac{1}{\pi} \int_{-\pi}^{\pi} f\left(\frac{cs}{\pi}\right) \cos ns \, ds \qquad (n = 0, 1, 2, \ldots) \tag{3}$$

and

$$b_n = \frac{1}{\pi} \int_{-\pi}^{\pi} f\left(\frac{cs}{\pi}\right) \sin ns \, ds \qquad (n = 1, 2, \ldots). \tag{4}$$

The function (1) is evidently also piecewise continuous, and we anticipate that correspondence (2) will become an equality when certain further conditions are imposed on f. Thus if we put

$$s = \frac{\pi x}{c}$$

in correspondence (2) and its conditions of validity, we arrive at the series

$$f(x) \sim \frac{a_0}{2} + \sum_{n=1}^{\infty} \left(a_n \cos \frac{n\pi x}{c} + b_n \sin \frac{n\pi x}{c}\right) \qquad (-c < x < c). \tag{5}$$

That same substitution in expressions (3) and (4), moreover, enables us to write

$$a_n = \frac{1}{c} \int_{-c}^{c} f(x) \cos \frac{n\pi x}{c} \, dx \qquad (n = 0, 1, 2, \ldots) \tag{6}$$

and

$$b_n = \frac{1}{c} \int_{-c}^{c} f(x) \sin \frac{n\pi x}{c} \, dx \qquad (n = 1, 2, \ldots). \tag{7}$$

Series (5) is a Fourier series on the fundamental interval $-c < x < c$ and becomes series (7) in Sec. 6 when $c = \pi$. Conditions on f ensuring that correspondence (5) is, in fact, an equality at points where f is continuous will be given in Chap. 2. Note that if the series does converge to $f(x)$ on $-c < x < c$, the graph of $y = f(x)$ is repeated every $2c$ units along the x axis.

Arguments similar to those used above lead to Fourier cosine and sine series on $0 < x < c$:

$$f(x) \sim \frac{a_0}{2} + \sum_{n=1}^{\infty} a_n \cos \frac{n\pi x}{c} \qquad (0 < x < c), \tag{8}$$

where

$$a_n = \frac{2}{c} \int_0^c f(x) \cos \frac{n\pi x}{c}\, dx \qquad (n = 0, 1, 2, \ldots), \tag{9}$$

and

$$f(x) \sim \sum_{n=1}^{\infty} b_n \sin \frac{n\pi x}{c} \qquad (0 < x < c), \tag{10}$$

where

$$b_n = \frac{2}{c} \int_0^c f(x) \sin \frac{n\pi x}{c}\, dx \qquad (n = 1, 2, \ldots). \tag{11}$$

The convergence of series (8) and (10) is also treated in Chap. 2.

The following example illustrates how Fourier series on intervals $-c < x < c$, as well as cosine and sine series on $0 < x < c$, can be obtained from known series on $-\pi < x < \pi$ and $0 < x < \pi$. Since we do not yet have theorems ensuring the convergence of Fourier series to the functions in question, we shall continue to use the tilde symbol $\sim$ to denote mere correspondence and not necessarily equality. Also, anticipating that the correspondences obtained will actually be equalities, we shall continue to include conditions of validity.

EXAMPLE. It is a simple matter to obtain the correspondence

$$x^2 \sim \frac{c^2}{3} + \frac{4c^2}{\pi^2} \sum_{n=1}^{\infty} \frac{(-1)^n}{n^2} \cos \frac{n\pi x}{c} \qquad (0 < x < c) \tag{12}$$

from the known one [Problem 3(*a*), Sec. 5]

$$x^2 \sim \frac{\pi^2}{3} + 4 \sum_{n=1}^{\infty} \frac{(-1)^n}{n^2} \cos nx \qquad (0 < x < \pi). \tag{13}$$

We let x be any number in the interval $0 < x < c$ and note how it follows that

$$0 < \frac{\pi x}{c} < \pi.$$

Hence it is legitimate to replace x by $\pi x/c$ in correspondence (13) and its condition of validity:

$$\frac{\pi^2 x^2}{c^2} \sim \frac{\pi^2}{3} + 4 \sum_{n=1}^{\infty} \frac{(-1)^n}{n^2} \cos \frac{n\pi x}{c} \qquad \left(0 < \frac{\pi x}{c} < \pi\right). \tag{14}$$

Then, multiplying each side by c^2/π^2 and multiplying through the new condition of validity by c/π, we arrive at correspondence (12).

Expression (9) could, of course, have been used to find the desired coefficients if correspondence (13) had not been available.

PROBLEMS

1. (*a*) Use the Fourier sine series in Example 1, Sec. 5, for

$$f(x) = x \qquad (0 < x < \pi)$$

to show that

$$x \sim \frac{2}{\pi}\sum_{n=1}^{\infty}\frac{(-1)^{n+1}}{n}\sin n\pi x \qquad (0 < x < 1).$$

(*b*) Obtain the correspondence in part (*a*) by using expression (11), Sec. 8, for the coefficients in a Fourier sine series on $0 < x < c$.

2. Show how it follows from the expansions obtained in Problem 1 and the example in Sec. 8 that

$$x(1+x) \sim \frac{1}{3} + \frac{2}{\pi}\sum_{n=1}^{\infty}(-1)^n\left(\frac{2}{n^2\pi}\cos n\pi x - \frac{1}{n}\sin n\pi x\right) \qquad (0 < x < 1).$$

3. Use the Fourier sine series found in Problem 3(*b*), Sec. 5, for

$$f(x) = x^2 \qquad (0 < x < \pi)$$

to obtain the correspondence

$$x^2 \sim 2c^2\sum_{n=1}^{\infty}\left[\frac{(-1)^{n+1}}{n\pi} - 2\,\frac{1-(-1)^n}{(n\pi)^3}\right]\sin\frac{n\pi x}{c} \qquad (0 < x < c).$$

4. (*a*) Use the Fourier sine series correspondence found in Example 3, Sec. 5, for the function

$$f(x) = x(\pi^2 - x^2) \qquad (0 < x < \pi)$$

to establish the correspondence

$$x(1-x^2) \sim \frac{12}{\pi^3}\sum_{n=1}^{\infty}\frac{(-1)^{n+1}}{n^3}\sin n\pi x \qquad (0 < x < 1).$$

(*b*) Replace x by $1 - x$ on each side of the correspondence in part (*a*) to show that

$$x(x-1)(x-2) \sim \frac{12}{\pi^3}\sum_{n=1}^{\infty}\frac{\sin n\pi x}{n^3} \qquad (0 < x < 1).$$

5. Show how it follows from the Fourier sine series obtained for

$$f(x) = x(\pi - x) \qquad (0 < x < \pi)$$

in Problem 5, Sec. 5, that

$$x(2c - x) \sim \frac{32c^2}{\pi^3}\sum_{n=1}^{\infty}\frac{1}{(2n-1)^3}\sin\frac{(2n-1)\pi x}{2c} \qquad (0 < x < 2c).$$

6. Use the Fourier series for

$$f(x) = e^{ax} \qquad (-\pi < x < \pi),$$

where $a \neq 0$, that was found in Problem 4, Sec. 7, to show that

$$e^x \sim \frac{\sinh c}{c} + 2\sinh c \sum_{n=1}^{\infty} \frac{(-1)^n}{c^2 + (n\pi)^2}\left(c \cos\frac{n\pi x}{c} - n\pi \sin\frac{n\pi x}{c}\right) \qquad (-c < x < c).$$

7. By starting with the Fourier cosine series correspondence obtained for the function

$$f(x) = \pi - x \qquad (0 < x < \pi)$$

in Problem 2(*a*), Sec. 5, show that

$$\frac{c}{4} - x \sim \frac{2c}{\pi^2} \sum_{n=1}^{\infty} \frac{1}{(2n-1)^2} \cos\frac{(4n-2)\pi x}{c} \qquad \left(0 < x < \frac{c}{2}\right).$$

8. Use expression (11), Sec. 8, for the coefficients in a Fourier sine series on $0 < x < c$ to obtain the correspondence

$$\cos \pi x \sim \frac{8}{\pi} \sum_{n=1}^{\infty} \frac{n}{4n^2 - 1} \sin 2n\pi x \qquad (0 < x < 1).$$

Suggestion: To evaluate the integrals that arise, recall the trigonometric identity

$$2 \sin A \cos B = \sin(A + B) + \sin(A - B).$$

9. Show that in Sec. 8 the Fourier series (5), with coefficients (6) and (7), can be written in the compact form

$$\frac{1}{2c}\int_{-c}^{c} f(s)\,ds + \frac{1}{c}\sum_{n=1}^{\infty}\int_{-c}^{c} f(s)\cos\left[\frac{n\pi}{c}(s - x)\right] ds.$$

(See Sec. 6, where this form was obtained when $c = \pi$.)

CHAPTER 2

CONVERGENCE OF FOURIER SERIES

In this chapter, we shall establish conditions on a function $f(x)$, defined on the interval $-\pi < x < \pi$, that ensure a valid Fourier series representation. Corresponding results for Fourier cosine and sine series representations will follow readily. It will be a simple matter to extend the theory to Fourier series on arbitrary intervals $-c < x < c$, as well as to Fourier cosine and sine series on intervals $0 < x < c$. Some further aspects of the theory of convergence of Fourier series will be touched on later in the chapter.

9. ONE-SIDED DERIVATIVES

In developing sufficient conditions on a function f such that its Fourier series on the interval $-\pi < x < \pi$ converges to $f(x)$ there, we need to generalize the concept of the derivative

$$f'(x_0) = \lim_{x \to x_0} \frac{f(x) - f(x_0)}{x - x_0} \tag{1}$$

of f at a point $x = x_0$.

Suppose that the right-hand limit $f(x_0+)$ exists at x_0 (see Sec. 1). The *right-hand derivative* of f at x_0 is defined as follows:

$$f'_R(x_0) = \lim_{\substack{x \to x_0 \\ x > x_0}} \frac{f(x) - f(x_0+)}{x - x_0}, \tag{2}$$

provided that the limit here exists. Note that although $f(x_0)$ need not exist, $f(x_0+)$ must exist if $f'_R(x_0)$ does. When the ordinary, or two-sided, derivative $f'(x_0)$ exists, f is continuous at x_0 and $f'_R(x_0) = f'(x_0)$.

Similarly, if $f(x_0-)$ exists, the *left-hand derivative* of f at x_0 is given by the equation

$$f'_L(x_0) = \lim_{\substack{x \to x_0 \\ x < x_0}} \frac{f(x) - f(x_0-)}{x - x_0} \tag{3}$$

when this limit exists; and if $f'(x_0)$ exists, $f'_L(x_0) = f'(x_0)$.

EXAMPLE 1. Let f denote the continuous function defined by the equations

$$f(x) = \begin{cases} x^2 & \text{when } x \le 0, \\ \sin x & \text{when } x > 0. \end{cases}$$

With the aid of l'Hôpital's rule, we see that

$$f'_R(0) = \lim_{\substack{x \to 0 \\ x > 0}} \frac{\sin x}{x} = 1;$$

furthermore,

$$f'_L(0) = \lim_{\substack{x \to 0 \\ x < 0}} \frac{x^2}{x} = \lim_{\substack{x \to 0 \\ x < 0}} x = 0.$$

Since these one-sided derivatives have different values, the ordinary derivative $f'(0)$ cannot exist.

The ordinary derivative $f'(x_0)$ can fail to exist even when $f(x_0)$ is defined and $f'_R(x_0)$ and $f'_L(x_0)$ have a common value.

EXAMPLE 2. If f is the step function

$$f(x) = \begin{cases} 0 & \text{when } x < 0, \\ 1 & \text{when } x \ge 0, \end{cases}$$

then $f'_R(0) = f'_L(0) = 0$. But the derivative $f'(0)$ does not exist since f is not continuous at $x = 0$.

As is the case with ordinary derivatives, the mere continuity of f at a point x_0 does *not* ensure the existence of either one-sided derivative there.

EXAMPLE 3. The function $f(x) = \sqrt{x}$ $(x \ge 0)$ has no right-hand derivative at the point $x = 0$, although it is continuous there.

A number of properties of ordinary derivatives remain valid for one-sided derivatives. Suppose, for instance, that the right-hand derivatives of two functions f and g exist at a point x_0. Let us find the right-hand derivative of the product

$$(fg)(x) = f(x)g(x)$$

at x_0. Since the difference quotient

$$\frac{(fg)(x) - (fg)(x_0+)}{x - x_0} = \frac{f(x)g(x) - f(x_0+)g(x_0+)}{x - x_0}$$

can be written

$$f(x)\,\frac{g(x) - g(x_0+)}{x - x_0} + \frac{f(x) - f(x_0+)}{x - x_0}\,g(x_0+),$$

it follows that

$$(fg)'_R(x_0) = f(x_0+)g'_R(x_0) + f'_R(x_0)g(x_0+).$$

Likewise, if $f'_L(x_0)$ and $g'_L(x_0)$ exist, the left-hand derivative of the product $(fg)(x)$ exists at x_0.

Finally, we turn to a property of one-sided derivatives that is particularly important in the theory of convergence of Fourier series. It concerns the subspace $C'_p(a, b)$ of $C_p(a, b)$ consisting of all piecewise continuous functions f on an interval $a < x < b$ whose derivatives f' are also piecewise continuous on that interval. Such a function is said to be *piecewise smooth* because, over the subintervals on which both f and f' are continuous, any tangents to the graph of $y = f(x)$ that turn do so continuously.

Theorem. *If a function f is piecewise smooth on an interval $a < x < b$, then at each point x_0 in the closed interval $a \leq x \leq b$ the one-sided derivatives of f, from the interior at the endpoints, exist and are the same as the corresponding one-sided limits of f':*

$$f'_R(x_0) = f'(x_0+), \qquad f'_L(x_0) = f'(x_0-). \tag{4}$$

To prove this, we assume for the moment that f and f' are actually continuous on the interval $a < x < b$ and that the one-sided limits of f and f' from the interior exist at the endpoints $x = a$ and $x = b$. If x_0 is a point in that open interval, $f'(x_0)$ exists. Hence $f'_R(x_0)$ and $f'_L(x_0)$ exist, and both are equal to $f'(x_0)$. Because f' is continuous at x_0, then, equations (4) hold.

The following argument shows that it is also true that $f'_R(a)$ exists and is equal to $f'(a+)$. If we let s denote any number in the interval $a < x < b$ and define $f(a)$ to be $f(a+)$, then f is continuous on the closed interval $a \leq x \leq s$ (Fig. 7). Since f' exists in the open interval $a < x < s$, the mean value theorem for derivatives applies. That is, there is a number c, where $a < c < s$, such that

$$\frac{f(s) - f(a+)}{s - a} = f'(c). \tag{5}$$

This is shown geometrically in Fig. 7, where the slopes of the secant line S and the tangent line T are the same. Letting s, and therefore c, tend to a in equation (5), we see that since $f'(a+)$ exists, the limit of $f'(c)$ exists and has that value. Consequently, the limit of the difference quotient on the left in equation (5) exists, its value being $f'_R(a)$. Thus $f'_R(a) = f'(a+)$. Similarly, $f'_L(b) = f'(b-)$.

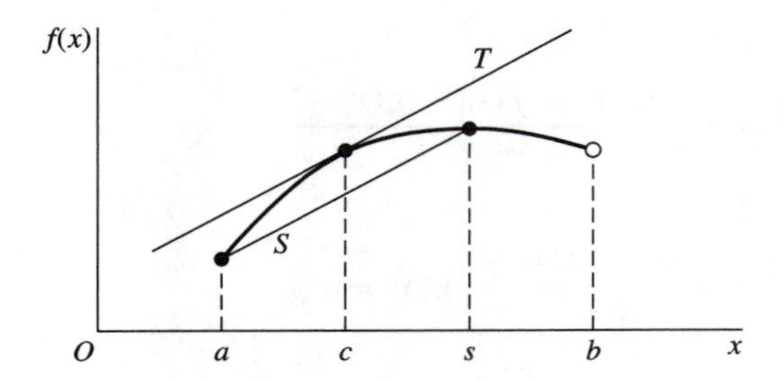

FIGURE 7

Now any piecewise smooth function f is continuous, along with its derivative f', on a finite number of subintervals at whose endpoints the one-sided limits of f and f' from the interior exist. If the results of the two preceding paragraphs are applied to each of those subintervals, the theorem is established.

Example 4 illustrates the distinction between one-sided derivatives and one-sided limits of derivatives.

EXAMPLE 4. Consider the function f whose values are

$$f(x) = \begin{cases} x^2 \sin(1/x) & \text{when } x \neq 0, \\ 0 & \text{when } x = 0. \end{cases}$$

Since $0 \le |x^2 \sin(1/x)| \le x^2$ when $x \neq 0$, both one-sided limits $f(0+)$ and $f(0-)$ exist and have value zero. Moreover, since $0 \le |x \sin(1/x)| \le |x|$ when $x \neq 0$,

$$f'_R(0) = \lim_{\substack{x \to 0 \\ x > 0}} \left(x \sin \frac{1}{x} \right) = 0 \qquad \text{and} \qquad f'_L(0) = \lim_{\substack{x \to 0 \\ x < 0}} \left(x \sin \frac{1}{x} \right) = 0.$$

But, from the expression

$$f'(x) = 2x \sin \frac{1}{x} - \cos \frac{1}{x} \qquad (x \neq 0),$$

we see that the one-sided limits $f'(0+)$ and $f'(0-)$ do not exist.

Note that although its one-sided derivatives exist everywhere, the function f is not piecewise smooth on any bounded interval containing the origin. Hence the above theorem is not applicable to this function on such an interval.

10. A PROPERTY OF FOURIER COEFFICIENTS

In treating the convergence of Fourier series, we shall find it useful to know that for a function f in $C_p(0, \pi)$, the coefficients a_n and b_n in the cosine and sine series always tend to zero as n tends to infinity.

To show that the coefficients

$$a_n = \frac{2}{\pi} \int_0^{\pi} f(x) \cos nx \, dx \qquad (n = 0, 1, 2, \ldots) \tag{1}$$

in a cosine series have this property, let $s_N(x)$ denote the partial sum consisting of the first $N+1$ $(N \geq 1)$ terms in such a series:

$$s_N(x) = \frac{a_0}{2} + \sum_{n=1}^{N} a_n \cos nx. \tag{2}$$

Then

$$\int_0^\pi [f(x) - s_N(x)]^2\,dx = \int_0^\pi [f(x)]^2\,dx - 2\int_0^\pi f(x)\,s_N(x)\,dx + \int_0^\pi [s_N(x)]^2\,dx. \tag{3}$$

We need to rewrite the last two integrals on the right here. From equation (2) we have

$$f(x)\,s_N(x) = \frac{a_0}{2} f(x) + \sum_{n=1}^{N} a_n f(x) \cos nx.$$

Hence

$$\int_0^\pi f(x)\,s_N(x)\,dx = \frac{a_0}{2}\int_0^\pi f(x)\,dx + \sum_{n=1}^{N} a_n \int_0^\pi f(x)\cos nx\,dx.$$

In view of expression (1), then,

$$\int_0^\pi f(x)\,s_N(x)\,dx = \frac{\pi}{2}\left(\frac{a_0^2}{2} + \sum_{n=1}^{N} a_n^2\right). \tag{4}$$

As for the integral on the far right in equation (3), we note from expression (2) that

$$\int_0^\pi s_N(x)\,dx = \frac{a_0}{2}\int_0^\pi dx + \sum_{n=1}^{N} a_n \int_0^\pi \cos nx\,dx = \frac{\pi}{2}a_0. \tag{5}$$

Also, using m as the index of summation in expression (2), we write

$$s_N(x)\cos nx = \frac{a_0}{2}\cos nx + \sum_{m=1}^{N} a_m \cos mx \cos nx,$$

where n has any one of the values $n = 1, 2, \ldots, N$. The integration formula (Sec. 2)

$$\int_0^\pi \cos mx \cos nx\,dx = \begin{cases} 0 & \text{when } m \neq n, \\ \pi/2 & \text{when } m = n, \end{cases}$$

where m and n are positive integers, now yields

$$\int_0^\pi s_N(x)\cos nx\,dx = \frac{a_0}{2}\int_0^\pi \cos nx\,dx + \sum_{m=1}^{N} a_m \int_0^\pi \cos mx \cos nx\,dx = \frac{\pi}{2}a_n \qquad (n = 1, 2, \ldots, N). \tag{6}$$

Consequently, by writing

$$[s_N(x)]^2 = \frac{a_0}{2} s_N(x) + \sum_{n=1}^{N} a_n s_N(x) \cos nx,$$

integrating each side from 0 to π, and then referring to expressions (5) and (6), we have

$$\int_0^\pi [s_N(x)]^2 \, dx = \frac{\pi}{2}\left(\frac{a_0^2}{2} + \sum_{n=1}^{N} a_n^2\right). \tag{7}$$

It now follows from equations (3), (4), and (7) that

$$\int_0^\pi [f(x) - s_N(x)]^2 \, dx = \int_0^\pi [f(x)]^2 \, dx - \frac{\pi}{2}\left(\frac{a_0^2}{2} + \sum_{n=1}^{N} a_n^2\right).$$

Since the value of the integral on the left here is nonnegative, we thus arrive at *Bessel's inequality* for the coefficients (1):

$$\frac{a_0^2}{2} + \sum_{n=1}^{N} a_n^2 \le \frac{2}{\pi} \int_0^\pi [f(x)]^2 \, dx \qquad (N = 1, 2, \ldots). \tag{8}$$

The desired result,

$$\lim_{n \to \infty} a_n = 0, \tag{9}$$

is an easy consequence of Bessel's inequality (8), as the following argument shows. We observe that the right-hand side of the inequality is independent of the positive integer N; and as N increases on the left-hand side, the sums of the squares there form a sequence that is bounded and nondecreasing. Since such a sequence must converge and since this particular sequence is the sequence of the partial sums of the series whose terms are $a_0^2/2$ and a_n^2 $(n = 1, 2, \ldots)$, that series must converge. Limit (9) then follows from the fact that the nth term of a convergent series always tends to zero as n tends to infinity.

A similar procedure can be used (Problem 3, Sec. 11) to show that the coefficients

$$b_n = \frac{2}{\pi} \int_0^\pi f(x) \sin nx \, dx \qquad (n = 1, 2, \ldots) \tag{10}$$

in the Fourier sine series for f satisfy the Bessel inequality

$$\sum_{n=1}^{N} b_n^2 \le \frac{2}{\pi} \int_0^\pi [f(x)]^2 \, dx \qquad (N = 1, 2, \ldots) \tag{11}$$

and that

$$\lim_{n \to \infty} b_n = 0. \tag{12}$$

Finally, we recall from Sec. 6 that the coefficients a_n and b_n in the Fourier series involving both cosines and sines for a piecewise continuous function f in $C_p(-\pi, \pi)$ are the same as the coefficients in the Fourier cosine and sine series, respectively, for certain related functions on $0 < x < \pi$. Hence those coefficients themselves tend to zero as n tends to infinity. (See also Problem 5, Sec. 11.)

11. TWO LEMMAS

We preface our theorem on the convergence of Fourier series with two lemmas, or preliminary theorems. The first is a special case of what is known as the *Riemann-Lebesgue lemma*. That lemma appears later in Chap. 6 (Sec. 46), where it is needed in full generality.

Lemma 1. *If a function $G(u)$ is piecewise continuous on the interval $0 < u < \pi$, then*

$$\lim_{N\to\infty} \int_0^{\pi} G(u) \sin\left(\frac{u}{2} + Nu\right) du = 0, \tag{1}$$

where N denotes positive integers.

Our proof starts with the trigonometric identity

$$\sin(A + B) = \sin A \cos B + \cos A \sin B,$$

which tells us that

$$\sin\left(\frac{u}{2} + Nu\right) = \sin\frac{u}{2}\cos Nu + \cos\frac{u}{2}\sin Nu.$$

This enables us to write

$$\int_0^{\pi} G(u) \sin\left(\frac{u}{2} + Nu\right) du = \frac{\pi}{2} a_N + \frac{\pi}{2} b_N, \tag{2}$$

where

$$a_N = \frac{2}{\pi}\int_0^{\pi}\left[G(u)\sin\frac{u}{2}\right]\cos Nu\, du$$

and

$$b_N = \frac{2}{\pi}\int_0^{\pi}\left[G(u)\cos\frac{u}{2}\right]\sin Nu\, du.$$

Now the a_N are coefficients in a Fourier cosine series on the interval $0 < u < \pi$, and the b_N are coefficients in a Fourier sine series on that interval. Thus, by limits (9) and (12) in Sec. 10,

$$\lim_{N\to\infty} a_N = 0 \qquad \text{and} \qquad \lim_{N\to\infty} b_N = 0. \tag{3}$$

With limits (3), we need only let N tend to infinity in equation (2) to see that Lemma 1 is true.

Our second lemma involves the *Dirichlet kernel*

$$D_N(u) = \frac{1}{2} + \sum_{n=1}^{N} \cos nu, \tag{4}$$

where N is any positive integer. Note that $D_N(u)$ is continuous, even, and periodic with period 2π. The Dirichlet kernel plays a central role in our theory, and two other properties will be useful:

$$\int_0^\pi D_N(u)\,du = \frac{\pi}{2}, \tag{5}$$

$$D_N(u) = \frac{\sin\left(\frac{u}{2} + Nu\right)}{2\sin\frac{u}{2}} \qquad (u \neq 0, \pm 2\pi, \pm 4\pi, \ldots). \tag{6}$$

Property (5) is obvious upon integrating each side of equation (4). Expression (6) can be derived with the aid of a certain trigonometric identity (Problem 6).

Lemma 2. *Suppose that a function $g(u)$ is piecewise continuous on the interval $0 < u < \pi$ and that the right-hand derivative $g'_R(0)$ exists. Then*

$$\lim_{N\to\infty} \int_0^\pi g(u)\,D_N(u)\,du = \frac{\pi}{2}\,g(0+), \tag{7}$$

where $D_N(u)$ is the Dirichlet kernel (4).

To prove this, we write

$$\int_0^\pi g(u)\,D_N(u)\,du = I_N + J_N, \tag{8}$$

where

$$I_N = \int_0^\pi [g(u) - g(0+)]D_N(u)\,du \tag{9}$$

and

$$J_N = \int_0^\pi g(0+)D_N(u)\,du. \tag{10}$$

In view of expression (6) for $D_N(u)$, integral (9) can be put in the form

$$I_N = \int_0^\pi \frac{g(u) - g(0+)}{2\sin\frac{u}{2}} \sin\left(\frac{u}{2} + Nu\right)du.$$

Thus

$$I_N = \int_0^\pi G(u)\sin\left(\frac{u}{2} + Nu\right)du, \tag{11}$$

where the function

$$G(u) = \frac{g(u) - g(0+)}{2\sin\frac{u}{2}} \tag{12}$$

is a quotient of two functions that are piecewise continuous on the interval $0 < u < \pi$. Although the denominator vanishes at the point $u = 0$, one can show that $G(u)$ is itself piecewise continuous on $0 < u < \pi$ by establishing the existence of $G(0+)$. This is done by referring to expression (12) and writing

$$\lim_{\substack{u\to 0\\ u>0}} G(u) = \lim_{\substack{u\to 0\\ u>0}} \frac{g(u)-g(0+)}{u-0} \lim_{\substack{u\to 0\\ u>0}} \frac{u}{2\sin\frac{u}{2}}.$$

The first of the limits on the right here is, of course, $g'_R(0)$; and an application of l'Hôpital's rule reveals that the second limit is unity. According to Lemma 1, then, the limit of the right-hand side of equation (11) is zero as N tends to infinity. That is,

$$\lim_{N\to\infty} I_N = 0. \tag{13}$$

With property (5) of the Dirichlet kernel, one can see from expression (10) for J_N that

$$J_N = g(0+)\int_0^{\pi} D_N(u)\,du = \frac{\pi}{2}\,g(0+).$$

Hence

$$\lim_{N\to\infty} J_N = \frac{\pi}{2}\,g(0+). \tag{14}$$

The desired result (7) now follows from equation (8) together with limits (13) and (14).

PROBLEMS

1. With the aid of l'Hôpital's rule, find $f(0+)$ and $f'_R(0)$ when

$$f(x) = \frac{e^x-1}{x} \qquad (x \neq 0).$$

Answers: $f(0+) = 1,\ f'_R(0) = \frac{1}{2}$.

2. Show that the function defined by the equations

$$f(x) = \begin{cases} x\sin(1/x) & \text{when } x \neq 0, \\ 0 & \text{when } x = 0 \end{cases}$$

is continuous at $x = 0$ but that neither $f'_R(0)$ nor $f'_L(0)$ exists. This provides another illustration (see Example 3, Sec. 9) of the fact that the continuity of a function f at a point x_0 is *not* a sufficient condition for the existence of one-sided derivatives of f at x_0.

3. By following the steps used in Sec. 10 to find Bessel's inequality for the coefficients a_n in the Fourier cosine series for a function f in $C_p(0,\pi)$, derive the Bessel inequality

$$\sum_{n=1}^{N} b_n^2 \le \frac{2}{\pi}\int_0^{\pi} [f(x)]^2\,dx \qquad (N = 1, 2, \ldots)$$

for the coefficients b_n in the sine series for f. Then use this result to show that

$$\lim_{n\to\infty} b_n = 0.$$

4. In Chap. 1 (Sec. 6) we expressed a function $f(x)$ in $C_p(-\pi,\pi)$ as a sum

$$f(x) = g(x) + h(x)$$

where

$$g(x) = \frac{f(x) + f(-x)}{2} \quad \text{and} \quad h(x) = \frac{f(x) - f(-x)}{2}.$$

We then saw that the coefficients a_n and b_n in the Fourier series

$$\frac{a_0}{2} + \sum_{n=1}^{\infty} (a_n \cos nx + b_n \sin nx)$$

for $f(x)$ on $-\pi < x < \pi$ are the same as the coefficients in the Fourier cosine and sine series for $g(x)$ and $h(x)$, respectively, on $0 < x < \pi$.

(*a*) By referring to the Bessel inequalities (8) and (11) in Sec. 10, write

$$\frac{a_0^2}{2} + \sum_{n=1}^{N} a_n^2 \leq \frac{2}{\pi} \int_0^{\pi} [g(x)]^2 \, dx \qquad (N = 1, 2, \ldots)$$

and

$$\sum_{n=1}^{N} b_n^2 \leq \frac{2}{\pi} \int_0^{\pi} [h(x)]^2 \, dx \qquad (N = 1, 2, \ldots).$$

Then point out how it follows that

$$\frac{a_0^2}{2} + \sum_{n=1}^{N} \left(a_n^2 + b_n^2\right) \leq \frac{1}{\pi} \left\{ \int_0^{\pi} [f(x)]^2 \, dx + \int_0^{\pi} [f(-s)]^2 \, ds \right\} \quad (N = 1, 2, \ldots).$$

(*b*) By making the substitution $x = -s$ in the last integral in part (*a*), obtain the Bessel inequality

$$\frac{a_0^2}{2} + \sum_{n=1}^{N} \left(a_n^2 + b_n^2\right) \leq \frac{1}{\pi} \int_{-\pi}^{\pi} [f(x)]^2 \, dx \qquad (N = 1, 2, \ldots).$$

5. Show how it follows from the Bessel inequality in Problem 4(*b*) that

$$\lim_{n \to \infty} a_n = 0 \quad \text{and} \quad \lim_{n \to \infty} b_n = 0,$$

where a_n and b_n are the coefficients in the Fourier series

$$\frac{a_0}{2} + \sum_{n=1}^{\infty} (a_n \cos nx + b_n \sin nx)$$

for a piecewise continuous function in $C_p(-\pi, \pi)$.

6. Derive the expression

$$D_N(u) = \frac{\sin\left(\dfrac{u}{2} + Nu\right)}{2 \sin \dfrac{u}{2}} \qquad (u \neq 0, \pm 2\pi, \pm 4\pi, \ldots)$$

for the Dirichlet kernel (Sec. 11)

$$D_N(u) = \frac{1}{2} + \sum_{n=1}^{N} \cos nu$$

by writing

$$A = \frac{u}{2} \qquad \text{and} \qquad B = nu$$

in the trigonometric identity

$$2\sin A\cos B = \sin(A + B) + \sin(A - B)$$

and then summing each side of the resulting equation from $n = 1$ to $n = N$.

Suggestion: Note that

$$\sum_{n=1}^{N} \sin\left(\frac{u}{2} - nu\right) = -\sum_{n=0}^{N-1} \sin\left(\frac{u}{2} + nu\right).$$

12. A FOURIER THEOREM

A theorem that gives conditions under which a Fourier series

$$\frac{a_0}{2} + \sum_{n=1}^{\infty}(a_n \cos nx + b_n \sin nx), \tag{1}$$

with coefficients

$$a_n = \frac{1}{\pi}\int_{-\pi}^{\pi} f(x)\cos nx\,dx \qquad (n = 0, 1, 2, \ldots) \tag{2}$$

and

$$b_n = \frac{1}{\pi}\int_{-\pi}^{\pi} f(x)\sin nx\,dx \qquad (n = 1, 2, \ldots), \tag{3}$$

converges to $f(x)$ is called a *Fourier theorem*. One such theorem will now be established. Although it is stated for periodic functions of period 2π, it also applies to functions defined only on the fundamental interval $-\pi < x < \pi$; for, as is done in the corollary following this theorem and its proof, we need only consider the periodic extensions, with period 2π, of such functions.

Theorem. *Let f denote a function that is piecewise continuous on the interval $-\pi < x < \pi$ and periodic, with period 2π, on the entire x axis. Its Fourier series* (1), *with coefficients* (2) *and* (3), *converges to the mean value*

$$\frac{f(x+) + f(x-)}{2} \tag{4}$$

of the one-sided limits of f at each point x $(-\infty < x < \infty)$ where both of the one-sided derivatives $f'_R(x)$ and $f'_L(x)$ exist.

Note that if f is actually continuous at x, the quotient (4) becomes $f(x)$. Hence

$$f(x) = \frac{a_0}{2} + \sum_{n=1}^{\infty}(a_n \cos nx + b_n \sin nx)$$

at x, provided that both $f'_R(x)$ and $f'_L(x)$ exist.

The fact that f is piecewise continuous on $-\pi < x < \pi$ ensures that the integrals (2) and (3) always exist; and we begin our proof of the theorem by writing series (1) as (see Sec. 6)

$$\frac{1}{2\pi}\int_{-\pi}^{\pi} f(s)\,ds + \frac{1}{\pi}\sum_{n=1}^{\infty}\int_{-\pi}^{\pi} f(s)\cos n(s-x)\,ds,$$

with those coefficients incorporated into it. Then, if $S_N(x)$ denotes the partial sum consisting of the sum of the first $N+1$ ($N \geq 1$) terms of the series,

$$S_N(x) = \frac{1}{2\pi}\int_{-\pi}^{\pi} f(s)\,ds + \frac{1}{\pi}\sum_{n=1}^{N}\int_{-\pi}^{\pi} f(s)\cos n(s-x)\,ds. \tag{5}$$

Using the Dirichlet kernel (Sec. 11)

$$D_N(u) = \frac{1}{2} + \sum_{n=1}^{N}\cos nu,$$

we can put equation (5) in the form

$$S_N(x) = \frac{1}{\pi}\int_{-\pi}^{\pi} f(s)D_N(s-x)\,ds.$$

The periodicity of the integrand here allows us to change the interval of integration to any interval of length 2π without altering the value of the integral (see Problem 9, Sec. 13). Thus

$$S_N(x) = \frac{1}{\pi}\int_{x-\pi}^{x+\pi} f(s)D_N(s-x)\,ds, \tag{6}$$

where point x is at the center of the interval we have chosen. It now follows from equation (6) that

$$S_N(x) = \frac{1}{\pi}[I_N(x) + J_N(x)], \tag{7}$$

where

$$I_N(x) = \int_{x}^{x+\pi} f(s)D_N(s-x)\,ds \tag{8}$$

and

$$J_N(x) = \int_{x-\pi}^{x} f(s)D_N(s-x)\,ds. \tag{9}$$

If we replace the variable of integration s in integral (8) by the new variable $u = s - x$, that integral becomes

$$I_N(x) = \int_{0}^{\pi} f(x+u)D_N(u)\,du. \tag{10}$$

Since f is piecewise continuous on the fundamental interval $-\pi < x < \pi$ and also periodic, it is piecewise continuous on any bounded interval of the x axis. So, for a fixed value of x, the function $g(u) = f(x+u)$ in expression (10) is piecewise continuous on any bounded interval of the u axis and, in particular, on the interval $0 < u < \pi$. Let us assume that the right-hand derivative $f'_R(x)$ exists. After observing that

$$g(0+) = \lim_{\substack{u \to 0 \\ u>0}} g(u) = \lim_{\substack{u \to 0 \\ u>0}} f(x+u) = \lim_{\substack{v \to x \\ v>x}} f(v) = f(x+),$$

one can show that the right-hand derivative of g at $u = 0$ exists:

$$g'_R(0) = \lim_{\substack{u \to 0 \\ u>0}} \frac{g(u) - g(0+)}{u - 0} = \lim_{\substack{u \to 0 \\ u>0}} \frac{f(x+u) - f(x+)}{u}$$

$$= \lim_{\substack{v \to x \\ v>x}} \frac{f(v) - f(x+)}{v - x} = f'_R(x).$$

According to Lemma 2 in Sec. 11, then,

$$\lim_{N \to \infty} I_N(x) = \frac{\pi}{2} g(0+) = \frac{\pi}{2} f(x+). \tag{11}$$

If, on the other hand, we make the substitution $u = x - s$ in integral (9) and recall from our discussion in Sec. 11 that $D_N(u)$ is an even function of u, we find that

$$J_N(x) = \int_0^\pi f(x-u) D_N(u)\, du. \tag{12}$$

This time, we assume that the left-hand derivative $f'_L(x)$ exists; and we note that the function $g(u) = f(x - u)$ in expression (12) is piecewise continuous on the interval $0 < u < \pi$. Furthermore,

$$g(0+) = \lim_{\substack{u \to 0 \\ u>0}} g(x) = \lim_{\substack{u \to 0 \\ u>0}} f(x-u) = \lim_{\substack{v \to x \\ v<x}} f(v) = f(x-)$$

and

$$g'_R(0) = \lim_{\substack{u \to 0 \\ u>0}} \frac{g(u) - g(0+)}{u - 0} = \lim_{\substack{u \to 0 \\ u>0}} \frac{f(x-u) - f(x-)}{u}$$

$$= -\lim_{\substack{v \to x \\ v<x}} \frac{f(v) - f(x-)}{v - x} = -f'_L(x).$$

So once again by Lemma 2 in Sec. 11,

$$\lim_{N \to \infty} J_N(x) = \frac{\pi}{2} g(0+) = \frac{\pi}{2} f(x-). \tag{13}$$

Finally, we may conclude from equation (7) and limits (11) and (13) that

$$\lim_{N \to \infty} S_N(x) = \frac{f(x+) + f(x-)}{2};$$

and the theorem is proved.

This theorem is especially suited to functions f that are piecewise smooth on the fundamental interval $-\pi < x < \pi$. We recall from Sec. 9 that f is piecewise smooth if *both* f and f' are piecewise continuous.

Corollary. *Let f denote a function that is piecewise smooth on the interval $-\pi < x < \pi$, and let F denote the periodic extension, with period 2π, of f. At each point x $(-\infty < x < \infty)$, the Fourier series for f on $-\pi < x < \pi$ converges to the mean value of the one-sided limits of $F(x+)$ and $F(x-)$, namely*

$$\frac{F(x+) + F(x-)}{2}. \tag{14}$$

The proof of this corollary relies on the theorem in Sec. 9, which tells us that when f is piecewise smooth on $-\pi < x < \pi$, its one-sided derivatives, from the interior at the endpoints $x = \pm\pi$, exist everywhere in the closed interval $-\pi \le x \le \pi$. Hence if F denotes the periodic extension of f, with period 2π, the one-sided derivatives of F exist at each point x $(-\infty < x < \infty)$. According to the theorem just proved, then, the Fourier series for f on $-\pi < x < \pi$ converges everywhere to the mean value of the one-sided limits of F.

13. DISCUSSION OF THE THEOREM AND ITS COROLLARY

It should be emphasized that the conditions in the theorem in Sec. 12, as well as the corollary there, are only sufficient, and there is no claim that they are *necessary* conditions. More general conditions are given in a number of the references listed in the Bibliography. Indeed, there are functions that even become unbounded at certain points but nevertheless have valid Fourier series representations.[†]

The corollary in Sec. 12 will be adequate for most of the applications in this book, where the functions are generally piecewise smooth. We note that if f and F denote the functions in the corollary, then

$$F(x+) = f(x+) \qquad \text{and} \qquad F(x-) = f(x-) \qquad \text{when } -\pi < x < \pi.$$

Consequently, when $-\pi < x < \pi$, the corollary tells us that the Fourier series for f on the interval $-\pi < x < \pi$ converges to the number

$$\frac{f(x+) + f(x-)}{2}, \tag{1}$$

which becomes $f(x)$ if x is a point of continuity of f.

At the endpoints $x = \pm\pi$, however, the series converges to

$$\frac{f(-\pi+) + f(\pi-)}{2}. \tag{2}$$

To see that this is so, consider first the point $x = \pi$. Since

$$F(\pi+) = f(-\pi+) \qquad \text{and} \qquad F(\pi-) = f(\pi-),$$

as is evident from Fig. 8, the quotient

$$\frac{F(x+) + F(x-)}{2}$$

[†]See, for instance, the book by Tolstov (1976, pp. 91–94), which is listed in the Bibliography.

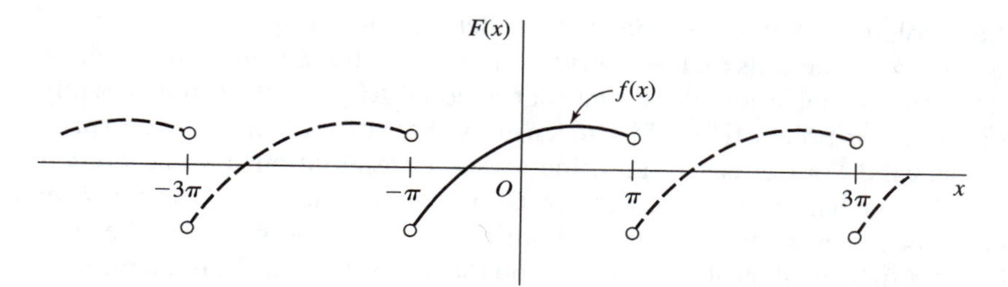

FIGURE 8

in the corollary becomes the quotient (2) when $x = \pi$. Because of the periodicity of the series, it also converges to the quotient (2) when $x = -\pi$.

EXAMPLE 1. In Example 1, Sec. 7, we obtained the Fourier series

$$\frac{\pi}{4} + \sum_{n=1}^{\infty} \left[\frac{(-1)^n - 1}{\pi n^2} \cos nx + \frac{(-1)^{n+1}}{n} \sin nx \right] \tag{3}$$

on the interval $-\pi < x < \pi$ for the function f defined by the equations

$$f(x) = \begin{cases} 0 & \text{when } -\pi < x \leq 0, \\ x & \text{when } \quad 0 < x < \pi. \end{cases}$$

Since

$$f'(x) = \begin{cases} 0 & \text{when } -\pi < x < 0, \\ 1 & \text{when } \quad 0 < x < \pi, \end{cases}$$

f is clearly piecewise smooth on the fundamental interval $-\pi < x < \pi$. In view of the continuity of f when $-\pi < x < \pi$, the series converges to $f(x)$ at each point x in that open interval. Since $f(-\pi +) = 0$ and $f(\pi -) = \pi$, it converges to $\pi/2$ at the endpoints $x = \pm\pi$. The series, in fact, converges to $\pi/2$ at each of the points $x = \pm\pi, \pm3\pi, \pm5\pi, \ldots$, as indicated in Fig. 5 (Sec. 7), where the sum of the series for all x is described graphically.

In particular, since series (3) converges to $\pi/2$ when $x = \pi$, we have the identity

$$\frac{\pi}{4} + \sum_{n=1}^{\infty} \frac{(-1)^n - 1}{\pi n^2} (-1)^n = \frac{\pi}{2},$$

which can be written

$$\sum_{n=1}^{\infty} \frac{1}{(2n-1)^2} = \frac{\pi^2}{8}.$$

This illustrates how Fourier series can sometimes be used to find the sums of convergent series encountered in calculus. Setting $x = 0$ in series (3) also yields this particular summation formula.

The corollary in Sec. 12 tells us that a function f in the space $C'_p(-\pi, \pi)$ of piecewise smooth functions on the interval $-\pi < x < \pi$ has a valid Fourier series representation on that interval, or one that is equal to $f(x)$ at all but possibly a finite number of points there. It also ensures that a function f in the space $C'_p(0, \pi)$ has valid Fourier cosine and sine series representations on the interval $0 < x < \pi$. This is because, according to Sec. 6, the cosine series for a function f on the interval $0 < x < \pi$ is the same as the Fourier series corresponding to the *even* extension of f on the interval $-\pi < x < \pi$ and the sine series for f on the interval $0 < x < \pi$ is the Fourier series for the *odd* extension of f. In view of the even periodic function represented by the cosine series, that series converges to $f(0+)$ at the point $x = 0$ and to $f(\pi-)$ at $x = \pi$. The sum of the sine series is, of course, zero when $x = 0$ and when $x = \pi$.

EXAMPLE 2. In Example 2, Sec. 3, we found the Fourier cosine series corresponding to the function $f(x) = \sin x$ on the interval $0 < x < \pi$:

$$\sin x \sim \frac{2}{\pi} - \frac{4}{\pi}\sum_{n=1}^{\infty}\frac{\cos 2nx}{4n^2 - 1}. \tag{4}$$

Since $\sin x$ is piecewise smooth on $0 < x < \pi$ and continuous on the closed interval $0 \le x \le \pi$, correspondence (4) is evidently an equality when $0 \le x \le \pi$.

Our final example here illustrates how the theorem in Sec. 12 can be useful when the corollary there fails to apply.

EXAMPLE 3. The odd function

$$f(x) = \sqrt[3]{x} \qquad (-\pi < x < \pi) \tag{5}$$

is piecewise continuous on the interval $-\pi < x < \pi$. But since

$$f'(x) = \frac{1}{3\sqrt[3]{x^2}}$$

when $x \ne 0$, it is clear that the one-sided limits $f'(0+)$ and $f'(0-)$ do not exist. Hence f is *not* piecewise smooth on $-\pi < x < \pi$, and the corollary in Sec. 12 does not apply.

If, however, F denotes the periodic extension, with period 2π, of the piecewise continuous function (5), the theorem in Sec. 12 can be applied to that extension. To be precise, since the one-sided derivatives of F exist everywhere in the interval $-\pi < x < \pi$ except at $x = 0$, we find that the Fourier series for F on $-\pi < x < \pi$ converges to $F(x)$ when $-\pi < x < 0$ and when $0 < x < \pi$. That series representation is also valid at $x = 0$ since F is odd and the series is actually a Fourier sine series on $0 < x < \pi$, which converges to zero when $x = 0$. Since $f(x) = F(x)$ when $-\pi < x < \pi$, we may conclude that the Fourier series for f on that interval is valid for all such x.

PROBLEMS

1. State why the Fourier sine series in Example 1, Sec. 5, for the function

$$f(x) = x \qquad (0 < x < \pi)$$

is a valid representation for x on the interval $-\pi < x < \pi$. Thus verify fully that the series converges for all x $(-\infty < x < \infty)$ to the function whose graph is shown in Fig. 4 (Sec. 5).

2. For each of the following functions, point out why its Fourier series on the interval $-\pi < x < \pi$ is convergent when $-\pi \leq x \leq \pi$, and state the sum of the series when $x = \pi$:
(*a*) the function

$$f(x) = \begin{cases} -\pi/2 & \text{when } -\pi < x < 0, \\ \pi/2 & \text{when } \quad 0 < x < \pi, \end{cases}$$

whose series was found in Problem 1, Sec. 7;
(*b*) the function

$$f(x) = e^{ax} \qquad (a \neq 0),$$

whose series was found in Problem 4, Sec. 7.
Answers: (*a*) sum = 0; (*b*) sum = $\cosh a\pi$.

3. By writing $x = 0$ and $x = \pi/2$ in the representation

$$\sin x = \frac{2}{\pi} - \frac{4}{\pi}\sum_{n=1}^{\infty}\frac{\cos 2nx}{4n^2 - 1} \qquad (0 \leq x \leq \pi),$$

established in Example 2, Sec. 13, obtain the following summations:

$$\sum_{n=1}^{\infty}\frac{1}{4n^2 - 1} = \frac{1}{2}, \qquad \sum_{n=1}^{\infty}\frac{(-1)^n}{4n^2 - 1} = \frac{1}{2} - \frac{\pi}{4}.$$

4. Point out why the Fourier series in Problem 7, Sec. 7, for the function

$$f(x) = \begin{cases} 0 & \text{when } -\pi \leq x \leq 0, \\ \sin x & \text{when } \quad 0 < x \leq \pi \end{cases}$$

converges to $f(x)$ everywhere in the interval $-\pi \leq x \leq \pi$.

5. State why the correspondence

$$x \sim \frac{\pi}{2} - \frac{4}{\pi}\sum_{n=1}^{\infty}\frac{\cos(2n-1)x}{(2n-1)^2} \qquad (0 < x < \pi),$$

obtained in Example 1, Sec. 3, is actually an equality on the closed interval $0 \leq x \leq \pi$. Thus show that

$$\sum_{n=1}^{\infty}\frac{1}{(2n-1)^2} = \frac{\pi^2}{8}.$$

(Compare with Example 1, Sec. 13.)

6. (*a*) Use the correspondence

$$x^2 \sim \frac{\pi^2}{3} + 4\sum_{n=1}^{\infty}\frac{(-1)^n}{n^2}\cos nx \qquad (0 < x < \pi),$$

found in Problem 3(a), Sec. 5, to show that

$$\sum_{n=1}^{\infty} \frac{(-1)^{n+1}}{n^2} = \frac{\pi^2}{12}, \qquad \sum_{n=1}^{\infty} \frac{1}{n^2} = \frac{\pi^2}{6}.$$

(b) By writing $x = \pi$ in the correspondence (Problem 6, Sec. 5)

$$x^4 \sim \frac{\pi^4}{5} + 8\sum_{n=1}^{\infty} (-1)^n \frac{(n\pi)^2 - 6}{n^4} \cos nx \qquad (0 < x < \pi)$$

and referring to the second summation obtained in part (a), show that

$$\sum_{n=1}^{\infty} \frac{1}{n^4} = \frac{\pi^4}{90}.$$

7. With the aid of the correspondence (Problem 6, Sec. 7)

$$\cos ax \sim \frac{2a \sin a\pi}{\pi} \left[\frac{1}{2a^2} + \sum_{n=1}^{\infty} \frac{(-1)^{n+1}}{n^2 - a^2} \cos nx \right] \qquad (-\pi < x < \pi),$$

where $a \neq 0, \pm 1, \pm 2, \ldots$, show that

$$\frac{a\pi}{\sin a\pi} = 1 + 2a^2 \sum_{n=1}^{\infty} \frac{(-1)^{n+1}}{n^2 - a^2} \qquad (a \neq 0, \pm 1, \pm 2, \ldots).$$

8. Without actually finding the Fourier series for the even function $f(x) = \sqrt[3]{x^2}$ on $-\pi < x < \pi$, point out how the theorem in Sec. 12 ensures the convergence of that series to $f(x)$ when $-\pi \leq x < 0$ and when $0 < x \leq \pi$ but not when $x = 0$.

9. Let f denote a function that is piecewise continuous on an interval $-c < x < c$ and periodic with period $2c$. Show that for any number a,

$$\int_{-c}^{c} f(x)\,dx = \int_{a-c}^{a+c} f(x)\,dx.$$

Suggestion: Write

$$\int_{-c}^{c} f(x)\,dx = \int_{-c}^{a+c} f(x)\,dx + \int_{a+c}^{c} f(s)\,ds$$

and then make the substitution $x = s - 2c$ in the second integral on the right-hand side of this equation.

14. CONVERGENCE ON OTHER INTERVALS

Section 8 began with a discussion of Fourier series corresponding to piecewise continuous functions f on arbitrary intervals $-c < x < c$. To treat the convergence of such series, we include here a few remarks about the function

$$g(s) = f\left(\frac{cs}{\pi}\right) \qquad (-\pi < s < \pi) \tag{1}$$

that was used in Sec. 8.

Let us write the function (1) as

$$g(s) = f(x) \qquad \text{where} \qquad x = \frac{cs}{\pi} \qquad (-\pi < s < \pi). \tag{2}$$

It is clear that the equation $x = cs/\pi$, or $s = \pi x/c$, establishes a one-to-one correspondence between points in the interval $-\pi < s < \pi$ and points in the interval $-c < x < c$. Suppose now that f is piecewise smooth on the interval $-c < x < c$ and that $f(x)$ at each point x where f is discontinuous is the mean value of the one-sided limits $f(x+)$ and $f(x-)$, as is the case when f is continuous at x.

One can see from equations (2) that if a specific point s_0 corresponds to a specific point x_0, then

$$g(s_0+) = f(x_0+), \qquad g(s_0-) = f(x_0-).$$

Since $f(x)$ is always the mean value of $f(x+)$ and $f(x-)$, it follows from these relations between one-sided limits that the number $g(s) = f(x)$ is always the mean value of $g(s+)$ and $g(s-)$. In particular, g is continuous at s when f is continuous at x. Since f is piecewise continuous on the interval $-c < x < c$, then, g is piecewise continuous on the interval $-\pi < s < \pi$. The derivative f' is also piecewise continuous, and a similar argument shows that g' is piecewise continuous. So g is piecewise smooth on the interval $-\pi < s < \pi$. According to the corollary in Sec. 12, the Fourier series for g on the interval $-\pi < s < \pi$ converges to $g(s)$ for each s in that interval. Moreover, since it was relation (1) that gave us the new correspondence for f on $-c < x < c$ in Sec. 8, the correspondence is, in fact, an equality. That is,

$$f(x) = \frac{a_0}{2} + \sum_{n=1}^{\infty} \left(a_n \cos \frac{n\pi x}{c} + b_n \sin \frac{n\pi x}{c} \right), \tag{3}$$

where

$$a_n = \frac{1}{c} \int_{-c}^{c} f(x) \cos \frac{n\pi x}{c} \, dx \qquad (n = 0, 1, 2, \ldots), \tag{4}$$

$$b_n = \frac{1}{c} \int_{-c}^{c} f(x) \sin \frac{n\pi x}{c} \, dx \qquad (n = 1, 2, \ldots). \tag{5}$$

We state this result as a theorem that is sufficient for our applications. The theorem and the one that follows it apply to any function f that has the following properties:

(*i*) The function f is piecewise smooth on the stated interval.
(*ii*) The value $f(x)$ at each point of discontinuity of f in that interval is the mean value of the one-sided limits $f(x+)$ and $f(x-)$.

Theorem 1. *Let f denote a function that has properties* (*i*) *and* (*ii*) *on an interval $-c < x < c$. The Fourier series representation* (3), *with coefficients* (4) *and* (5), *is valid for each x in that interval.*

Note that series (3) also represents the *periodic extension, with period* $2c$, of the function f. That is, it converges to a function $F(x)$ whose graph coincides with the graph of $f(x)$ on $-c < x < c$ and is repeated every $2c$ units along the x axis. The series has the expected sums at the endpoints $x = \pm c$. The sum at $x = c$ is, for instance, the mean value of $F(c+)$ and $F(c-)$.

If we restrict function (1) to the interval $0 < s < \pi$, Fourier cosine and sine series representations on $0 < x < c$ follow from representations on $0 < s < \pi$ that involve only cosines and sines, respectively.

Theorem 2. *Let f denote a function that has properties (i) and (ii) on an interval $0 < x < c$. The Fourier cosine series representation*

$$f(x) = \frac{a_0}{2} + \sum_{n=1}^{\infty} a_n \cos \frac{n\pi x}{c}, \tag{6}$$

with coefficients

$$a_n = \frac{2}{c} \int_0^c f(x) \cos \frac{n\pi x}{c}\, dx \qquad (n = 0, 1, 2, \ldots), \tag{7}$$

is valid for all x in that interval. The same is true of the Fourier sine series representation

$$f(x) = \sum_{n=1}^{\infty} b_n \sin \frac{n\pi x}{c}, \tag{8}$$

with coefficients

$$b_n = \frac{2}{c} \int_0^c f(x) \sin \frac{n\pi x}{c}\, dx \qquad (n = 1, 2, \ldots). \tag{9}$$

Series (6) represents, of course, the *even* periodic extension, with period $2c$, of f; and series (8) represents the *odd* periodic extension, with period $2c$, of f.

PROBLEMS

1. Use formulas (4) and (5), Sec. 14, as well as Theorem 1 in that section, to show that if

$$f(x) = \begin{cases} 0 & \text{when } -3 < x < 0, \\ 1 & \text{when } \quad 0 < x < 3, \end{cases}$$

and if $f(0) = \dfrac{1}{2}$, then

$$f(x) = \frac{1}{2} + \frac{2}{\pi} \sum_{n=1}^{\infty} \frac{1}{2n-1} \sin \frac{(2n-1)\pi x}{3} \qquad (-3 < x < 3).$$

Describe graphically the function that is represented by this series for all values of x $(-\infty < x < \infty)$.

2. Let f denote the function whose values are

$$f(x) = \begin{cases} 0 & \text{when } -2 < x < 1, \\ 1 & \text{when } \quad 1 < x < 2, \end{cases}$$

and

$$f(-2) = f(1) = f(2) = \frac{1}{2}.$$

Use formulas (4) and (5) in Sec. 14, together with Theorem 1 there, to show that

$$f(x) = \frac{1}{4} - \frac{1}{\pi}\sum_{n=1}^{\infty}\frac{1}{n}\left[\sin\frac{n\pi}{2}\cos\frac{n\pi x}{2} + \left(\cos n\pi - \cos\frac{n\pi}{2}\right)\sin\frac{n\pi x}{2}\right]$$

for each x in the closed interval $-2 \le x \le 2$.

3. Let $M(c, t)$ denote the square wave (Fig. 9) defined by the equations

$$M(c,t) = \begin{cases} 1 & \text{when } 0 < t < c, \\ -1 & \text{when } c < t < 2c, \end{cases}$$

and $M(c, t + 2c) = M(c, t)$ when $t > 0$. Show that

$$M(c,t) = \frac{4}{\pi}\sum_{n=1}^{\infty}\frac{1}{2n-1}\sin\frac{(2n-1)\pi t}{c} \qquad (t \neq c, 2c, 3c, \ldots).$$

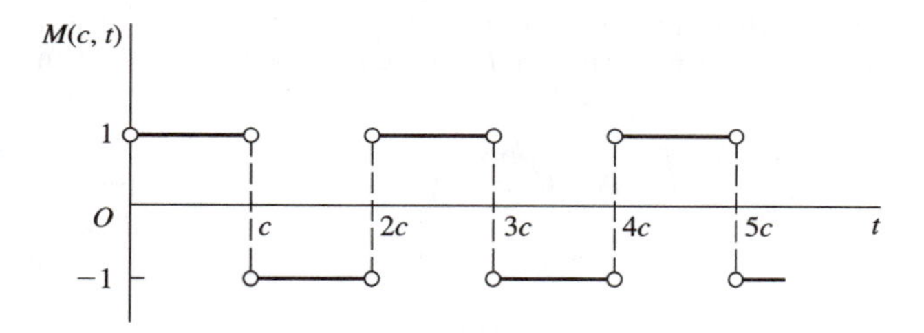

FIGURE 9

4. Let F denote the periodic function, of period c, where

$$F(x) = \begin{cases} \dfrac{c}{4} - x & \text{when } 0 \le x \le \dfrac{c}{2}, \\ x - \dfrac{3c}{4} & \text{when } \dfrac{c}{2} < x \le c. \end{cases}$$

(*a*) Describe the function $F(x)$ graphically, and show that it is, in fact, the even periodic extension, with period c, of the function

$$f(x) = \frac{c}{4} - x \qquad \left(0 \le x \le \frac{c}{2}\right).$$

(*b*) Use the result in part (*a*) and the Fourier cosine series correspondence found in Problem 7, Sec. 8, to show that

$$F(x) = \frac{2c}{\pi^2}\sum_{n=1}^{\infty}\frac{1}{(2n-1)^2}\cos\frac{(4n-2)\pi x}{c} \qquad (-\infty < x < \infty).$$

5. Let f denote the periodic function, of period 2, where

$$f(x) = \begin{cases} \cos \pi x & \text{when } 0 < x < 1, \\ 0 & \text{when } 1 < x < 2 \end{cases}$$

and where

$$f(0) = \frac{1}{2} \quad \text{and} \quad f(1) = -\frac{1}{2}.$$

By referring to the correspondence

$$\cos \pi x \sim \frac{8}{\pi} \sum_{n=1}^{\infty} \frac{n}{4n^2 - 1} \sin 2n\pi x \qquad (0 < x < 1),$$

obtained in Problem 7, Sec. 8, show that

$$f(x) = \frac{1}{2} \cos \pi x + \frac{4}{\pi} \sum_{n=1}^{\infty} \frac{n}{4n^2 - 1} \sin 2n\pi x \qquad (-\infty < x < \infty).$$

6. Suppose that a function f is piecewise smooth on an interval $0 < x < c$, and let F denote this extension of f on the interval $0 < x < 2c$:

$$F(x) = \begin{cases} f(x) & \text{when } 0 < x < c, \\ f(2c - x) & \text{when } c < x < 2c. \end{cases}$$

[The graph of $y = F(x)$ is evidently symmetric with respect to the line $x = c$.] Show that the coefficients B_n in the Fourier sine series for F on the interval $0 < x < 2c$ can be written

$$B_n = \frac{1 - (-1)^n}{c} \int_0^c f(x) \sin \frac{n\pi x}{2c}\, dx \qquad (n = 1, 2, \ldots).$$

Thus show that

$$f(x) = \sum_{n=1}^{\infty} b_n \sin \frac{(2n-1)\pi x}{2c},$$

where

$$b_n = \frac{2}{c} \int_0^c f(x) \sin \frac{(2n-1)\pi x}{2c}\, dx \qquad (n = 1, 2, \ldots),$$

for each point x $(0 < x < c)$ at which f is continuous.

Suggestion: Write

$$B_n = \frac{1}{c} \left[\int_0^c f(x) \sin \frac{n\pi x}{2c}\, dx + \int_c^{2c} f(2c - s) \sin \frac{n\pi s}{2c}\, ds \right]$$

and make the substitution $x = 2c - s$ in the second of these integrals.

7. Use the result in Problem 6 to establish the representation

$$x = \frac{8c}{\pi^2} \sum_{n=1}^{\infty} \frac{(-1)^{n+1}}{(2n-1)^2} \sin \frac{(2n-1)\pi x}{2c} \qquad (-c \le x \le c).$$

8. After writing the Fourier series representation (3), Sec. 14, as

$$f(x) = \frac{a_0}{2} + \lim_{N \to \infty} \sum_{n=1}^{N} \left(a_n \cos \frac{n\pi x}{c} + b_n \sin \frac{n\pi x}{c} \right),$$

use the exponential forms[†]

$$\cos\theta = \frac{e^{i\theta} + e^{-i\theta}}{2}, \qquad \sin\theta = \frac{e^{i\theta} - e^{-i\theta}}{2i}$$

of the cosine and sine functions to put that representation in exponential form:

$$f(x) = \lim_{N\to\infty} \sum_{n=-N}^{N} A_n \exp\left(i\frac{n\pi x}{c}\right),$$

where

$$A_0 = \frac{a_0}{2}, \qquad A_n = \frac{a_n - ib_n}{2}, \qquad A_{-n} = \frac{a_n + ib_n}{2}. \qquad (n = 1, 2, \ldots).$$

Then use expressions (4) and (5), Sec. 14, for the coefficients a_n and b_n to obtain the single formula

$$A_n = \frac{1}{2c}\int_{-c}^{c} f(x)\exp\left(-i\frac{n\pi x}{c}\right)dx \qquad (n = 0, \pm 1, \pm 2, \ldots).$$

15. A LEMMA

We prove here an important lemma to be used in Sec. 16. That section, regarding the absolute and uniform convergence of Fourier series, and Secs. 17 and 18, dealing with differentiation and integration of such series, will be used only occasionally later on and will be specifically cited as needed. Hence the reader may at this time pass directly to Chap. 3 without serious disruption.

For convenience, we treat only Fourier series for which the fundamental interval is $-\pi < x < \pi$. Adaptations of our results to series on any fundamental interval $-c < x < c$ can be made by the method used in Sec. 8.

Lemma. *Let f denote a function such that*

(i) *f is continuous on the interval $-\pi \le x \le \pi$;*
(ii) *$f(-\pi) = f(\pi)$;*
(iii) *its derivative f' is piecewise continuous on the interval $-\pi < x < \pi$.*

If a_n and b_n are the Fourier coefficients

$$a_n = \frac{1}{\pi}\int_{-\pi}^{\pi} f(x)\cos nx\,dx, \qquad b_n = \frac{1}{\pi}\int_{-\pi}^{\pi} f(x)\sin nx\,dx \tag{1}$$

for f, the series

$$\sum_{n=1}^{\infty}\sqrt{a_n^2 + b_n^2} \tag{2}$$

converges.

[†]For background on these forms and an introduction to series and integrals involving complex-valued functions, see the authors' book (2004), listed in the Bibliography.

The class of functions satisfying conditions (*i*) through (*iii*) here is, of course, a subspace of the space of piecewise smooth functions on the interval $-\pi < x < \pi$.

We begin the proof of the lemma with the observation that the Fourier coefficients

$$(3) \qquad \alpha_n = \frac{1}{\pi}\int_{-\pi}^{\pi} f'(x)\cos nx\,dx, \qquad \beta_n = \frac{1}{\pi}\int_{-\pi}^{\pi} f'(x)\sin nx\,dx$$

for f' exist because of the piecewise continuity of f'. Note that

$$\alpha_0 = \frac{1}{\pi}\int_{-\pi}^{\pi} f'(x)\,dx = \frac{1}{\pi}[f(\pi) - f(-\pi)] = 0.$$

Also, since f is continuous and $f(-\pi) = f(\pi)$, integration by parts reveals that when $n = 1, 2, \ldots,$

$$\begin{aligned}\alpha_n &= \frac{1}{\pi}\int_{-\pi}^{\pi} (\cos nx)\, f'(x)\,dx\\ &= \frac{1}{\pi}\left\{[(\cos nx)\, f(x)]_{-\pi}^{\pi} + n\int_{-\pi}^{\pi} f(x)\sin nx\,dx\right\} = nb_n.\end{aligned}$$

Likewise,

$$\begin{aligned}\beta_n &= \frac{1}{\pi}\int_{-\pi}^{\pi} (\sin nx)\, f'(x)\,dx\\ &= \frac{1}{\pi}\left\{[(\sin nx)\, f(x)]_{-\pi}^{\pi} - n\int_{-\pi}^{\pi} f(x)\cos nx\,dx\right\} = -na_n;\end{aligned}$$

and we find that

$$(4) \qquad a_n = -\frac{\beta_n}{n}, \qquad b_n = \frac{\alpha_n}{n} \qquad (n = 1, 2, \ldots).$$

In view of relations (4), the sum s_N of the first N terms of the infinite series (2) becomes

$$(5) \qquad s_N = \sum_{n=1}^{N} \sqrt{a_n^2 + b_n^2} = \sum_{n=1}^{N} \frac{1}{n}\sqrt{\alpha_n^2 + \beta_n^2}.$$

Cauchy's inequality

$$\left(\sum_{n=1}^{N} p_n q_n\right)^2 \le \left(\sum_{n=1}^{N} p_n^2\right)\left(\sum_{n=1}^{N} q_n^2\right),$$

which applies to any two sets of real numbers p_n $(n = 1, 2, \ldots, N)$ and q_n $(n = 1, 2, \ldots, N)$ (see Problem 6, Sec. 18, for a derivation), can now be used to write

$$(6) \qquad s_N^2 \le \left(\sum_{n=1}^{N} \frac{1}{n^2}\right)\left[\sum_{n=1}^{N} (\alpha_n^2 + \beta_n^2)\right] \qquad (N = 1, 2, \ldots).$$

The sequence of sums

$$\sum_{n=1}^{N} \frac{1}{n^2} \qquad (N = 1, 2, \ldots)$$

here is clearly bounded since each sum is a partial sum of the convergent series whose terms are $1/n^2$ [see Problem 6(*a*), Sec. 13]. The sequence

$$\sum_{n=1}^{N} \left(\alpha_n^2 + \beta_n^2\right) \qquad (N = 1, 2, \ldots)$$

is also bounded since α_n $(n = 0, 1, 2, \ldots)$ and β_n $(n = 1, 2, \ldots)$ are the Fourier coefficients for f' on the interval $-\pi < x < \pi$ and must, therefore, satisfy Bessel's inequality:

$$\sum_{n=1}^{N} \left(\alpha_n^2 + \beta_n^2\right) \le \frac{1}{\pi} \int_{-\pi}^{\pi} [f'(x)]^2 \, dx \qquad (N = 1, 2, \ldots).$$

[See Problem 4(*b*), Sec. 11.] It now follows from inequality (6) that the sequence s_N^2 $(N = 1, 2, \ldots)$ is both bounded and nondecreasing. Hence it converges, and this means that the sequence s_N $(N = 1, 2, \ldots)$ converges. Thus series (2) converges.

16. ABSOLUTE AND UNIFORM CONVERGENCE OF FOURIER SERIES

We turn now to the absolute and uniform convergence of Fourier series. We begin by recalling some facts about uniformly convergent series of functions.[†]

Let $s(x)$ denote the sum of an infinite series of functions $f_n(x)$, where the series is convergent for all x in some interval $a \le x \le b$. Thus

$$(1) \qquad s(x) = \sum_{n=1}^{\infty} f_n(x) = \lim_{N \to \infty} s_N(x) \qquad (a \le x \le b),$$

where $s_N(x)$ is the partial sum consisting of the sum of the first N terms of the series. The series converges *uniformly* with respect to x if the absolute value of its remainder $r_N(x) = s(x) - s_N(x)$ can be made arbitrarily small for all x in the interval by taking N sufficiently large; that is, for each positive number ε, there exists a positive integer N_ε, *independent of* x, such that

$$(2) \qquad |s(x) - s_N(x)| < \varepsilon \quad \text{whenever} \quad N > N_\varepsilon \qquad (a \le x \le b).$$

A sufficient condition for uniform convergence is given by the *Weierstrass M-test*. Namely, if there is a convergent series

$$(3) \qquad \sum_{n=1}^{\infty} M_n$$

[†]See, for instance, the book by Kaplan (2003, chap. 6) or the one by Taylor and Mann (1983, chap. 20), both listed in the Bibliography.

of positive constants such that

$$|f_n(x)| \leq M_n \qquad (a \leq x \leq b) \tag{4}$$

for each n, then series (1) is uniformly convergent on the stated interval.

We include here a few properties of uniformly convergent series that are often useful. If the functions f_n are continuous and if series (1) is uniformly convergent, then the sum $s(x)$ of that series is a continuous function. Also, the series can be integrated term by term over the interval $a \leq x \leq b$ to give the integral of $s(x)$ from $x = a$ to $x = b$. If the functions f_n and their derivatives f_n' are continuous, if series (1) converges, and if the series whose terms are $f_n'(x)$ is uniformly convergent, then $s'(x)$ is found by differentiating series (1) term by term.

Theorem. *Let f denote a function such that*

(i) f is continuous on the interval $-\pi \leq x \leq \pi$;

(ii) $f(-\pi) = f(\pi)$;

(iii) its derivative f' is piecewise continuous on the interval $-\pi < x < \pi$.

The Fourier series

$$\frac{a_0}{2} + \sum_{n=1}^{\infty}(a_n \cos nx + b_n \sin nx) \tag{5}$$

for f, with coefficients

$$a_n = \frac{1}{\pi}\int_{-\pi}^{\pi} f(x)\cos nx\,dx, \qquad b_n = \frac{1}{\pi}\int_{-\pi}^{\pi} f(x)\sin nx\,dx, \tag{6}$$

converges absolutely and uniformly to $f(x)$ on the interval $-\pi \leq x \leq \pi$.

To prove this, we first note that the conditions on f ensure the continuity of the periodic extension of f for all x. Hence it follows from the corollary in Sec. 12 that series (5) converges to $f(x)$ everywhere in the interval $-\pi \leq x \leq \pi$. Observe how it follows from the inequalities

$$|a_n| \leq \sqrt{a_n^2 + b_n^2} \quad \text{and} \quad |b_n| \leq \sqrt{a_n^2 + b_n^2}$$

that

$$|a_n \cos nx + b_n \sin nx| \leq |a_n| + |b_n| \leq 2\sqrt{a_n^2 + b_n^2} \qquad (n = 1, 2, \ldots).$$

Since the series

$$\sum_{n=1}^{\infty} \sqrt{a_n^2 + b_n^2}$$

converges, according to the lemma in Sec. 15, the comparison test and the Weierstrass M-test thus apply to show that the convergence of series (5) is absolute and uniform on the interval $-\pi \leq x \leq \pi$, as stated.

Modifications of the statements in both the lemma in Sec. 15 and the above theorem are apparent. For instance, it follows from the theorem that the Fourier cosine series on $0 < x < \pi$ for a function f that is continuous on the closed interval $0 \le x \le \pi$ converges absolutely and uniformly to $f(x)$ when $0 \le x \le \pi$ if f' is piecewise continuous on $0 < x < \pi$. For the sine series, however, the additional conditions $f(0) = f(\pi) = 0$ are needed.

Since a uniformly convergent series of continuous functions always converges to a continuous function, a Fourier series for a function f cannot converge uniformly on an interval that contains a point at which f is discontinuous. Hence the continuity of f, assumed in the theorem, is necessary for the series there to converge uniformly.

Suppose that x_0 is a point at which a piecewise smooth function f is discontinuous. The nature of the deviation near x_0 of the values of the partial sums of a Fourier series for f from the values of f is commonly referred to as the *Gibbs phenomenon* and is illustrated below.†

EXAMPLE. Consider the piecewise smooth function defined by the equations

$$f(x) = \begin{cases} -\pi/2 & \text{when} \quad -\pi < x < 0, \\ \pi/2 & \text{when} \quad 0 < x < \pi, \end{cases}$$

and $f(0) = 0$. According to Problem 1, Sec. 7, and Theorem 1 in Sec. 14, the Fourier (sine) series

$$2\sum_{n=1}^{\infty} \frac{\sin(2n-1)x}{2n-1} \qquad (-\pi < x < \pi)$$

for f converges to $f(x)$ everywhere in the interval $-\pi < x < \pi$.

Let $S_N(x)$ denote the sum of the first N terms of this series. The sequence $S_N(x)$ $(N=1, 2, \ldots)$ thus converges to $f(x)$ when $-\pi < x < \pi$. In particular, it converges to the number $\pi/2 = 1.57\cdots$ when $0 < x < \pi$. But, as shown in Problem 7, Sec. 18, there is a fixed number $\sigma = 1.85\cdots$ such that

$$S_N\left(\frac{\pi}{2N}\right)$$

tends to σ. See Fig. 10, which indicates how "spikes" in the graphs of the partial sums $y = S_N(x)$, moving to the left as N increases, are formed, their tips tending to the point σ on the y axis. The behavior of the partial sums is similar on the interval $-\pi < x < 0$.

This illustrates that special care must be taken when a function is approximated by a partial sum of its Fourier series near a point of discontinuity.

†For a detailed analysis of this phenomenon, see the book by Carslaw (1952, chap. 9), which is listed in the Bibliography.

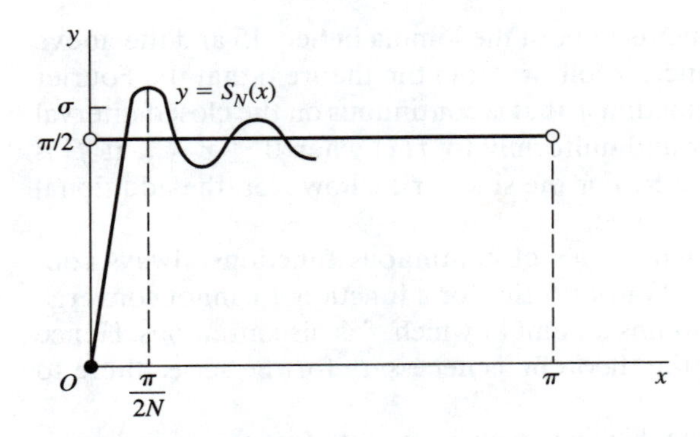

FIGURE 10

17. DIFFERENTIATION OF FOURIER SERIES

Not all Fourier series are differentiable, as Example 1 illustrates.

EXAMPLE 1. According to Theorem 1 in Sec. 14, the Fourier series in Example 3, Sec. 7, for the function $f(x) = x\ (-\pi < x < \pi)$ converges to $f(x)$ at each point in the interval $-\pi < x < \pi$:

$$x = 2\sum_{n=1}^{\infty} \frac{(-1)^{n+1}}{n} \sin nx \qquad (-\pi < x < \pi). \tag{1}$$

But the differentiated series

$$2\sum_{n=1}^{\infty} (-1)^{n+1} \cos nx$$

does not converge since its nth term fails to approach zero as n tends to infinity.

Sufficient conditions for differentiability can be stated as follows, where the conditions on f are the same as those in the theorem in Sec. 16, as well as in the lemma in Sec. 15.

Theorem. *Let f denote a function such that*

(i) f is continuous on the interval $-\pi \le x \le \pi$;

(ii) $f(-\pi) = f(\pi)$;

(iii) its derivative f' is piecewise continuous on the interval $-\pi < x < \pi$.

The Fourier series representation

$$f(x) = \frac{a_0}{2} + \sum_{n=1}^{\infty} (a_n \cos nx + b_n \sin nx) \qquad (-\pi \le x \le \pi), \tag{2}$$

where

$$a_n = \frac{1}{\pi}\int_{-\pi}^{\pi} f(x)\cos nx\,dx, \qquad b_n = \frac{1}{\pi}\int_{-\pi}^{\pi} f(x)\sin nx\,dx,$$

is differentiable at each point x in the interval $-\pi < x < \pi$ *at which the second-order derivative* f'' *exists:*

$$f'(x) = \sum_{n=1}^{\infty} n(-a_n \sin nx + b_n \cos nx). \tag{3}$$

Our proof of this theorem is especially brief. To start, we consider a point x $(-\pi < x < \pi)$ at which f'' exists; and we note that f' is therefore continuous at x. Hence an application of the Fourier theorem in Sec. 12 to the function f' shows that

$$f'(x) = \frac{\alpha_0}{2} + \sum_{n=1}^{\infty}(\alpha_n \cos nx + \beta_n \sin nx), \tag{4}$$

where

$$\alpha_n = \frac{1}{\pi}\int_{-\pi}^{\pi} f'(x)\cos nx\,dx, \qquad \beta_n = \frac{1}{\pi}\int_{-\pi}^{\pi} f'(x)\sin nx\,dx.$$

But since f and f' satisfy all the conditions stated in the lemma in Sec. 15, we know from the proof there that

$$\alpha_0 = 0, \qquad \alpha_n = nb_n, \qquad \beta_n = -na_n \qquad (n = 1, 2, \ldots). \tag{5}$$

When these substitutions are made, equation (4) takes the form (3); and the proof is complete.

At a point x where $f''(x)$ does not exist, but where f' has one-sided derivatives, differentiation is still valid in the sense that the series in equation (3) converges to the mean of the values $f'(x+)$ and $f'(x-)$. This is also true for the periodic extension of f.

The theorem applies, with obvious changes, to other Fourier series. For instance, if f is continuous when $0 \le x \le \pi$ and f' is piecewise continuous on the interval $0 < x < \pi$, then the Fourier cosine series for f on $0 < x < \pi$ is differentiable at each point x $(0 < x < \pi)$ where $f''(x)$ exists.

EXAMPLE 2. We know from Problem 5, Sec. 7, and the corollary in Sec. 12 that when $a \neq 0$,

$$\cosh ax = \frac{\sinh a\pi}{a\pi}\left[1 + 2a^2 \sum_{n=1}^{\infty} \frac{(-1)^n}{a^2+n^2}\cos nx\right]$$

on the closed interval $-\pi \le x \le \pi$. Inasmuch as the hypothesis in the theorem here is satisfied when $f(x) = \cosh ax$, it follows that

$$a \sinh ax = \frac{2a \sinh a\pi}{\pi} \sum_{n=1}^{\infty} \frac{(-1)^n}{a^2+n^2}(-n \sin nx)$$

on the interval $-\pi < x < \pi$. That is, when $a \neq 0$,

$$\sinh ax = \frac{2\sinh a\pi}{\pi}\sum_{n=1}^{\infty}(-1)^{n+1}\frac{n}{a^2+n^2}\sin nx \qquad (-\pi < x < \pi). \tag{6}$$

Note that equation (6) is, in fact, valid when the condition $a \neq 0$ is dropped.

18. INTEGRATION OF FOURIER SERIES

Integration of a Fourier series is possible under much more general conditions than those for differentiation. This is to be expected because an integration of the series in the correspondence

$$f(x) \sim \frac{a_0}{2} + \sum_{n=1}^{\infty}(a_n \cos nx + b_n \sin nx) \qquad (-\pi < x < \pi), \tag{1}$$

where

$$a_n = \frac{1}{\pi}\int_{-\pi}^{\pi} f(x)\cos nx\,dx, \qquad b_n = \frac{1}{\pi}\int_{-\pi}^{\pi} f(x)\sin nx\,dx, \tag{2}$$

introduces a factor n in the denominator of the general term. In the following theorem, it is not even essential that the original series converge in order that the integrated series converge to the integral of the function.

Theorem. *Let f be a function that is piecewise continuous on the interval $-\pi < x < \pi$. Regardless of whether series* (1) *converges or not, the following equation is valid when $-\pi \le x \le \pi$:*

$$\int_{-\pi}^{x} f(s)\,ds = \frac{a_0}{2}(x+\pi) + \sum_{n=1}^{\infty}\frac{1}{n}\{a_n \sin nx - b_n[\cos nx + (-1)^{n+1}]\}. \tag{3}$$

Series (3) is, of course, obtained by replacing x by s in series (1) and then integrating term by term from $s = -\pi$ to $s = x$. Observe that if $a_0 \neq 0$, the first term on the right in equation (3) is not of the type encountered in a Fourier series. Hence integrating a Fourier series does not always yield a Fourier series.

Our proof starts with the fact that since f is piecewise continuous, the function

$$F(x) = \int_{-\pi}^{x} f(s)\,ds - \frac{a_0}{2}x \qquad (-\pi \le x \le \pi) \tag{4}$$

is continuous; moreover,

$$F'(x) = f(x) - \frac{a_0}{2} \qquad (-\pi < x < \pi), \tag{5}$$

except at points where f is discontinuous. Hence F' is piecewise continuous on the interval $-\pi < x < \pi$. Since F is piecewise smooth, then, it follows from Theorem 1 in Sec. 14 that

$$F(x) = \frac{A_0}{2} + \sum_{n=1}^{\infty}(A_n \cos nx + B_n \sin nx) \qquad (-\pi < x < \pi), \tag{6}$$

where

$$A_n = \frac{1}{\pi}\int_{-\pi}^{\pi} F(x)\cos nx\,dx, \qquad B_n = \frac{1}{\pi}\int_{-\pi}^{\pi} F(x)\sin nx\,dx. \tag{7}$$

We note from expression (4) and the first of expressions (2) when $n = 0$ that

$$F(-\pi) = \frac{a_0}{2}\pi \quad \text{and} \quad F(\pi) = \int_{-\pi}^{\pi} f(s)\,ds - \frac{a_0}{2}\pi = a_0\pi - \frac{a_0}{2}\pi = \frac{a_0}{2}\pi; \tag{8}$$

hence

$$F(-\pi) = F(\pi). \tag{9}$$

This shows that representation (6) is also valid at the endpoints of the open interval $-\pi < x < \pi$ (see Sec. 13) and, therefore, at each point of the closed interval $-\pi \le x \le \pi$.

Let us now write the coefficients A_n and B_n in terms of a_n and b_n. When $n \ge 1$, we may integrate integrals (7) by parts, using the fact that F is continuous and F' is piecewise continuous. Thus

$$\begin{aligned} A_n &= \frac{1}{\pi}\left\{\left[F(x)\frac{\sin nx}{n}\right]_{-\pi}^{\pi} - \int_{-\pi}^{\pi} \frac{\sin nx}{n} F'(x)\,dx\right\} \\ &= -\frac{1}{n\pi}\int_{-\pi}^{\pi} F'(x)\sin nx\,dx; \end{aligned}$$

and, in view of expression (5) for $F'(x)$, we have

$$\begin{aligned} A_n &= -\frac{1}{n\pi}\int_{-\pi}^{\pi}\left[f(x) - \frac{a_0}{2}\right]\sin nx\,dx \\ &= -\frac{1}{n}\cdot\frac{1}{\pi}\int_{-\pi}^{\pi} f(x)\sin nx\,dx + \frac{a_0}{2n\pi}\int_{-\pi}^{\pi}\sin nx\,dx = -\frac{b_n}{n}. \end{aligned}$$

Likewise, keeping relation (9) in mind, we find that

$$B_n = \frac{1}{n\pi}\int_{-\pi}^{\pi} F'(x)\cos nx\,dx;$$

and, using expression (5) once again, we can see that

$$B_n = \frac{1}{n}\cdot\frac{1}{\pi}\int_{-\pi}^{\pi} f(x)\cos nx\,dx - \frac{a_0}{2n\pi}\int_{-\pi}^{\pi}\cos nx\,dx = \frac{a_n}{n}.$$

As for A_0, from the final value for $F(\pi)$ shown in the second of relations (8) and the fact that representation (6) is valid when $x = \pi$, we know that

$$\frac{a_0}{2}\pi = \frac{A_0}{2} + \sum_{n=1}^{\infty} A_n(-1)^n.$$

So, by solving for A_0 here and then using and relation $A_n = -b_n/n$ found above, we arrive at

$$A_0 = a_0\pi - 2\sum_{n=1}^{\infty} A_n(-1)^n = a_0\pi - 2\sum_{n=1}^{\infty}\frac{(-1)^{n+1}}{n}b_n.$$

With these expressions for A_n and B_n, including A_0, representation (6) takes the form

$$F(x) = \frac{a_0}{2}\pi + \sum_{n=1}^{\infty} \frac{1}{n}\{a_n \sin nx - b_n[\cos nx + (-1)^{n+1}]\}.$$

Finally, if we use expression (4) to substitute for $F(x)$ here, we arrive at the desired result (3).

The theorem can be written for the integral from x_0 to x, where $-\pi \le x_0 \le \pi$ and $-\pi \le x \le \pi$, by noting that

$$\int_{x_0}^{x} f(s)\,ds = \int_{-\pi}^{x} f(s)\,ds - \int_{-\pi}^{x_0} f(s)\,ds.$$

PROBLEMS

1. Show that the function

$$f(x) = \begin{cases} 0 & \text{when } -\pi \le x \le 0, \\ \sin x & \text{when } \quad 0 < x \le \pi, \end{cases}$$

satisfies all the conditions in the theorem in Sec. 16. Then, with the aid of the Weierstrass M-test (Sec. 16), verify that the Fourier series

$$\frac{1}{\pi} + \frac{1}{2}\sin x - \frac{2}{\pi}\sum_{n=1}^{\infty} \frac{\cos 2nx}{4n^2 - 1} \qquad (-\pi < x < \pi)$$

for f, found in Problem 7, Sec. 7, converges uniformly on the interval $-\pi \le x \le \pi$, as the theorem in Sec. 16 tells us. Also, state why this series is differentiable in the interval $-\pi < x < \pi$, except at the point $x = 0$, and describe graphically the function that is represented by the differentiated series for all x.

2. We know from Example 1, Sec. 3, that the series

$$\frac{\pi}{2} - \frac{4}{\pi}\sum_{n=1}^{\infty} \frac{\cos(2n-1)x}{(2n-1)^2}$$

is the Fourier cosine series for the function $f(x) = x$ on the interval $0 < x < \pi$. Differentiate this series term by term to obtain a representation for the function $f'(x) = 1$ on that interval. State why the procedure is reliable here.

3. State the theorem in Sec. 17 as it applies to Fourier sine series. Point out, in particular, why the conditions $f(0) = f(\pi) = 0$ are present in this case.

4. Let a_n and b_n denote the Fourier coefficients in the lemma in Sec. 15. Using the fact that the coefficients in the Fourier series for a function in $C_p(-\pi, \pi)$ always tend to zero as n tends to infinity (Problem 5, Sec. 11), show why

$$\lim_{n\to\infty} na_n = 0 \qquad \text{and} \qquad \lim_{n\to\infty} nb_n = 0.$$

5. Integrate from $s = 0$ to $s = x$ $(-\pi \le x \le \pi)$ the Fourier series

$$2\sum_{n=1}^{\infty} \frac{(-1)^{n+1}}{n} \sin ns$$

in Example 1, Sec. 17, and the one

$$2\sum_{n=1}^{\infty}\frac{\sin(2n-1)s}{2n-1}$$

appearing in the example in Sec. 16. In each case, describe graphically the function that is represented by the new series.

6. Let p_n $(n = 1, 2, \dots, N)$ and q_n $(n = 1, 2, \dots, N)$ denote real numbers, where at least one of the numbers p_n, say p_m, is nonzero. By writing the quadratic equation

$$x^2\sum_{n=1}^{N}p_n^2 + 2x\sum_{n=1}^{N}p_nq_n + \sum_{n=1}^{N}q_n^2 = 0$$

in the form

$$\sum_{n=1}^{N}(p_nx + q_n)^2 = 0,$$

show that the number $x_0 = -q_m/p_m$ is the only possible real root. Conclude that since *there cannot be two distinct real roots,* the discriminant

$$\left(2\sum_{n=1}^{N}p_nq_n\right)^2 - 4\left(\sum_{n=1}^{N}p_n^2\right)\left(\sum_{n=1}^{N}q_n^2\right)$$

of this quadratic equation is negative or zero. Thus derive Cauchy's inequality (Sec. 15)

$$\left(\sum_{n=1}^{N}p_nq_n\right)^2 \le \left(\sum_{n=1}^{N}p_n^2\right)\left(\sum_{n=1}^{N}q_n^2\right),$$

which is clearly valid even if all the numbers p_n are zero.

7. As in the example in Sec. 16, let $S_N(x)$ denote the partial sum consisting of the sum of the first N terms of the Fourier series

$$2\sum_{n=1}^{\infty}\frac{\sin(2n-1)x}{2n-1} \qquad (-\pi < x < \pi)$$

for the function f there.

(*a*) By writing $A = x$ and $B = (2n-1)x$ in the trigonometric identity

$$2\sin A\cos B = \sin(A+B) + \sin(A-B)$$

and then summing each side of the resulting equation from $n = 1$ to $n = N$, derive the summation formula

$$2\sum_{n=1}^{N}\cos(2n-1)x = \frac{\sin 2Nx}{\sin x} \qquad (x \neq 0, \pm\pi, \pm 2\pi, \dots).$$

Use this formula to write the derivative of $S_N(x)$ on the interval $0 < x < \pi$ as a simple quotient:

$$S_N'(x) = \frac{\sin 2Nx}{\sin x} \qquad (0 < x < \pi).$$

(*b*) With the aid of the expression for the derivative $S_N'(x)$ in part (*a*), show that the first extremum of $S_N(x)$ in the interval $0 < x < \pi$ is a relative maximum occurring when $x = \pi/(2N)$.

(*c*) By integrating each side of the summation formula in part (*a*) from $x = 0$ to $x = \pi/(2N)$, show that

$$S_N\left(\frac{\pi}{2N}\right) = I_1 + I_2$$

where

$$I_1 = \int_0^{\pi/(2N)} \frac{x - \sin x}{x \sin x} \sin 2Nx \, dx \qquad \text{and} \qquad I_2 = \int_0^{\pi/(2N)} \frac{\sin 2Nx}{x} \, dx.$$

Verify that the integrands of these two integrals are piecewise continuous on the interval $0 < x < \pi/(2N)$ and hence that the integrals actually exist.

(*d*) Using the fact that the integrand of the integral I_1 in part (*c*) is bounded (see Sec. 1), show that the value of I_1 tends to zero as N tends to infinity. Then conclude that

$$\lim_{N\to\infty} S_N\left(\frac{\pi}{2N}\right) = \int_0^{\pi} \frac{\sin t}{t} \, dt.$$

The value of this last integral is the number σ in the example in Sec. 16.[†]

[†]The integral occurs as a particular value of the *sine integral function* $\text{Si}(x)$, which is tabulated in, for instance, the handbook edited by Abramowitz and Stegun (1972, p. 244), listed in the Bibliography. Approximation methods for evaluating definite integrals can also be used to find σ.

CHAPTER 3

PARTIAL DIFFERENTIAL EQUATIONS OF PHYSICS

This chapter is devoted mainly to the derivation of certain partial differential equations arising in studies of heat conduction, electrostatics, and mechanical vibrations. Solutions of these equations, to be obtained in subsequent chapters, will involve not only Fourier series of the type treated in Chaps. 1 and 2 but also other kinds of series to be developed in the later chapters.

19. LINEAR BOUNDARY VALUE PROBLEMS

In the theory and application of partial differential equations, the dependent variable, denoted here by u, is usually required to satisfy some conditions on the boundary of the domain on which the differential equation is defined. The equations that represent those boundary conditions may involve values of derivatives of u, as well as values of u itself, at points on the boundary. In addition, some conditions on the continuity of u and its derivatives within the domain and on the boundary may be required. Such a set of requirements constitutes a *boundary value problem* in the function u. We use that terminology whenever the differential equation is accompanied by some boundary conditions.

A boundary value problem is correctly set if it has one and only one solution within a given class of functions. Physical interpretations often suggest boundary conditions under which a problem may be correctly set. In fact, it is sometimes helpful to interpret a problem physically in order to judge whether the boundary conditions may be adequate. This is a prominent reason for associating such problems with their physical applications, aside from the opportunity to illustrate connections between mathematical analysis and the physical sciences.

The theory of partial differential equations gives results on the existence and uniqueness of solutions of boundary value problems. But such results are necessarily limited by the great variety of types of differential equations and domains on which they are defined, as well as types of boundary conditions. Instead of appealing to general theory in treating a specific problem, our approach will be to actually find a solution, which can often be verified and shown to be the only one possible.

Frequently, it is convenient to indicate partial differentiation by writing independent variables as subscripts. If, for instance, u is a function of x and y, we may write

$$u_x \text{ or } u_x(x, y) \text{ for } \frac{\partial u}{\partial x}, \qquad u_{xx} \text{ for } \frac{\partial^2 u}{\partial x^2}, \qquad u_{xy} \text{ for } \frac{\partial^2 u}{\partial y\,\partial x},$$

etc. We shall always assume that the partial derivatives of u satisfy continuity conditions allowing us to write $u_{yx} = u_{xy}$. Also, we shall be free to use the symbol $u_x(c, y)$, for example, to denote values of the function $\partial u/\partial x$ on the line $x = c$.

EXAMPLE. The problem consisting of the differential equation

$$u_{xx}(x, y) + u_{yy}(x, y) = 0 \qquad (x > 0, y > 0) \tag{1}$$

and the two boundary conditions

$$u(0, y) = u_x(0, y) \qquad (y > 0), \tag{2}$$

$$u(x, 0) = \sin x + \cos x \qquad (x > 0) \tag{3}$$

is a boundary value problem in partial differential equations. The differential equation is defined in the first quadrant of the xy plane. As the reader can readily verify, the function

$$u(x, y) = e^{-y}(\sin x + \cos x) \tag{4}$$

is a solution of this problem. The function (4) and its partial derivatives of the first and second orders are continuous in the region $x \geq 0$, $y \geq 0$.

A differential equation in a function u, or a boundary condition on u, is *linear* if it is an equation of the first degree in u and derivatives of u. Thus the terms of the equation are either prescribed functions of the independent variables alone, including constants, or such functions multiplied by u or a derivative of u. Note that the general linear partial differential equation of the second order in $u = u(x, y)$ has the form

$$Au_{xx} + Bu_{xy} + Cu_{yy} + Du_x + Eu_y + Fu = G, \tag{5}$$

where the letters A through G denote either constants or functions of the independent variables x and y only. The differential equation and the boundary conditions in the example above are all linear. The differential equation

$$zu_{xx} + xy^2u_{yy} - e^xu_z = f(y, z) \tag{6}$$

is linear in $u = u(x, y, z)$; but the equation $u_{xx} + uu_y = x$ is nonlinear in $u = u(x, y)$ because the term uu_y is not of the first degree as an algebraic expression in the two variables u and u_y, in accordance with equation (5).

As expected, a boundary value problem is said to be linear if its differential equation and all its boundary conditions are linear. The boundary value problem in our example is, therefore, linear. The method of solution presented in this book does not apply to nonlinear problems.

A linear differential equation or boundary condition in u is *homogeneous* if each of its terms, other than zero itself, is of the first degree in the function u and its derivatives. Homogeneity will play a central role in our treatment of linear boundary value problems. Observe that equation (1) and condition (2) are homogeneous but that condition (3) is not. Equation (5) is homogeneous in a domain of the xy plane only when the function G is identically equal to zero ($G \equiv 0$) throughout that domain; and equation (6) is nonhomogeneous unless $f(y, z) \equiv 0$ for all values of y and z being considered.

20. ONE-DIMENSIONAL HEAT EQUATION

Thermal energy is transferred from warmer to cooler regions interior to a solid body by means of conduction. It is convenient to refer to that transfer as a *flow of heat*, as if heat were a fluid or gas that diffused through the body from regions of high concentration into regions of low concentration.

Let P_0 denote a point (x_0, y_0, z_0) interior to the body and S a smooth surface through P_0. Also, let $\mathbf{n}$ be a unit vector that is normal to S at the point P_0 (Fig. 11). At time t, the *flux of heat* $\Phi(x_0, y_0, z_0, t)$ across S at P_0 in the direction of $\mathbf{n}$ is the quantity of heat per unit area per unit time that is being conducted across S at P_0 in that direction. Flux is, therefore, measured in such units as calories per square centimeter per second.

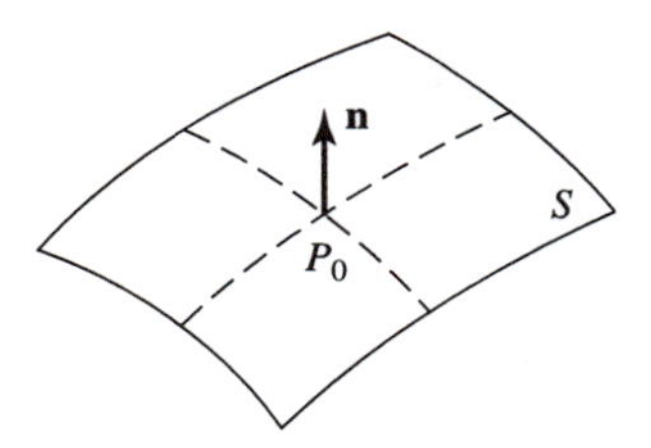

FIGURE 11

If $u(x, y, z, t)$ denotes temperatures at points (x, y, z) of the body at time t and if n is a coordinate that represents distance in the direction of $\mathbf{n}$, the flux $\Phi(x_0, y_0, z_0, t)$ is positive when the directional derivative du/dn is negative at P_0

and negative when du/dn is positive there. A fundamental postulate, known as *Fourier's law,* in the mathematical theory of heat conduction states that the magnitude of the flux $\Phi(x_0, y_0, z_0, t)$ is proportional to the magnitude of the directional derivative du/dn at P_0 at time t. That is, there is a coefficient K, known as the *thermal conductivity* of the material, such that

$$\Phi = -K\frac{du}{dn} \qquad (K > 0) \tag{1}$$

at P_0 and time t.

Another thermal coefficient of the material is its *specific heat* σ. This is the quantity of heat required to raise the temperature of a unit mass of the material one unit on the temperature scale. Unless otherwise stated, we shall always assume that the coefficients K and σ are constant throughout the solid body and that the same is true of δ, the mass per unit volume of the material. With these assumptions, a second postulate in the mathematical theory is that conduction leads to a temperature function u which, together with its derivative u_t and those of the first or second order with respect to x, y, and z, is continuous throughout each domain interior to a solid body in which no heat is generated or lost.

Suppose now that heat flows only parallel to the x axis in the body, so that flux Φ and temperatures u depend on only x and t. Thus $\Phi = \Phi(x, t)$ and $u = u(x, t)$. We assume at present that heat is neither generated nor lost within the body and hence that heat enters or leaves only through its surface. We then construct a small rectangular parallelepiped, lying in the interior of the body, with one vertex at a point (x, y, z) and with faces parallel to the coordinate planes. The lengths of the edges are Δx, Δy, and Δz, as shown in Fig. 12. Observe that since the parallelepiped is small, the continuous function u_t varies little in that region and has approximately the value $u_t(x, t)$ throughout it. This approximation improves, of course, as Δx tends to zero.

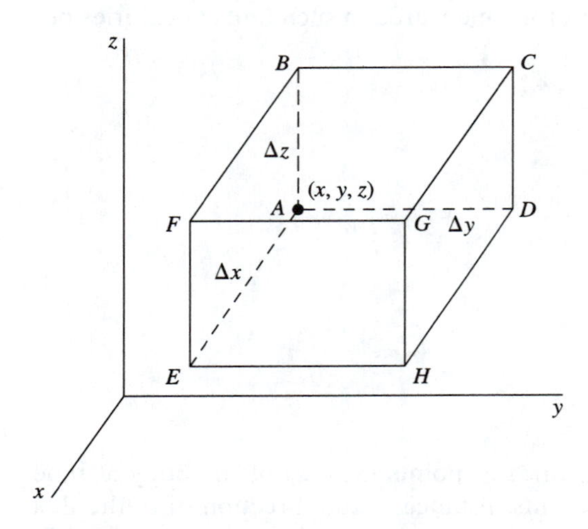

FIGURE 12

The mass of the element of material occupying the parallelepiped is $\delta\,\Delta x \Delta y\,\Delta z$. So, in view of the definition of specific heat σ stated above, we know that one measure of the quantity of heat entering that element per unit time at time t is approximately

$$\sigma\delta\,\Delta x\,\Delta y\,\Delta z\,u_t(x,t). \tag{2}$$

Another way to measure that quantity is to observe that since the flow of heat is parallel to the x axis, heat crosses only the surfaces $ABCD$ and $EFGH$ of the element, which are parallel to the yz plane. If the direction of the flux $\Phi(x,t)$ is in the *positive* direction of the x axis, it follows that the quantity of heat per unit time crossing the surface $ABCD$ into the element at time t is $\Phi(x,t)\Delta y\,\Delta z$. Because of the heat leaving the element through the face $EFGH$, the net quantity of heat entering the element per unit time is, then,

$$\Phi(x,t)\,\Delta y\,\Delta z - \Phi(x+\Delta x,t)\,\Delta y\,\Delta z.$$

In view of Fourier's law (1), this expression can be written

$$K\,[u_x(x+\Delta x,t) - u_x(x,t)]\,\Delta y\,\Delta z. \tag{3}$$

Equating expressions (2) and (3) for the quantity of heat entering the element per unit time and then dividing by $\sigma\delta\,\Delta x\,\Delta y\,\Delta z$, we have

$$u_t(x,t) = \frac{K}{\sigma\delta}\cdot\frac{u_x(x+\Delta x,t) - u_x(x,\,t)}{\Delta x}.$$

Letting Δx tend to zero here, we find that temperatures in a solid body, when heat flows only parallel to the x axis, satisfy the one-dimensional *heat equation*

$$u_t(x,\,t) = ku_{xx}(x,\,t), \tag{4}$$

where

$$k = \frac{K}{\sigma\delta}.$$

The constant k here is called the *thermal diffusivity* of the material.

In the derivation of equation (4), we assumed that there is no source (or sink) of heat within the solid body, only heat transfer by conduction. If there is a uniform source throughout the body that generates heat at a constant rate Q per unit volume, it is easy to modify the derivation to obtain the nonhomogeneous heat equation

$$u_t(x,\,t) = ku_{xx}(x,\,t) + q_0, \tag{5}$$

where

$$q_0 = \frac{Q}{\sigma\delta}.$$

This is accomplished by simply adding the term $Q\,\Delta x\,\Delta y\,\Delta z$ to expression (3) and proceeding in the same way as before. The rate Q per unit volume at which heat is generated may, in fact, be any continuous function of x and t, in which case the term q_0 in equation (5) is replaced by a function $q(x,\,t)$.

The heat equation describing flow in two and three dimensions is discussed in Sec. 21.

21. RELATED EQUATIONS

When the direction of heat flow in a solid body is not restricted to be simply parallel to the x axis, temperatures u in the body depend, in general, on all the space variables, as well as t. By considering the rate of heat passing through each of the six faces of the element in Fig. 12 (Sec. 20), one can derive (see Problem 1) the *three-dimensional* heat equation, satisfied by $u = u(x, y, z, t)$:

$$u_t = k(u_{xx} + u_{yy} + u_{zz}). \tag{1}$$

The constant k is the thermal diffusivity of the material, appearing in equation (4), Sec. 20. When the *laplacian*

$$\nabla^2 u = u_{xx} + u_{yy} + u_{zz} \tag{2}$$

is used, equation (1) takes the compact form

$$u_t = k\nabla^2 u. \tag{3}$$

Note that when there is no flow of heat parallel to the z axis, so that $u_{zz} = 0$ and $u = u(x, y, t)$, equation (1) reduces to the heat equation for *two-dimensional* flow parallel to the xy plane:

$$u_t = k(u_{xx} + u_{yy}). \tag{4}$$

The one-dimensional heat equation $u_t = ku_{xx}$ that was obtained in Sec. 20 for temperatures $u = u(x, t)$ follows, of course, from this when there is, in addition, no flow parallel to the y axis. If temperatures are in a *steady state,* in which case u does not vary with time, equation (1) becomes *Laplace's equation*

$$u_{xx} + u_{yy} + u_{zz} = 0. \tag{5}$$

Equation (5) is often written $\nabla^2 u = 0$.

The derivation of equation (1) in Problem 1 takes into account the possibility that heat may be generated in the solid body at a constant rate Q per unit volume, and the generalization

$$u_t = k\nabla^2 u + q_0 \tag{6}$$

of equation (5), Sec. 20, is obtained. If the rate Q is a continuous function of the space variables x, y, and z and if temperatures are in a steady state, equation (6) becomes *Poisson's equation*

$$\nabla^2 u = f(x, y, z), \tag{7}$$

where $f(x, y, z) = -q(x, y, z)/k$.

It should be emphasized that the various partial differential equations in this section are important in other areas of applied mathematics. In simple diffusion problems, for example, Fourier's law $\Phi = -K\,du/dn$ applies to the flux Φ of a substance that is diffusing within a porous solid. In that case, Φ represents the mass of the substance that is diffused per unit area per unit time through a surface, u denotes *concentration* (the mass of the diffusing substance per unit volume of the solid), and K is the *coefficient of diffusion*. Since the mass of the substance entering the element of volume in Fig. 12 per unit time is $\Delta x \, \Delta y \, \Delta z \, u_t$, one can write $Q = 0$ in the derivation of equation (6) and replace the product $\sigma\delta$ in that derivation by unity to see that the concentration u satisfies the *diffusion equation*

$$u_t = K\nabla^2 u. \tag{8}$$

A function $u = u(x, y, z)$ that is continuous, together with its partial derivatives of the first and second orders, and satisfies Laplace's equation (5) is called a *harmonic function*. We have seen in this section that the steady-state temperatures at points interior to a solid body in which no heat is generated are represented by a harmonic function. The steady-state concentration of a diffusing substance is also represented by such a function.

Among the many physical examples of harmonic functions, the velocity potential for the steady-state irrotational motion of an incompressible fluid is prominent in hydrodynamics and aerodynamics. An important harmonic function in electric field theory is the electrostatic potential $V(x, y, z)$ in a region of space that is free of electric charges. The potential may be caused by a static distribution of electric charges outside that region. The fact that V is harmonic is a consequence of the inverse-square law of attraction or repulsion between charges. Likewise, gravitational potential is a harmonic function in regions of space not occupied by matter.

In this book, the physical problems involving the laplacian, and Laplace's equation in particular, are limited mostly to those for which the differential equations are derived in this chapter. Derivations of such differential equations in other areas of applied mathematics can be found in books on hydrodynamics, elasticity, vibrations and sound, electric field theory, potential theory, and other branches of continuum mechanics. A number of such books are listed in the Bibliography at the back of this book.

PROBLEMS

1. Let $u = u(x, y, z, t)$ denote temperatures in a solid body throughout which there is a uniform heat source. Derive the heat equation

$$u_t = k\nabla^2 u + q_0$$

for those temperatures, where the constants k and q_0 are the same ones as in equation (5), Sec. 20.

Suggestion: Modify the derivation of equation (5), Sec. 20, by also considering the net rate of heat entering the element in Fig. 12 (Sec. 20) through the faces parallel to the xz and xy planes. Since the faces are small, one may consider the needed flux at points on a given face to be constant over that face. Thus, for instance, the net rate of heat entering the element through the faces parallel to the xy plane is to be taken as

$$K[u_z(x, y, z + \Delta z, t) - u_z(x, y, z, t)]\,\Delta x\,\Delta y.$$

2. Suppose that the thermal coefficients K and σ (Sec. 20) are functions of x, y, and z. Modify the derivation in Problem 1 to show that the heat equation takes the form

$$\sigma \delta u_t = (K u_x)_x + (K u_y)_y + (K u_z)_z$$

in a domain where all functions and derivatives involved are continuous and where there is no heat generated.

3. Show that the substitution $\tau = kt$ can be used to write the two-dimensional heat equation

$$u_t = k(u_{xx} + u_{yy})$$

in the form

$$u_\tau = u_{xx} + u_{yy},$$

where the thermal diffusivity is unity.

Suggestion: Note that

$$\frac{\partial u}{\partial t} = \frac{\partial u}{\partial \tau}\frac{d\tau}{dt}.$$

4. Show that the physical dimensions of thermal diffusivity k (Sec. 20) are L^2T^{-1}, where L denotes length and T time.

Suggestion: Observe first that the dimensions of thermal conductivity K and specific heat σ are $AL^{-1}T^{-1}B^{-1}$ and $AM^{-1}B^{-1}$, respectively, where M denotes mass, A is quantity of heat, and B is temperature. Then recall that $k = K/(\sigma\delta)$, where δ is density (ML^{-3}).

22. LAPLACIAN IN CYLINDRICAL AND SPHERICAL COORDINATES

We recall (Sec. 21) that the heat equation, and also Laplace's equation, can be written in terms of the laplacian

$$\nabla^2 u = u_{xx} + u_{yy} + u_{zz}. \tag{1}$$

Often, because of the geometric configuration of the physical problem, it is more convenient to use the laplacian in other than the rectangular coordinates x, y, and z. In this section, we give expressions for $\nabla^2 u$ in two different coordinate systems already encountered in calculus. These alternative forms of $\nabla^2 u$ will then be derived in Sec. 23, where it is assumed that u possesses continuous partial derivatives of the first and second orders with respect to the independent variables.

The *cylindrical coordinates* ρ, ϕ, and z determine a point $P(\rho, \phi, z)$, shown in Fig. 13. Cylindrical and rectangular coordinates evidently share the coordinate z. Also, ρ and ϕ are the polar coordinates in the xy plane of the projection Q of P onto that plane.†

---

†In calculus, the symbols r and θ are often used instead of ρ and ϕ, but the notation used here is common in physics and engineering. The notation for spherical coordinates, appearing later in this section, may also differ somewhat from that learned in calculus.

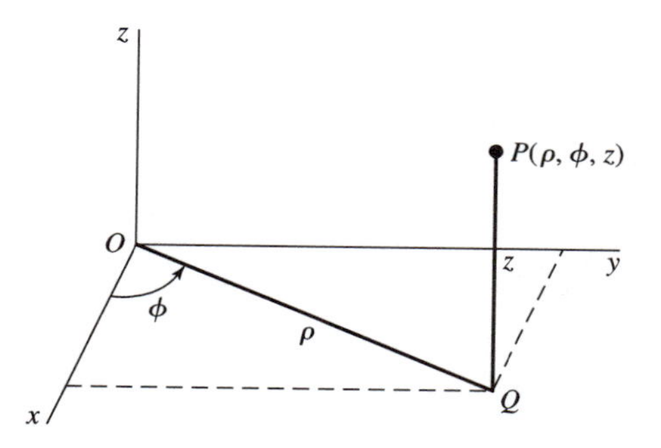

FIGURE 13

We shall show in Sec. 23 that the *laplacian of u in cylindrical coordinates* is

$$\nabla^2 u = u_{\rho\rho} + \frac{1}{\rho} u_\rho + \frac{1}{\rho^2} u_{\phi\phi} + u_{zz}. \tag{2}$$

Note how one can group the first two terms in this expression to write

$$\nabla^2 u = \frac{1}{\rho} (\rho u_\rho)_\rho + \frac{1}{\rho^2} u_{\phi\phi} + u_{zz}. \tag{3}$$

When u is independent of z, so that $u = u(\rho, \phi)$, expression (2) becomes the two-dimensional laplacian of u in polar coordinates:

$$\nabla^2 u = u_{\rho\rho} + \frac{1}{\rho} u_\rho + \frac{1}{\rho^2} u_{\phi\phi}. \tag{4}$$

Laplace's equation $\nabla^2 u = 0$ in polar coordinates can, therefore, be written in the form

$$\rho^2 u_{\rho\rho} + \rho u_\rho + u_{\phi\phi} = 0. \tag{5}$$

Also, it follows from expression (2) that when temperatures u in a solid body vary only with ρ and time t, and not with the space variables ϕ and z, the heat equation $u_t = k\nabla^2 u$ becomes

$$u_t = k\left(u_{\rho\rho} + \frac{1}{\rho} u_\rho\right). \tag{6}$$

Equations (5) and (6) will be of particular interest in the applications.

The *spherical coordinates* r, ϕ, and θ of a point $P(r, \phi, \theta)$ are shown in Fig. 14. Note that the coordinate ϕ is common to spherical and cylindrical coordinates.

It will also be shown in Sec. 23 that the *laplacian of u in spherical coordinates* is

$$\nabla^2 u = u_{rr} + \frac{2}{r} u_r + \frac{1}{r^2 \sin^2 \theta} u_{\phi\phi} + \frac{1}{r^2} u_{\theta\theta} + \frac{\cot\theta}{r^2} u_\theta. \tag{7}$$

Other forms of this expression are

$$\nabla^2 u = \frac{1}{r} (ru)_{rr} + \frac{1}{r^2 \sin^2 \theta} u_{\phi\phi} + \frac{1}{r^2 \sin\theta} (\sin\theta \, u_\theta)_\theta, \tag{8}$$

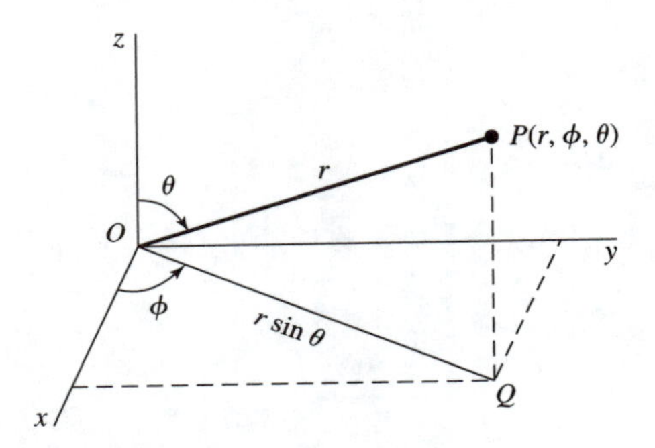

FIGURE 14

$$\nabla^2 u = \frac{1}{r^2}\left(r^2 u_r\right)_r + \frac{1}{r^2 \sin^2\theta}\, u_{\phi\phi} + \frac{1}{r^2 \sin\theta}\,(\sin\theta\, u_\theta)_\theta. \tag{9}$$

Some of our applications later on will involve Laplace's equation $\nabla^2 u = 0$ in spherical coordinates when u is independent of ϕ. According to expression (8), that equation can then be written

$$r(ru)_{rr} + \frac{1}{\sin\theta}\,(\sin\theta\, u_\theta)_\theta = 0. \tag{10}$$

The reader who wishes to accept expressions (2) and (7) without derivation can pass directly to Sec. 24 without disruption.

23. DERIVATIONS

One can see from Fig. 13 in Sec. 22 that the cylindrical coordinates ρ, ϕ, and z are related to the rectangular coordinates x, y, and z by the equations

$$x = \rho\cos\phi, \qquad y = \rho\sin\phi, \qquad z = z, \tag{1}$$

as well as the ones

$$\rho = \sqrt{x^2 + y^2}, \qquad \phi = \tan^{-1}\frac{y}{x}, \qquad z = z, \tag{2}$$

where the quadrant to which the angle ϕ belongs is determined by the signs of x and y, not by the ratio y/x alone.

Let u denote a function of x, y, and z. Then, in view of relations (1), it is also a function of the three independent variables ρ, ϕ, and z. If u is continuous and possesses continuous partial derivatives of the first and second orders, the following method, based on the chain rule for differentiating composite functions, can be used to express the laplacian

$$\nabla^2 u = \frac{\partial^2 u}{\partial x^2} + \frac{\partial^2 u}{\partial y^2} + \frac{\partial^2 u}{\partial z^2} \tag{3}$$

in terms of ρ, ϕ, and z.

Since

$$\frac{\partial u}{\partial x} = \frac{\partial u}{\partial \rho}\frac{\partial \rho}{\partial x} + \frac{\partial u}{\partial \phi}\frac{\partial \phi}{\partial x} + \frac{\partial u}{\partial z}\frac{\partial z}{\partial x},$$

it follows from relations (2) that

$$\frac{\partial u}{\partial x} = \frac{\partial u}{\partial \rho}\frac{x}{\sqrt{x^2+y^2}} - \frac{\partial u}{\partial \phi}\frac{y}{x^2+y^2} + \frac{\partial u}{\partial z}0$$
$$= \frac{x}{\rho}\frac{\partial u}{\partial \rho} - \frac{y}{\rho^2}\frac{\partial u}{\partial \phi}.$$

Hence, by relations (1),

$$\frac{\partial u}{\partial x} = \cos\phi\,\frac{\partial u}{\partial \rho} - \frac{\sin\phi}{\rho}\frac{\partial u}{\partial \phi}. \tag{4}$$

Replacing the function u in equation (4) by $\partial u/\partial x$, we see that

$$\frac{\partial^2 u}{\partial x^2} = \cos\phi\,\frac{\partial}{\partial \rho}\left(\frac{\partial u}{\partial x}\right) - \frac{\sin\phi}{\rho}\frac{\partial}{\partial \phi}\left(\frac{\partial u}{\partial x}\right). \tag{5}$$

We may now use expression (4) to substitute for the derivative $\partial u/\partial x$ appearing on the right-hand side of equation (5):

$$\frac{\partial^2 u}{\partial x^2} = \cos\phi\,\frac{\partial}{\partial \rho}\left(\cos\phi\,\frac{\partial u}{\partial \rho} - \frac{\sin\phi}{\rho}\frac{\partial u}{\partial \phi}\right) - \frac{\sin\phi}{\rho}\frac{\partial}{\partial \phi}\left(\cos\phi\,\frac{\partial u}{\partial \rho} - \frac{\sin\phi}{\rho}\frac{\partial u}{\partial \phi}\right).$$

By applying rules for differentiating differences and products of functions and using the relation

$$\frac{\partial^2 u}{\partial \rho \partial \phi} = \frac{\partial^2 u}{\partial \phi\,\partial \rho},$$

which is ensured by the continuity of the partial derivatives, we find that

$$\frac{\partial^2 u}{\partial x^2} = \cos^2\phi\,\frac{\partial^2 u}{\partial \rho^2} - \frac{2\sin\phi\cos\phi}{\rho}\frac{\partial^2 u}{\partial \phi\,\partial \rho} + \frac{\sin^2\phi}{\rho^2}\frac{\partial^2 u}{\partial \phi^2}$$
$$+ \frac{\sin^2\phi}{\rho}\frac{\partial u}{\partial \rho} + \frac{2\sin\phi\cos\phi}{\rho^2}\frac{\partial u}{\partial \phi}. \tag{6}$$

In the same way, one can show that

$$\frac{\partial u}{\partial y} = \frac{y}{\rho}\frac{\partial u}{\partial \rho} + \frac{x}{\rho^2}\frac{\partial u}{\partial \phi}, \tag{7}$$

or

$$\frac{\partial u}{\partial y} = \sin\phi\,\frac{\partial u}{\partial \rho} + \frac{\cos\phi}{\rho}\frac{\partial u}{\partial \phi}, \tag{8}$$

and also that

$$\frac{\partial^2 u}{\partial y^2} = \sin^2\phi\,\frac{\partial^2 u}{\partial \rho^2} + \frac{2\sin\phi\cos\phi}{\rho}\frac{\partial^2 u}{\partial \phi\,\partial \rho} + \frac{\cos^2\phi}{\rho^2}\frac{\partial^2 u}{\partial \phi^2}$$
$$+ \frac{\cos^2\phi}{\rho}\frac{\partial u}{\partial \rho} - \frac{2\sin\phi\cos\phi}{\rho^2}\frac{\partial u}{\partial \phi}. \tag{9}$$

By adding corresponding sides of equations (6) and (9), we arrive at

$$\frac{\partial^2 u}{\partial x^2} + \frac{\partial^2 u}{\partial y^2} = \frac{\partial^2 u}{\partial \rho^2} + \frac{1}{\rho}\frac{\partial u}{\partial \rho} + \frac{1}{\rho^2}\frac{\partial^2 u}{\partial \phi^2}. \tag{10}$$

Since cylindrical and rectangular coordinates share the coordinate z, it follows that equation (3) becomes

$$\nabla^2 u = \frac{\partial^2 u}{\partial \rho^2} + \frac{1}{\rho}\frac{\partial u}{\partial \rho} + \frac{1}{\rho^2}\frac{\partial^2 u}{\partial \phi^2} + \frac{\partial^2 u}{\partial z^2} \tag{11}$$

in cylindrical coordinates. This is, of course, the same as expression (2) in Sec. 22.

As for the spherical coordinates r, ϕ, and θ, Fig. 14 in Sec. 22 shows that they are related to the rectangular coordinates x, y, and z as follows:

$$x = r\sin\theta\cos\phi, \qquad y = r\sin\theta\sin\phi, \qquad z = r\cos\theta. \tag{12}$$

Spherical and cylindrical coordinates are, moreover, related by the equations

$$z = r\cos\theta, \qquad \rho = r\sin\theta, \qquad \phi = \phi. \tag{13}$$

Expression (11) for $\nabla^2 u$ in cylindrical coordinates can be transformed into spherical coordinates quite readily by means of the proper interchange of letters, *without any further application of the chain rule*. This is accomplished in three steps, described below.

First, we observe that except for the names of the variables involved, relations (13) are the same as relations (1) connecting cylindrical and rectangular coordinates. Since relations (1) gave us expressions (8) and (10), it follows immediately, then, that

$$\frac{\partial u}{\partial \rho} = \sin\theta\,\frac{\partial u}{\partial r} + \frac{\cos\theta}{r}\frac{\partial u}{\partial \theta} \tag{14}$$

and

$$\frac{\partial^2 u}{\partial z^2} + \frac{\partial^2 u}{\partial \rho^2} = \frac{\partial^2 u}{\partial r^2} + \frac{1}{r}\frac{\partial u}{\partial r} + \frac{1}{r^2}\frac{\partial^2 u}{\partial \theta^2}. \tag{15}$$

Next, we note how equation (14) and the second of relations (13) enable us to write

$$\frac{1}{\rho}\frac{\partial u}{\partial \rho} + \frac{1}{\rho^2}\frac{\partial^2 u}{\partial \phi^2} = \frac{1}{r}\frac{\partial u}{\partial r} + \frac{\cot\theta}{r^2}\frac{\partial u}{\partial \theta} + \frac{1}{r^2\sin^2\theta}\frac{\partial^2 u}{\partial \phi^2}. \tag{16}$$

Finally, we rewrite expression (11) for $\nabla^2 u$ in cylindrical coordinates as

$$\nabla^2 u = \left(\frac{\partial^2 u}{\partial z^2} + \frac{\partial^2 u}{\partial \rho^2}\right) + \left(\frac{1}{\rho}\frac{\partial u}{\partial \rho} + \frac{1}{\rho^2}\frac{\partial^2 u}{\partial \phi^2}\right). \tag{17}$$

Using equations (15) and (16) to substitute for the sums in parentheses here, we arrive at the expression

$$\nabla^2 u = \frac{\partial^2 u}{\partial r^2} + \frac{2}{r}\frac{\partial u}{\partial r} + \frac{1}{r^2\sin^2\theta}\frac{\partial^2 u}{\partial \phi^2} + \frac{1}{r^2}\frac{\partial^2 u}{\partial \theta^2} + \frac{\cot\theta}{r^2}\frac{\partial u}{\partial \theta}, \tag{18}$$

which is the same as expression (7), Sec. 22, for $\nabla^2 u$ in spherical coordinates.

24. BOUNDARY CONDITIONS

Equations that describe thermal conditions on the surfaces of a solid body and initial temperatures throughout the body must accompany the heat equation if we are to determine the temperature function u. The conditions on the surfaces may be other than just prescribed temperatures. Suppose, for example, that the flux Φ into the solid at points on a surface S is some constant Φ_0. That is, at each point P on S, we know that Φ_0 units of heat per unit area per unit time flow across S in the direction *opposite to* an outward unit normal vector $\mathbf{n}$ at P. From Fourier's law (1) in Sec. 20, we know that if du/dn is the directional derivative of u at P in the direction of $\mathbf{n}$, the flux into the solid across S at P is the value of $K du/dn$ there. Hence

$$K\frac{du}{dn} = \Phi_0 \tag{1}$$

on the surface S. Observe that if S is perfectly insulated, $\Phi_0 = 0$ at points on S and condition (1) becomes

$$\frac{du}{dn} = 0. \tag{2}$$

On the other hand, there may be surface heat transfer between a boundary surface S and a medium whose temperature is a constant T. The inward flux Φ, which can be negative, may then vary from point to point on S; and we assume that at each point P, the flux is proportional to the difference between the temperature of the medium and the temperature at P. Under this assumption, which is sometimes called *Newton's law of cooling*, there is a positive constant H, known as the *surface conductance* of the material, such that $\Phi = H(T - u)$ at points on S. Condition (1) is then replaced by the condition

$$K\frac{du}{dn} = H(T - u), \tag{3}$$

or

$$\frac{du}{dn} = h(T - u) \qquad \left(h = \frac{H}{K}\right). \tag{4}$$

EXAMPLE 1. Consider a semi-infinite slab occupying the region $0 \leq x \leq c$, $y \geq 0$ of three-dimensional space. Figure 15 shows the cross section of the slab in

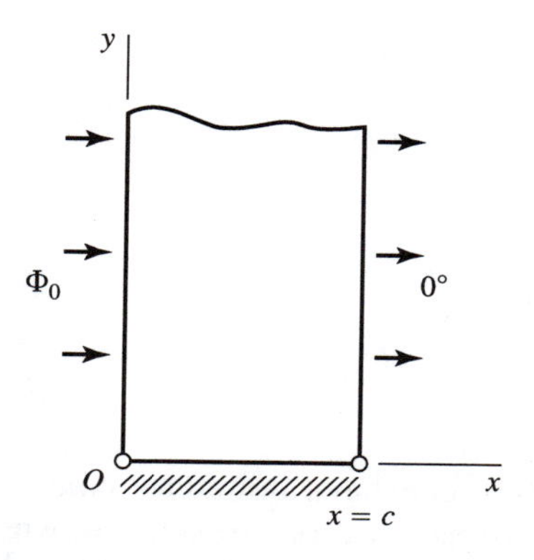

FIGURE 15

the xy plane. Suppose that there is a constant flux Φ_0 into the slab at points on the face in the plane $x = 0$ and that there is surface heat transfer (possibly inward) between the face in the plane $x = c$ and a medium at temperature zero. Also, the surface in the plane $y = 0$ is insulated. Since

$$\frac{du}{dn} = -\frac{\partial u}{\partial x} \quad \text{and} \quad \frac{du}{dn} = \frac{\partial u}{\partial x}$$

on the faces in the planes $x = 0$ and $x = c$, respectively, a temperature function $u = u(x, y, z, t)$ evidently satisfies the boundary conditions

$$-Ku_x(0, y, z, t) = \Phi_0, \qquad u_x(c, y, z, t) = -hu(c, y, z, t).$$

The insulated surface gives rise to the boundary condition $u_y(x, 0, z, t) = 0$.

EXAMPLE 2. Let u denote temperatures in a long rod, parallel to the z axis, whose cross section in the xy plane is the sector $0 \le \rho \le 1, 0 \le \phi \le \pi/2$ of a disk (Fig. 16). We assume that u is independent of the cylindrical coordinate z, so that $u = u(\rho, \phi)$, and that the rod is insulated on its planar surfaces, where $\phi = 0$ and $\phi = \pi/2$. Evidently,

$$\frac{\partial u}{\partial y} = 0 \quad \text{when } y = 0 \quad \text{and} \quad \frac{\partial u}{\partial x} = 0 \quad \text{when } x = 0.$$

Also, in view of the transformation equations $x = \rho \cos\phi$ and $y = \rho \sin\phi$ (Sec. 23), the chain rule tells us that

$$\frac{\partial u}{\partial \phi} = \frac{\partial u}{\partial x}\frac{\partial x}{\partial \phi} + \frac{\partial u}{\partial y}\frac{\partial y}{\partial \phi} = -y\frac{\partial u}{\partial x} + x\frac{\partial u}{\partial y}.$$

Consequently, u must satisfy the boundary conditions

$$u_\phi(\rho, 0) = 0, \qquad u_\phi\left(\rho, \frac{\pi}{2}\right) = 0 \qquad (0 < \rho < 1).$$

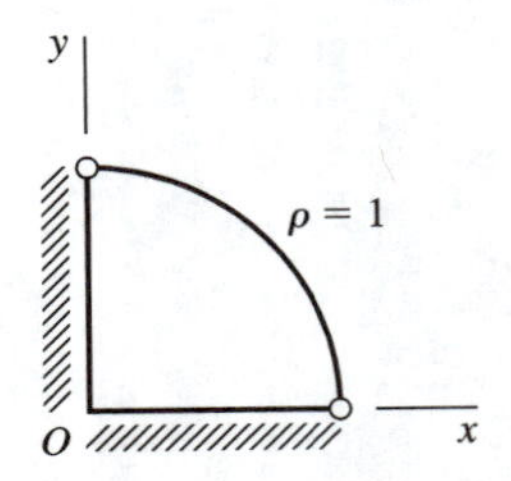

FIGURE 16

Boundary value problems in heat conduction can have boundary conditions that are time-dependent. We state here a special case of a result, known as *Duhamel's theorem,* that is often useful in solving boundary value problems involving such conditions.† Suppose that a function $u = u(P, t)$ satisfies the heat equation $u_t = k\nabla^2 u$ at points P in a region throughout which the initial temperatures are zero. Also, suppose that $u = F(t)$ on a portion S of the boundary of the region and that $u = 0$ on the remaining part S' (see Fig. 17). Thus

$$u(P, t) = F(t) \quad \text{on } S, \qquad u(P, t) = 0 \quad \text{on } S', \qquad \text{and} \qquad u(P, 0) = 0.$$

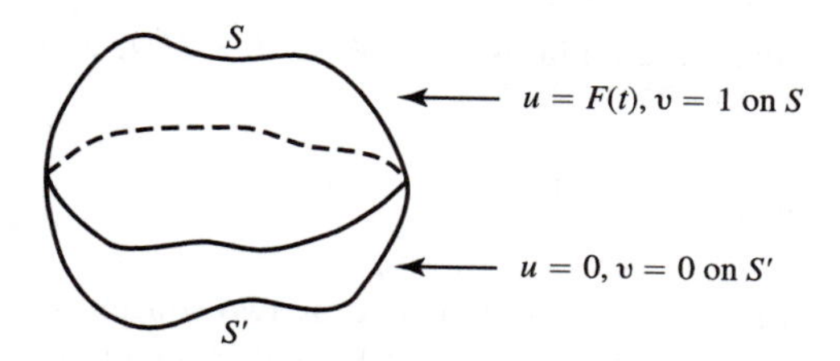

FIGURE 17

For simplicity here, we assume that $F(t)$ is continuous and differentiable when $t \geq 0$ and that $F(0) = 0$. Our special case of Duhamel's theorem tells us that if $v(P, t)$ satisfies the boundary value problem consisting of the heat equation $v_t = k\nabla^2 v$ and the conditions

$$v(P, t) = 1 \quad \text{on } S, \qquad v(P, t) = 0 \quad \text{on } S', \qquad \text{and} \qquad v(P, 0) = 0,$$

then

$$u(P, t) = \int_0^t F(\tau)\, v_t(P, t - \tau)\, d\tau. \tag{5}$$

Note that the boundary value problem for $v(P, t)$ is obtained from the one for $u(P, t)$ by replacing $F(t)$ by unity.

PROBLEMS

1. Let $u(x)$ denote the steady-state temperatures in a slab bounded by the planes $x = 0$ and $x = c$ when those faces are kept at fixed temperatures $u = 0$ and $u = u_0$, respectively. Set up the boundary value problem for $u(x)$ and solve it to show that

$$u(x) = \frac{u_0}{c}\, x \qquad \text{and} \qquad \Phi_0 = K\, \frac{u_0}{c},$$

where Φ_0 is the flux of heat to the left across each plane $x = x_0$ $(0 \leq x_0 \leq c)$.

2. A slab occupies the region $0 \leq x \leq c$. There is a constant flux of heat Φ_0 into the slab through the face $x = 0$. The face $x = c$ is kept at temperature $u = 0$. Set up and solve the boundary value problem for the steady-state temperatures $u(x)$ in the slab.

Answer: $u(x) = \dfrac{\Phi_0}{K}\,(c - x)$.

†The book by Churchill (1972) uses Laplace transforms to establish a more general form of Duhamel's theorem, which includes other types of time-dependent boundary conditions. It also adapts Duhamel's theorem to boundary value problems in mechanical vibrations.

3. Let a slab $0 \le x \le c$ be subjected to surface heat transfer, according to Newton's law of cooling, at its faces $x = 0$ and $x = c$, the surface conductance H being the same on each face. Show that if the medium $x < 0$ has temperature zero and the medium $x > c$ has the constant temperature T, then the boundary value problem for steady-state temperatures $u(x)$ in the slab is

$$u''(x) = 0 \qquad (0 < x < c),$$

$$Ku'(0) = Hu(0), \qquad Ku'(c) = H[T - u(c)],$$

where K is the thermal conductivity of the material in the slab. Write $h = H/K$ and derive the expression

$$u(x) = \frac{T}{hc + 2}(hx + 1)$$

for those temperatures.

4. Let $u(r)$ denote the steady-state temperatures in a solid bounded by two concentric spheres $r = a$ and $r = b$ $(a < b)$ when the inner surface $r = a$ is kept at temperature zero and the outer surface $r = b$ is maintained at a constant temperature u_0. Show why Laplace's equation for $u = u(r)$ reduces to

$$\frac{d^2}{dr^2}(ru) = 0,$$

and then derive the expression

$$u(r) = \frac{bu_0}{b - a}\left(1 - \frac{a}{r}\right) \qquad (a \le r \le b).$$

Sketch the graph of $u(r)$ versus r.

5. In Problem 4, replace the condition on the outer surface $r = b$ with the condition that there be surface heat transfer into a medium at constant temperature T according to Newton's law of cooling. Then obtain the expression

$$u(r) = \frac{hb^2 T}{a + hb(b - a)}\left(1 - \frac{a}{r}\right) \qquad (a \le r \le b)$$

for the steady-state temperatures, where h is the ratio of the surface conductance H to the thermal conductivity K of the material.

6. A slender wire lies along the x axis, and surface heat transfer takes place along the wire into the surrounding medium at a fixed temperature T. Modify the procedure in Sec. 20 to show that if $u = u(x, t)$ denotes temperatures in the wire, then

$$u_t = ku_{xx} + b(T - u),$$

where b is a positive constant.

Suggestion: Let r denote the radius of the wire, and apply Newton's law of cooling to see that the quantity of heat entering the element in Fig. 18 through its cylindrical surface per unit time is approximately $H[T - u(x, t)]2\pi r\,\Delta x$.

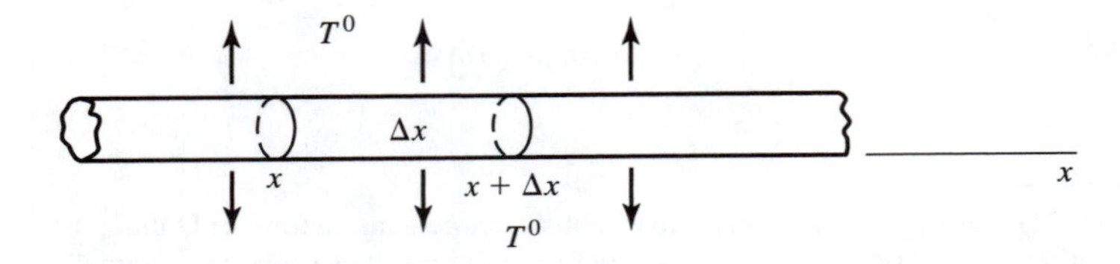

FIGURE 18

7. In Sec. 22, show how expressions (8) and (9) for $\nabla^2 u$ in spherical coordinates follow from expression (7).
8. Derive expressions (8) and (9) in Sec. 23 for

$$\frac{\partial u}{\partial y} \quad \text{and} \quad \frac{\partial^2 u}{\partial y^2}$$

in cylindrical coordinates.
9. Suppose that temperatures u in a solid hemisphere $r \leq 1, 0 \leq \theta \leq \pi/2$ are independent of the spherical coordinate ϕ, so that $u = u(r, \theta)$, and that the base of the hemisphere is insulated (Fig. 19). Use transformation (13), Sec. 23, which relates spherical and cylindrical coordinates, to show that

$$\frac{\partial u}{\partial \theta} = -\rho \frac{\partial u}{\partial z} + z \frac{\partial u}{\partial \rho}.$$

Thus show that u must satisfy the boundary condition

$$u_\theta\left(r, \frac{\pi}{2}\right) = 0.$$

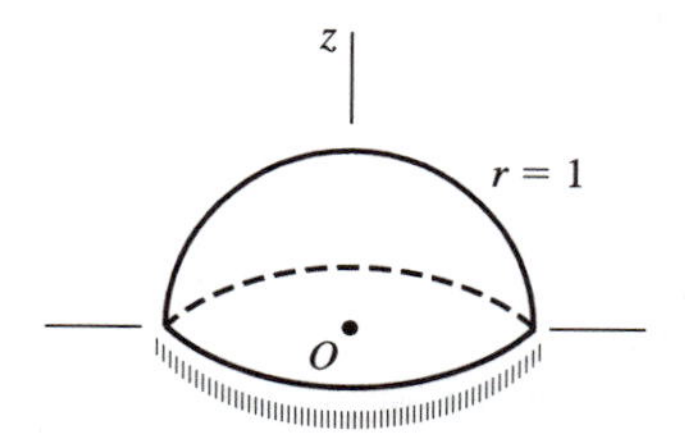

FIGURE 19

25. A VIBRATING STRING

A tightly stretched string, whose position of equilibrium is some interval on the x axis, is vibrating in the xy plane. Each point of the string, with coordinates $(x, 0)$ in the equilibrium position, has a transverse displacement $y = y(x, t)$ at time t. We assume that the displacements y are small relative to the length of the string, that slopes are small, and that other conditions are such that the movement of each point is parallel to the y axis. Then, at time t, a point on the string has coordinates (x, y), where $y = y(x, t)$.

Let the tension of the string be great enough that the string behaves as if it were perfectly flexible. That is, at a point (x, y) on the string, the part of the string to the left of that point exerts a force $\mathbf{T}$, in the tangential direction, on the part to the right (see Fig. 20); and any resistance to bending at the point is to be neglected. The magnitude of the x component of the tensile force $\mathbf{T}$ is denoted by H. Our final assumption here is that H is constant. That is, the variation of H with respect to x and t can be neglected.

These idealizing assumptions are severe, but they are justified in many applications. They are adequately satisfied, for instance, by strings of musical

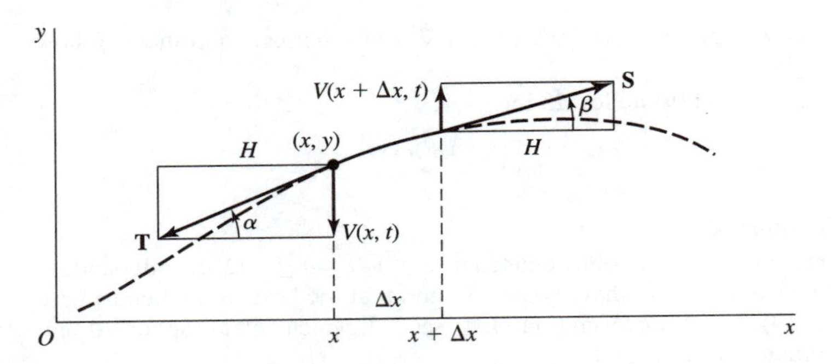

FIGURE 20

instruments under ordinary conditions of operation. Mathematically, the assumptions will lead us to a partial differential equation in $y(x, t)$ that is linear.

Now let $V(x, t)$ denote the y component of the tensile force $\mathbf{T}$ exerted by the left-hand portion of the string on the right-hand portion at the point (x, y). We take the positive sense of V to be that of the y axis. It is straightforward to show that *the y component $V(x, t)$ of the force exerted at time t by the part of the string to the left of a point (x, y) on the part to the right is given by the equation*

$$V(x, t) = -Hy_x(x, t) \qquad (H > 0), \tag{1}$$

which is basic for deriving the equation of motion of the string. Equation (1) is also used in setting up certain types of boundary conditions.

To establish expression (1), we note that if α is the angle of inclination of the string at the point (x, y) at time t, then

$$\frac{-V(x, t)}{H} = \tan\alpha = y_x(x, t).$$

This is indicated in Fig. 20, where $V(x, t) < 0$ and $y_x(x, t) > 0$. If $V(x, t) > 0$, then $\pi/2 < \alpha < \pi$ and $y_x(x, t) < 0$; and a similar sketch shows that

$$\frac{V(x, t)}{H} = \tan(\pi - \alpha) = -\tan\alpha = -y_x(x, t).$$

Hence expression (1) continues to hold. Note, too, that $V(x, t) = 0$ when $y_x(x, t) = 0$.

Suppose that all external forces such as the weight of the string and resistance forces, other than forces at the endpoints, can be neglected. Consider a segment of the string not containing an endpoint and whose projection onto the x axis has length Δx. Since x components of displacements are negligible, the mass of the segment is $\delta\Delta x$, where the constant δ is the mass per unit length of the string. At time t, the y component of the force exerted by the string on the segment at the left-hand end (x, y) is $V(x, t)$, given by equation (1). The tangential force $\mathbf{S}$ exerted on the other end of the segment by the part of the string to the right is also indicated in Fig. 20. Its y component $V(x + \Delta x, t)$ evidently satisfies the relation

$$\frac{V(x + \Delta x, t)}{H} = \tan\beta,$$

where β is the angle of inclination of the string at that other end. That is,

$$V(x + \Delta x, t) = Hy_x(x + \Delta x, t) \qquad (H > 0). \tag{2}$$

Note that except for a minus sign, this is equation (1) when the argument x there is replaced by $x + \Delta x$.

Now the acceleration of the end (x, y) in the y direction is $y_{tt}(x, t)$. Consequently, by Newton's second law of motion (mass times acceleration equals force), it follows from equations (1) and (2) that

$$\delta \Delta x \, y_{tt}(x, t) = -Hy_x(x, t) + Hy_x(x + \Delta x, t), \tag{3}$$

approximately, when Δx is small. That is,

$$y_{tt}(x, t) = \frac{H}{\delta} \cdot \frac{y_x(x + \Delta x, t) - y_x(x, t)}{\Delta x};$$

and as Δx tends to zero,

$$y_{tt}(x, t) = a^2 y_{xx}(x, t) \qquad \left(a^2 = \frac{H}{\delta}\right) \tag{4}$$

whenever these partial derivatives exist. This is the one-dimensional *wave equation,* satisfied by the transverse displacements $y(x, t)$ in a stretched string under the conditions stated above. The constant a has the physical dimensions of velocity.

One can choose units for the time variable so that $a = 1$ in the wave equation. More precisely, if we make the substitution $\tau = at$, the chain rule shows that

$$\frac{\partial y}{\partial t} = a \frac{\partial y}{\partial \tau} \qquad \text{and} \qquad \frac{\partial^2 y}{\partial t^2} = a \frac{\partial}{\partial \tau}\left(a \frac{\partial y}{\partial \tau}\right) = a^2 \frac{\partial^2 y}{\partial \tau^2}.$$

Equation (4) then becomes $y_{\tau\tau} = y_{xx}$. (A similar observation was made in Problem 3, Sec. 21, with regard to the heat equation.)

When external forces parallel to the y axis act along the string, we let F denote the force per unit length of the string, the positive sense of F being that of the y axis. Then a term $F \, \Delta x$ must be added on the right-hand side of equation (3), and the equation of motion is

$$y_{tt}(x, t) = a^2 y_{xx}(x, t) + \frac{F}{\delta}. \tag{5}$$

In particular, with the y axis vertical and its positive sense upward, suppose that the external force consists of the weight of the string. Then $F \Delta x = -\delta \, \Delta x \, g$, where the positive constant g is acceleration due to gravity; and equation (5) becomes the linear nonhomogeneous equation

$$y_{tt}(x, t) = a^2 y_{xx}(x, t) - g. \tag{6}$$

In equation (5), F may be a function of x, t, y, or derivatives of y. If the external force per unit length is a damping force proportional to the velocity in the y direction, for example, F is replaced by $-By_t$, where the positive constant B is a damping coefficient. Then the equation of motion is linear and homogeneous:

$$y_{tt}(x,t) = a^2 y_{xx}(x,t) - by_t(x,t) \qquad \left(b = \frac{B}{\delta}\right). \tag{7}$$

If an end $x = 0$ of the string is kept fixed at the origin at all times $t \geq 0$, the boundary condition there is clearly

$$y(0,t) = 0 \qquad (t \geq 0). \tag{8}$$

But if the end is permitted to slide along the y axis and is moved along that axis with a displacement $f(t)$, the boundary condition is the linear nonhomogeneous one

$$y(0,t) = f(t) \qquad (t \geq 0). \tag{9}$$

Suppose that the left-hand end is attached to a ring which can slide along the y axis. When a force $F(t)$ $(t > 0)$ in the y direction is applied to that end, $F(t)$ is the limit, as x tends to zero through positive values, of the force $V(x,t)$ described earlier in this section. According to equation (1), the boundary condition at $x = 0$ is then

$$-Hy_x(0,t) = F(t) \qquad (t > 0).$$

The minus sign disappears, however, if $x = 0$ is the *right-hand* end, in view of equation (2).

PROBLEMS

1. A stretched string, with its ends fixed at the points 0 and $2c$ on the x axis, hangs at rest under its own weight. The y axis is directed vertically upward. Point out how it follows from the nonhomogeneous wave equation (6), Sec. 25, that the static displacements $y(x)$ of points on the string must satisfy the differential equation

$$a^2 y''(x) = g \qquad \left(a^2 = \frac{H}{\delta}\right)$$

on the interval $0 < x < 2c$, in addition to the boundary conditions

$$y(0) = 0, \qquad y(2c) = 0.$$

By solving this boundary value problem, show that the string hangs in the parabolic arc

$$(x-c)^2 = \frac{2a^2}{g}\left(y + \frac{gc^2}{2a^2}\right) \qquad (0 \leq x \leq 2c)$$

and that the depth of the vertex of the arc varies directly with c^2 and δ and inversely with H.

2. Use expression (1), Sec. 25, for the vertical force V and the equation of the arc in which the string in Problem 1 lies to show that the vertical force exerted on that string by each support is δcg, one-half the weight of the string.

3. Give the needed details in the derivation of equation (5), Sec. 25, for the forced vibrations of a stretched string.

4. The physical dimensions of the magnitude H of the x component of the tensile force in a string are those of mass times acceleration: MLT^{-2}, where M denotes mass, L length, and T time. Show that since $a^2 = H/\delta$, the constant a has the dimensions of velocity: LT^{-1}.
5. A strand of wire 1 ft long, stretched between the origin and the point 1 on the x axis, weighs 0.032 lb ($\delta g = 0.032, g = 32$ ft/s^2) and $H = 10$ lb. At the instant $t = 0$, the strand lies along the x axis but has a velocity of 1 ft/s in the direction of the y axis, perhaps because the supports were in motion and were brought to rest at that instant. Assuming that no external forces act along the wire, state why the displacements $y(x, t)$ should satisfy this boundary value problem:

$$y_{tt}(x,t) = 10^4 y_{xx}(x,t) \qquad (0 < x < 1, t > 0),$$

$$y(0,t) = y(1,t) = 0, \qquad y(x,0) = 0, \qquad y_t(x,0) = 1.$$

26. VIBRATIONS OF BARS AND MEMBRANES

We describe here two other types of vibrations for which the displacements satisfy wave equations. We continue to limit our attention to fairly simple phenomena.

Let the coordinate x denote distances from one end of an elastic bar, in the shape of a cylinder or prism, to other cross sections when the bar is unstrained. Displacements of the ends or initial displacements or velocities in the bar, all directed lengthwise along it and uniform over each cross section involved, cause the sections to move parallel to the x axis. At time t, the longitudinal displacement of the section at a point x is denoted by $y(x, t)$. Thus the origin of the displacement y of the section at x is in a fixed coordinate system outside of the bar, in the plane of the unstrained position of that section (Fig. 21).

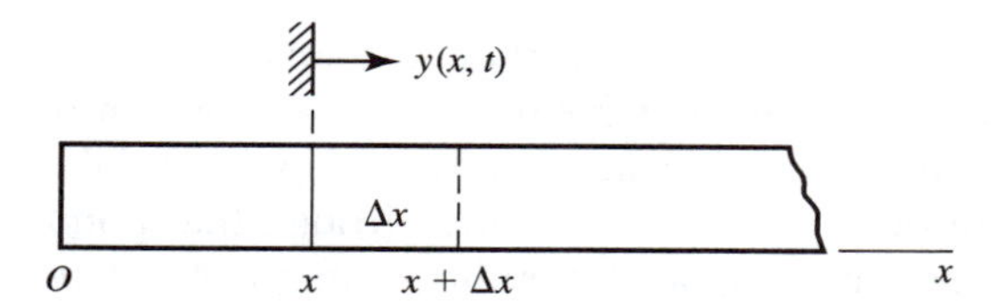

FIGURE 21

At the same time, a neighboring section, labeled $x + \Delta x$ in Fig. 21 and to the right of the section at x, has a displacement $y(x + \Delta x, t)$. The element of the bar with natural length Δx is, then, stretched or compressed by the amount $y(x + \Delta x, t) - y(x, t)$. We assume that such an extension or compression of the element satisfies Hooke's law and that the effect of the inertia of the moving element is negligible. Hence the force exerted on the section at x by the part of the bar to the left of that section is

$$-AE\frac{y(x+\Delta x, t) - y(x,t)}{\Delta x},$$

where A is the area of each cross section, the positive constant E is *Young's modulus of elasticity,* and the ratio shown represents the relative change in the length of the element. When Δx tends to zero, it follows that the longitudinal force $F(x,t)$ exerted on the element at its left-hand end is given by the basic equation

$$F(x,t) = -AEy_x(x,t). \tag{1}$$

Similarly, the force on the right-hand end is

$$F(x+\Delta x,t) = AEy_x(x+\Delta x,t). \tag{2}$$

Let the constant δ denote the mass per unit volume of the material. Then, applying Newton's second law to the motion of an element of the bar of length Δx, we may write

$$\delta A\,\Delta x\, y_{tt}(x,t) = -AEy_x(x,t) + AEy_x(x+\Delta x,t). \tag{3}$$

We find, after dividing by $\delta A\,\Delta x$ and letting Δx tend to zero, that

$$y_{tt}(x,t) = a^2 y_{xx}(x,t) \qquad \left(a^2 = \frac{E}{\delta}\right). \tag{4}$$

Thus the longitudinal displacements $y(x,t)$ in an elastic bar satisfy the wave equation (4) when no external longitudinal forces act on the bar, other than at the ends. We have assumed that displacements are small enough that Hooke's law applies and that sections remain planar after being displaced. The elastic bar here may be replaced by a column of air, in which case equation (4) has applications in the theory of sound.

The boundary condition $y(0,t) = 0$ signifies that the end $x = 0$ of the bar is held fixed. If, instead, the end $x = 0$ is free when $t > 0$, then no force acts at that end; that is, $F(0,t) = 0$ and, in view of equation (1),

$$y_x(0,t) = 0 \qquad (t > 0). \tag{5}$$

Turning to another type of vibration, we let $z(x,y,t)$ denote small displacements in the z direction, at time t, of points on a flexible membrane that is tightly stretched over a horizontal frame. In the equilibrium position, a point on the membrane has coordinates (x,y) in the xy plane. The plane through that point and parallel to the xz plane intersects the displaced membrane in a curve containing the points labeled A and B in Fig. 22. By making similar constructions,

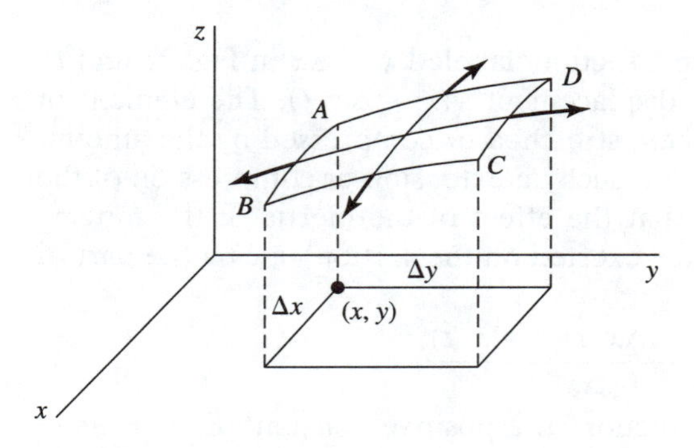

FIGURE 22

we can form the element $ABCD$ of the membrane that is also shown in Fig. 22. The projection of the element onto the xy plane is a small rectangle with edges of length Δx and Δy.

We now examine the internal tensile forces that are exerted on the element at points of the curve AB, those forces being tangent to the element and normal to AB. In analyzing such a force, we let H denote the magnitude *per unit length* along AB of the component parallel to the xy plane. We assume that H is constant, regardless of what point or curve on the membrane is being discussed. In view of expressions (1) and (2) in Sec. 25 for the forces V on the ends of a segment of a vibrating string, we know that the force in the z direction exerted over the curve AB is approximately $-Hz_y(x, y, t)\Delta x$ and that the corresponding force over the curve DC is approximately $Hz_y(x, y + \Delta y, t)\Delta x$. Similar expressions are found for the vertical forces exerted over AD and BC when the tensile forces on those curves are considered. It then follows that the sum of the vertical forces exerted over the entire boundary of the element is approximately

$$(6)\qquad \begin{aligned} -Hz_y(x, y, t)\,\Delta x + Hz_y(x, y + \Delta y, t)\,\Delta x \\ -Hz_x(x, y, t)\,\Delta y + Hz_x(x + \Delta x, y, t)\,\Delta y. \end{aligned}$$

If Newton's second law is applied to the motion of the element in the z direction and if δ denotes the mass per unit area of the membrane, it follows from expression (6) for the total force on the element that $z(x, y, t)$ satisfies the *two-dimensional* wave equation

$$(7)\qquad z_{tt} = a^2(z_{xx} + z_{yy}) \qquad \left(a^2 = \frac{H}{\delta}\right).$$

Details of these final steps are left to the problems, where it is also shown that if an external transverse force $F(x, y, t)$ per unit area acts over the membrane, the equation of motion takes the form

$$(8)\qquad z_{tt} = a^2(z_{xx} + z_{yy}) + \frac{F}{\delta}.$$

Equation (8) arises, for example, when the z axis is directed vertically upward and the weight of the membrane is taken into account in the derivation of equation (7). Then $F/\delta = -g$, where g is acceleration due to gravity.

From equation (7), one can see that the *static* transverse displacements $z(x, y)$ of a stretched membrane satisfy Laplace's equation (Sec. 21) in two dimensions. Here the displacements are the result of displacements, perpendicular to the xy plane, of parts of the frame that support the membrane when no external forces are exerted except at the boundary.

PROBLEMS

1. Let $z(\rho)$ represent static transverse displacements in a membrane, stretched between the two circles $\rho = 1$ and $\rho = \rho_0$ $(\rho_0 > 1)$ in the plane $z = 0$, after the outer support $\rho = \rho_0$ is displaced by a distance $z = z_0$. State why the boundary value problem in $z(\rho)$

can be written

$$\frac{d}{d\rho}\left(\rho\,\frac{dz}{d\rho}\right) = 0 \qquad (1 < \rho < \rho_0),$$

$$z(1) = 0, \qquad z(\rho_0) = z_0,$$

and obtain the solution

$$z(\rho) = z_0\,\frac{\ln\rho}{\ln\rho_0} \qquad (1 \le \rho \le \rho_0).$$

2. Show that the steady-state temperatures $u(\rho)$ in an infinitely long hollow cylinder $1 \le \rho \le \rho_0, -\infty < z < \infty$ also satisfy the boundary value problem written in Problem 1 if $u = 0$ on the inner cylindrical surface and $u = z_0$ on the outer one. Thus show that Problem 1 is a *membrane analogy* for this temperature problem. Soap films have been used to display such analogies.

3. The end $x = 0$ of a cylindrical elastic bar is kept fixed, and a constant compressive force of magnitude F_0 units per unit area is exerted at all times $t > 0$ over the end $x = c$. The bar is initially unstrained and at rest, with no external forces acting along it. State why the function $y(x, t)$ representing the longitudinal displacements of cross sections should satisfy this boundary value problem, where $a^2 = E/\delta$:

$$y_{tt}(x,t) = a^2 y_{xx}(x,t) \qquad (0 < x < c, t > 0),$$

$$y(0,t) = 0, \qquad Ey_x(c,t) = -F_0, \qquad y(x,0) = y_t(x,0) = 0.$$

4. The left-hand end $x = 0$ of a horizontal elastic bar is elastically supported in such a way that the longitudinal force per unit area exerted on the bar at that end is proportional to the displacement of the end, but opposite in sign. State why the end condition there has the form

$$y_x(0,t) = by(0,t) \qquad (b > 0).$$

5. Use expression (6), Sec. 26, to derive the nonhomogeneous wave equation (8), Sec. 26, for a membrane when there is an external transverse force $F(x, y, t)$ per unit area acting on it. [Note that if this force is zero ($F \equiv 0$), the equation reduces to equation (7), Sec. 26.]

6. Let $z(x, y)$ denote the static transverse displacements in a membrane over which an external transverse force $F(x, y)$ per unit area acts. Show how it follows from the nonhomogeneous wave equation (8), Sec. 26, that $z(x, y)$ satisfies *Poisson's equation*:

$$z_{xx} + z_{yy} + f = 0 \qquad \left(f = \frac{F}{H}\right).$$

[Compare with equation (7), Sec. 21.]

7. A uniform transverse force of F_0 units per unit area acts over a membrane, stretched between the two circles $\rho = 1$ and $\rho = \rho_0$ ($\rho_0 > 1$) in the plane $z = 0$. From Problem 6, show that the static transverse displacements $z(\rho)$ satisfy the equation

$$(\rho z')' + f_0\rho = 0 \qquad \left(f_0 = \frac{F_0}{H}\right),$$

and derive the expression

$$z(\rho) = \frac{f_0}{4}(\rho_0^2 - 1)\left(\frac{\ln\rho}{\ln\rho_0} - \frac{\rho^2 - 1}{\rho_0^2 - 1}\right) \qquad (1 \le \rho \le \rho_0).$$

27. GENERAL SOLUTION OF THE WAVE EQUATION

In this section, we shall find a general solution of the wave equation (Secs. 25 and 26)

$$y_{tt}(x,t) = a^2 y_{xx}(x,t) \qquad (-\infty < x < \infty, t > 0) \tag{1}$$

that can be useful in special cases.

The differential equation (1) can be simplified as follows by introducing the new independent variables

$$u = x + at, \qquad v = x - at. \tag{2}$$

According to the chain rule for differentiating composite functions,

$$\frac{\partial y}{\partial t} = \frac{\partial y}{\partial u}\frac{\partial u}{\partial t} + \frac{\partial y}{\partial v}\frac{\partial v}{\partial t}.$$

That is,

$$\frac{\partial y}{\partial t} = a\frac{\partial y}{\partial u} - a\frac{\partial y}{\partial v}. \tag{3}$$

Replacing the function y by $\partial y/\partial t$ in equation (3) yields the expression

$$\frac{\partial^2 y}{\partial t^2} = a\frac{\partial}{\partial u}\left(\frac{\partial y}{\partial t}\right) - a\frac{\partial}{\partial v}\left(\frac{\partial y}{\partial t}\right);$$

and using equation (3) again, this time to substitute for $\partial y/\partial t$ on the right here, we see that

$$\frac{\partial^2 y}{\partial t^2} = a\frac{\partial}{\partial u}\left(a\frac{\partial y}{\partial u} - a\frac{\partial y}{\partial v}\right) - a\frac{\partial}{\partial v}\left(a\frac{\partial y}{\partial u} - a\frac{\partial y}{\partial v}\right),$$

or

$$\frac{\partial^2 y}{\partial t^2} = a^2\left(\frac{\partial^2 y}{\partial u^2} - 2\frac{\partial^2 y}{\partial v\,\partial u} + \frac{\partial^2 y}{\partial v^2}\right). \tag{4}$$

We have, of course, assumed that

$$\frac{\partial^2 y}{\partial u\,\partial v} = \frac{\partial^2 y}{\partial v\,\partial u}.$$

In like manner, one can show that

$$\frac{\partial^2 y}{\partial x^2} = \frac{\partial^2 y}{\partial u^2} + 2\frac{\partial^2 y}{\partial v\,\partial u} + \frac{\partial^2 y}{\partial v^2}. \tag{5}$$

In view of expressions (4) and (5), then, equation (1) becomes

$$y_{uv} = 0 \tag{6}$$

with the change of variables (2).

Equation (6) can be solved by successive integrations to give $y_u = \phi'(u)$ and

$$y = \phi(u) + \psi(v),$$

where the arbitrary functions ϕ and ψ are twice differentiable. The *general solution* of the wave equation (1) is, therefore,

$$y(x,t) = \phi(x+at) + \psi(x-at). \tag{7}$$

EXAMPLE. Suppose that the function $y(x,t)$ in the wave equation (1) is subject to the boundary conditions

$$y(x,0) = f(x), \qquad y_t(x,0) = 0 \qquad (-\infty < x < \infty). \tag{8}$$

Physically, $y(x,t)$ represents transverse displacements in a stretched string of infinite length, initially released at rest from the position $y = f(x)$.

The boundary conditions here are simple enough that we can actually determine the functions ϕ and ψ in expression (7). Observe that the solution (7) satisfies conditions (8) when

$$\phi(x) + \psi(x) = f(x) \qquad \text{and} \qquad a\phi'(x) - a\psi'(x) = 0.$$

The second of these equations tells us that $\psi(x) = \phi(x) + C$, where C is some constant. It then follows from the first equation that

$$\phi(x) = \frac{1}{2}[f(x) - C] \qquad \text{and} \qquad \psi(x) = \frac{1}{2}[f(x) + C].$$

Hence expression (7) becomes

$$y(x,t) = \frac{1}{2}[f(x+at) + f(x-at)]. \tag{9}$$

The solution (9) of the boundary value problem consisting of equations (1) and (8) is known as *d'Alembert's solution*. It is easily verified under the assumption that $f'(x)$ and $f''(x)$ exist for all x.

Note how solution (9) can be used to display the instantaneous position of the string at time t graphically by adding ordinates of two curves, one obtained by translating the curve

$$y = \frac{1}{2}f(x) \tag{10}$$

to the right through the distance at and the other by translating it to the left through the same distance. As t varies, the curve (10) moves in each direction as a wave with velocity a.

PROBLEMS

1. Use the general solution (7), Sec. 27, of the wave equation to solve the boundary value problem

$$y_{tt}(x,t) = a^2 y_{xx}(x,t) \qquad (-\infty < x < \infty, t > 0),$$

$$y(x,0) = 0, \qquad y_t(x,0) = g(x) \qquad (-\infty < x < \infty).$$

Suggestion: Note that one can write

$$\int g(x)\,dx = \int_0^x g(s)\,ds + C.$$

Answer: $y(x,t) = \dfrac{1}{2a}\displaystyle\int_{x-at}^{x+at} g(s)\,ds.$

2. Let $Y(x,t)$ denote d'Alembert's solution (9), Sec. 27, of the boundary value problem solved in the example in that section, and let $Z(x,t)$ denote the solution found in Problem 1 for a related boundary value problem. Verify directly that the sum

$$y(x,t) = Y(x,t) + Z(x,t)$$

is a solution of the boundary value problem

$$y_{tt}(x,t) = a^2 y_{xx}(x,t) \qquad (-\infty < x < \infty, t > 0),$$

$$y(x,0) = f(x), \qquad y_t(x,0) = g(x) \qquad (-\infty < x < \infty).$$

Thus show that

$$y(x,t) = \frac{1}{2}[f(x+at) + f(x-at)] + \frac{1}{2a}\int_{x-at}^{x+at} g(s)\,ds$$

is a solution of the problem here.

3. Let $y(x,t)$ represent transverse displacements in a long stretched string one end of which is attached to a ring that can slide along the y axis. The other end is so far out on the positive x axis that it may be considered to be infinitely far from the origin. The ring is initially at the origin and is then moved along the y axis (Fig. 23) so that $y = f(t)$ when $x = 0$ and $t \geq 0$, where f is a prescribed continuous function and $f(0) = 0$. We assume that the string is initially at rest on the x axis; thus $y(x,t) \to 0$ as $x \to \infty$. The boundary value problem for $y(x,t)$ is

$$y_{tt}(x,t) = a^2 y_{xx}(x,t) \qquad (x > 0, t > 0),$$

$$y(x,0) = 0, \qquad y_t(x,0) = 0 \qquad (x \geq 0),$$

$$y(0,t) = f(t) \qquad (t \geq 0).$$

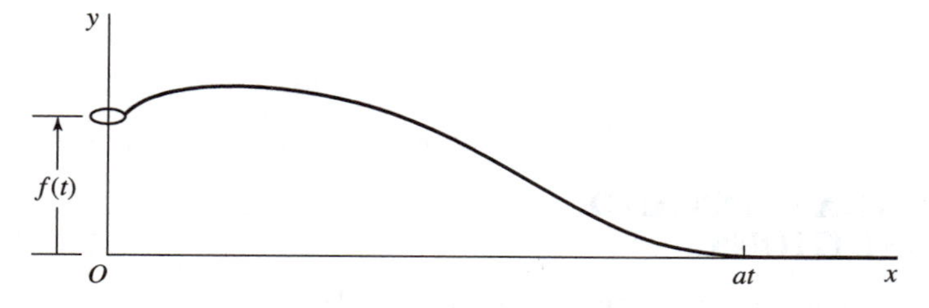

FIGURE 23

(*a*) Apply the first two of these boundary conditions to the general solution (Sec. 27)

$$y(x,t) = \phi(x+at) + \psi(x-at)$$

of the wave equation to show that there is a constant C such that

$$\phi(x) = C \qquad \text{and} \qquad \psi(x) = -C \qquad (x \geq 0).$$

Then apply the third boundary condition to show that

$$\psi(-x) = f\left(\frac{x}{a}\right) - C \qquad (x \geq 0),$$

where C is the same constant.

(*b*) With the aid of the results in part (*a*), derive the solution

$$y(x,t) = \begin{cases} 0 & \text{when } x \geq at, \\ f\left(t - \dfrac{x}{a}\right) & \text{when } x \leq at. \end{cases}$$

Note that the part of the string to the right of the point $x = at$ on the x axis is unaffected by the movement of the ring prior to time t, as shown in Fig. 23.

4. Use the solution obtained in Problem 3 to show that if the ring at the left-hand end of the string in that problem is moved according to the function

$$f(t) = \begin{cases} \sin \pi t & \text{when } 0 \leq t \leq 1, \\ 0 & \text{when } \quad t \geq 1, \end{cases}$$

then

$$y(x,t) = \begin{cases} 0 & \text{when } x \leq a(t-1) \text{ or } x \geq at, \\ \sin\left[\pi\left(t - \dfrac{x}{a}\right)\right] & \text{when } \quad a(t-1) \leq x \leq at. \end{cases}$$

Observe that the ring is lifted up 1 unit and then returned to the origin, where it remains after time $t = 1$. The expression for $y(x,t)$ here shows that when $t > 1$, the string coincides with the x axis except on an interval of length a, where it forms one arch of a sine curve (Fig. 24). Furthermore, as t increases, the arch moves to the right with speed a.

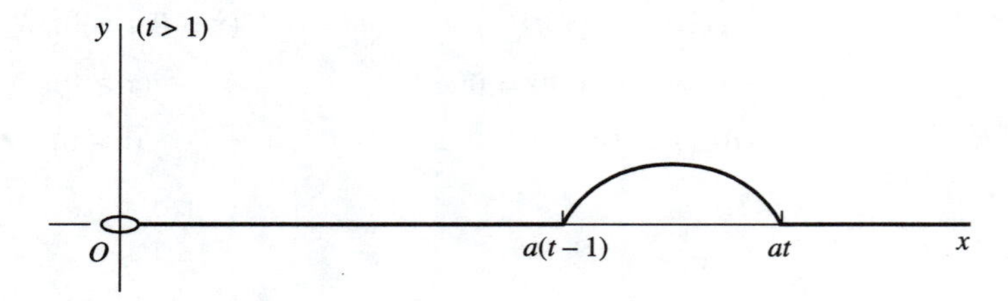

FIGURE 24

28. TYPES OF EQUATIONS AND BOUNDARY CONDITIONS

The second-order linear partial differential equation (Sec. 19)

$$\text{(1)} \qquad Au_{xx} + Bu_{xy} + Cu_{yy} + Du_x + Eu_y + Fu = G$$

in $u = u(x,y)$, where $A, B, \ldots, G$ are constants or functions of x and y, is classified in any given region of the xy plane according to whether $B^2 - 4AC$ is positive, negative, or zero throughout that region. Specifically, equation (1) is

(*i*) *hyperbolic* if $B^2 - 4AC > 0$;
(*ii*) *elliptic* if $B^2 - 4AC < 0$;
(*iii*) *parabolic* if $B^2 - 4AC = 0$.

For each of these categories, equation (1) and its solutions have distinct features. Some indication of this is given in Problems 2 and 3. The terminology here is suggested by the equation

$$Ax^2 + Bxy + Cy^2 + Dx + Ey + F = 0, \tag{2}$$

where $A, B, \ldots, F$ are constants. From analytic geometry, we recall that equation (2) represents a conic section in the xy plane and that the different types of conic sections arising are similarly determined by $B^2 - 4AC$.

EXAMPLES. Laplace's equation

$$u_{xx} + u_{yy} = 0$$

is a special case of equation (1) in which $A = C = 1$ and $B = 0$. Hence it is elliptic throughout the xy plane. Poisson's equation (Sec. 21)

$$u_{xx} + u_{yy} = f(x, y)$$

in two dimensions is elliptic in any region of the xy plane where $f(x, y)$ is defined. The one-dimensional heat equation

$$-ku_{xx} + u_t = 0$$

in $u = u(x, t)$ is parabolic in the xt plane, and the one-dimensional wave equation

$$-a^2 y_{xx} + y_{tt} = 0$$

in $y = y(x, t)$ is hyperbolic there.

As indicated below, the three types of second-order linear equations just described require, in general, different types of boundary conditions to determine a solution.

Let u denote the dependent variable in a boundary value problem. A condition that prescribes the values of u itself along a portion of the boundary is known as a *Dirichlet condition.* The problem of determining a harmonic function on a domain such that the function assumes prescribed values over the entire boundary of that domain is called a *Dirichlet problem.* In that case, the values of the function can be interpreted as steady-state temperatures. Such a physical interpretation leads us to expect that a Dirichlet problem may have a unique solution if the functions considered satisfy certain requirements as to their regularity.

A *Neumann condition* prescribes the values of normal derivatives du/dn on a part of the boundary. Another type of boundary condition is a *Robin condition.*

It prescribes values of $hu + du/dn$ at boundary points, where h is either a constant or a function of the independent variables.[†]

If a partial differential equation in y is of the second order with respect to one of the independent variables t and if the values of both y and y_t are prescribed when $t = 0$, the boundary condition is one of *Cauchy type* with respect to t. In the case of the wave equation $y_{tt} = a^2 y_{xx}$, such a condition corresponds physically to that of prescribing the initial values of the transverse displacements y and velocities y_t in a stretched string. Initial values for both y and y_t appear to be needed if the displacements $y(x, t)$ are to be determined.

When the equation is Laplace's equation $u_{xx} + u_{yy} = 0$ or the heat equation $ku_{xx} = u_t$, however, conditions of Cauchy type on u with respect to x cannot be imposed without severe restrictions. This is suggested by interpreting u physically as a temperature function. When the temperatures u in a slab $0 \le x \le c$ are prescribed on the face $x = 0$, for example, the flux Ku_x to the left through that face is ordinarily determined by the values of u there and by other conditions in the problem. Conversely, if the flux Ku_x is prescribed at $x = 0$, the temperatures there are affected.

PROBLEMS

1. Whether a second-order linear partial differential equation in $u = u(x, y)$ is hyperbolic, elliptic, or parabolic can vary from region to region in the xy plane when at least one of the coefficients is a nonconstant function of x and y. Classify each of the following differential equations in various regions, and sketch those regions.

(*a*) $yu_{xx} + u_{yy} = 0$; (*b*) $u_{xx} + 2x^2 u_{xy} + yu_{yy} = 0$;

(*c*) $xu_{xx} + yu_{yy} - 3u_y = 2$; (*d*) $u_{xx} - 2xu_{xy} + (1 - y^2)u_{yy} = 0$.

Answers: (*a*) Parabolic on the x axis, elliptic above it, and hyperbolic below it; (*b*) parabolic on the curve $y = x^4$, elliptic above it, and hyperbolic below it; (*d*) parabolic on the circle $x^2 + y^2 = 1$, elliptic inside it, and hyperbolic outside of it.

2. Consider the partial differential equation

$$Ay_{xx} + By_{xt} + Cy_{tt} = 0 \qquad (A \ne 0,\ C \ne 0),$$

where A, B, and C are constants, and assume that it is *hyperbolic*, so that $B^2 - 4AC > 0$.

(*a*) Use the transformation

$$u = x + \alpha t, \qquad v = x + \beta t \qquad (\alpha \ne \beta)$$

to obtain the new differential equation

$$(A + B\alpha + C\alpha^2)y_{uu} + [2A + B(\alpha + \beta) + 2C\alpha\beta]y_{uv} + (A + B\beta + C\beta^2)y_{vv} = 0.$$

[†]When such a condition is prescribed on the entire boundary of a region throughout which u is harmonic, the boundary value problem is sometimes referred to as a *Churchill problem*. See pp. 154–156 of the book by Sneddon (1957) that is listed in the Bibliography.

(*b*) Show that when α and β have the values

$$\alpha_0 = \frac{-B + \sqrt{B^2 - 4AC}}{2C} \quad \text{and} \quad \beta_0 = \frac{-B - \sqrt{B^2 - 4AC}}{2C},$$

respectively, the differential equation in part (*a*) reduces to $y_{uv} = 0$.

(*c*) Conclude from the result in part (*b*) that the general solution of the original differential equation is

$$y = \phi(x + \alpha_0 t) + \psi(x + \beta_0 t),$$

where ϕ and ψ are arbitrary functions that are twice differentiable. Then show how the general solution (7), Sec. 27, of the wave equation

$$-a^2 y_{xx} + y_{tt} = 0$$

follows as a special case.

3. Show that under the transformation

$$u = x, \qquad v = \alpha x + \beta t \qquad (\beta \neq 0),$$

the given differential equation in Problem 2 becomes

$$A y_{uu} + (2A\alpha + B\beta) y_{uv} + (A\alpha^2 + B\alpha\beta + C\beta^2) y_{vv} = 0.$$

Then show that this new equation reduces to

(*a*) $y_{uu} + y_{vv} = 0$ when the original equation is *elliptic* ($B^2 - 4AC < 0$) and

$$\alpha = \frac{-B}{\sqrt{4AC - B^2}}, \qquad \beta = \frac{2A}{\sqrt{4AC - B^2}};$$

(*b*) $y_{uu} = 0$ when the original equation is *parabolic* ($B^2 - 4AC = 0$) and

$$\alpha = -B, \qquad \beta = 2A.$$

CHAPTER 4

THE FOURIER METHOD

We turn now to a careful presentation of the Fourier method for solving boundary value problems involving partial differential equations. Once the basics of the method have been developed, we shall, in Chap. 5, use it to solve a variety of boundary value problems whose solutions entail Fourier series. Then, in subsequent chapters, we shall apply the method to problems with solutions involving other, but closely related, types of representations.

29. LINEAR OPERATORS

If u_1 and u_2 are functions and c_1 and c_2 are constants, the function $c_1u_1 + c_2u_2$ is called a *linear combination* of u_1 and u_2. Note that $u_1 + u_2$ and c_1u_1, as well as the constant function 0, are special cases. A linear space of functions, or *function space,* is a class of functions, all with a common domain of definition, such that each linear combination of any two functions in that class remains in it; that is, if u_1 and u_2 are in the class, then so is $c_1u_1 + c_2u_2$. An example is the function space $C_p(a, b)$, introduced in Sec. 1.

A *linear operator* L on a given function space transforms each function u of that space into a function Lu, which need not be in the space, and has the property that for each pair of functions u_1 and u_2,

$$L(c_1u_1 + c_2u_2) = c_1Lu_1 + c_2Lu_2 \tag{1}$$

whenever c_1 and c_2 are constants. In particular,

$$L(u_1 + u_2) = Lu_1 + Lu_2 \quad \text{and} \quad L(c_1u_1) = c_1Lu_1. \tag{2}$$

The function Lu may be a constant function; in particular,

$$L(0) = L(0 \cdot 0) = 0L(0) = 0.$$

If u_3 is a third function in the space, then

$$\begin{aligned} L(c_1u_1 + c_2u_2 + c_3u_3) &= L(c_1u_1 + c_2u_2) + L(c_3u_3) \\ &= c_1 Lu_1 + c_2 Lu_2 + c_3 Lu_3. \end{aligned}$$

Proceeding by induction, we find that L transforms linear combinations of N functions in this manner:

$$L\left(\sum_{n=1}^{N} c_n u_n\right) = \sum_{n=1}^{N} c_n Lu_n. \tag{3}$$

EXAMPLE 1. Suppose that both u_1 and u_2 are functions of the independent variables x and y. According to elementary properties of derivatives, a derivative of any linear combination of the two functions can be written as the same linear combination of the individual derivatives. Thus

$$\frac{\partial}{\partial x}(c_1u_1 + c_2u_2) = c_1\frac{\partial u_1}{\partial x} + c_2\frac{\partial u_2}{\partial x}, \tag{4}$$

provided that $\partial u_1/\partial x$ and $\partial u_2/\partial x$ exist. In view of property (4), the class of functions of x and y that have partial derivatives of the first order with respect to x in the xy plane is a function space. The operator $\partial/\partial x$ is a linear operator on that space. It is naturally classified as a linear *differential operator*.

EXAMPLE 2. Consider a space of functions $u(x, y)$ defined on the xy plane. If $f(x, y)$ is a fixed function, also defined on the xy plane, then the operator L that multiplies each function $u(x, y)$ by $f(x, y)$ is a linear operator, where $Lu = fu$.

If linear operators L and M, distinct or not, are such that M transforms each function u of some function space into a function Mu to which L applies and if u_1 and u_2 are functions in that space, it follows from equation (1) that

$$LM(c_1u_1 + c_2u_2) = L(c_1 Mu_1 + c_2 Mu_2) = c_1 LMu_1 + c_2 LMu_2. \tag{5}$$

That is, the *product* LM of linear operators is itself a linear operator.

The *sum* of two linear operators L and M is defined by the equation

$$(L + M)u = Lu + Mu \tag{6}$$

and is found to be linear by writing

$$\begin{aligned} (L + M)(c_1u_1 + c_2u_2) &= L(c_1u_1 + c_2u_2) + M(c_1u_1 + c_2u_2) \\ &= c_1(Lu_1 + Mu_1) + c_2(Lu_2 + Mu_2) \\ &= c_1(L + M)u_1 + c_2(L + M)u_2. \end{aligned}$$

The sum of any finite number of linear operators is, in fact, linear.

EXAMPLE 3. Let L denote the linear operator $\partial^2/\partial x^2$ defined on the space of functions $u(x, y)$ whose derivatives of the first and second orders with respect to x exist in a given domain of the xy plane. The product $M = f\,\partial/\partial x$ of the linear operators in Examples 1 and 2 is linear on the same space, and the sum

$$L + M = \frac{\partial^2}{\partial x^2} + f\frac{\partial}{\partial x}$$

is therefore linear.

30. PRINCIPLE OF SUPERPOSITION

Each nonzero term of a linear homogeneous differential equation in u consists of a constant or a function of the independent variables alone times one of the derivatives of u or u itself. Hence every linear homogeneous differential equation has the form

$$Lu = 0, \tag{1}$$

where L is a linear differential operator.

In particular, we recall from Sec. 19 that

$$Au_{xx} + Bu_{xy} + Cu_{yy} + Du_x + Eu_y + Fu = 0, \tag{2}$$

where the letters A through F denote constants or functions of x and y only, is the general second-order linear homogeneous partial differential equation in $u(x, y)$. It can be written in form (1) when

$$L = A\frac{\partial^2}{\partial x^2} + B\frac{\partial^2}{\partial y\,\partial x} + C\frac{\partial^2}{\partial y^2} + D\frac{\partial}{\partial x} + E\frac{\partial}{\partial y} + F. \tag{3}$$

Linear homogeneous boundary conditions also have the form (1). Then the variables appearing as arguments of u and as arguments of functions that serve as coefficients in the linear operator L are restricted so that they represent points on the boundary of the domain.

We now state a *principle of superposition*, which is fundamental to the Fourier method for solving linear boundary value problems.

Theorem. *Suppose that each function of an infinite set $u_1, u_2, \ldots$ satisfies a linear homogeneous differential equation or boundary condition $Lu = 0$. Then the infinite series*

$$u = \sum_{n=1}^{\infty} c_n u_n, \tag{4}$$

where the c_n are constants, also satisfies $Lu = 0$, provided that the series converges and is differentiable for all derivatives involved in L and provided that any required continuity condition at the boundary is satisfied by Lu when $Lu = 0$ is a boundary condition.

Superposition is also useful in the theory of ordinary differential equations. For example, from the two solutions $y = e^x$ and $y = e^{-x}$ of the linear homogeneous equation $y'' - y = 0$, we know that $y = c_1 e^x + c_2 e^{-x}$ is also a solution. In this book, we shall be concerned mainly with applying the principle of superposition to solutions of *partial* differential equations.

To prove the theorem, we must deal with the convergence and differentiability of infinite series. Suppose that functions u_n and constants c_n are such that series (4) converges to u throughout some domain of the independent variables, and let x represent one of those variables. The series is *differentiable,* or termwise differentiable, with respect to x if the derivatives $\partial u_n/\partial x$ and $\partial u/\partial x$ exist and if the series of functions $c_n \partial u_n/\partial x$ converges to $\partial u/\partial x$:

$$\frac{\partial u}{\partial x} = \sum_{n=1}^{\infty} c_n \frac{\partial u_n}{\partial x}. \tag{5}$$

Note that a series must be convergent if it is to be differentiable. If, in addition, series (5) is differentiable with respect to x, then series (4) is differentiable twice with respect to x.

Let L be a linear operator where Lu is a product of a function f of the independent variables by u or by a derivative of u, or where Lu is a sum of a finite number of such terms. We now show that if series (4) is differentiable for all the derivatives involved in L and if each of the functions u_n in series (4) satisfies the linear homogeneous differential equation $Lu = 0$, then series (4) satisfies it.

To accomplish this, we first note that according to the definition of the sum of an infinite series,

$$f \frac{\partial u}{\partial x} = f \lim_{N\to\infty} \sum_{n=1}^{N} c_n \frac{\partial u_n}{\partial x}$$

when series (4) is differentiable with respect to x. Thus

$$f \frac{\partial u}{\partial x} = \lim_{N\to\infty} f \frac{\partial}{\partial x} \sum_{n=1}^{N} c_n u_n. \tag{6}$$

Here the operator $\partial/\partial x$ can be replaced by other derivatives if the series is so differentiable. Then, by adding corresponding sides of equations similar to equation (6), including one that may not have any derivative at all, we find that

$$Lu = \lim_{N\to\infty} L\left(\sum_{n=1}^{N} c_n\, u_n\right). \tag{7}$$

The sum on the right-hand side of equation (7) is a linear combination of the functions $u_1, u_2, \ldots, u_N$; and if $Lu_n = 0$ $(n = 1, 2, \ldots)$, it follows, with the aid of property (3), Sec. 29, that

$$Lu = \lim_{N\to\infty} \sum_{n=1}^{N} c_n L u_n = \lim_{N\to\infty} 0 = 0.$$

This is, of course, the desired result.

The above discussion applies as well to linear homogeneous boundary conditions $Lu = 0$. In that case, we may require the function Lu to satisfy a condition of continuity at points on the boundary so that its values there will represent limiting values as those points are approached from the interior of the domain. This completes the proof of the theorem.

We now illustrate how the superposition theorem is to be used in solving boundary value problems. In our discussion, we shall assume that needed conditions for convergence and differentiability of series are satisfied. We assume, moreover, that any continuity requirements involving boundary conditions are satisfied. The examples here will be used in sections immediately following this one, where two boundary value problems will be completely solved by the Fourier method.

EXAMPLE 1. Consider the linear homogeneous heat equation (Sec. 20)

$$u_t(x, t) = k u_{xx}(x, t) \qquad (0 < x < c, t > 0), \tag{8}$$

together with the linear homogeneous boundary conditions

$$u_x(0, t) = 0, \qquad u_x(c, t) = 0 \qquad (t > 0). \tag{9}$$

Equation (8) takes the form $Lu = 0$ when

$$L = k\frac{\partial^2}{\partial x^2} - \frac{\partial}{\partial t};$$

and it is straightforward to verify that if

$$u_0 = 1, \qquad u_n = \exp\left(-\frac{n^2\pi^2 k}{c^2}t\right)\cos\frac{n\pi x}{c} \qquad (n = 1, 2, \ldots), \tag{10}$$

then $Lu_0 = 0$ and $Lu_n = 0\,(n = 1, 2, \ldots)$. Thus, by the superposition theorem, $Lu = 0$ if u denotes the infinite series

$$u = A_0\,u_0 + \sum_{n=1}^{\infty} A_n\,u_n, \tag{11}$$

where A_n $(n = 0, 1, 2, \ldots)$ are constants. In view of expressions (10), then, the series

$$u(x, t) = A_0 + \sum_{n=1}^{\infty} A_n \exp\left(-\frac{n^2\pi^2 k}{c^2}t\right)\cos\frac{n\pi x}{c} \tag{12}$$

satisfies the heat equation (8).

Conditions (9) can, moreover, be written in terms of the operator $L = \partial/\partial x$ as $Lu = 0$, where Lu is to be evaluated at $x = 0$ and at $x = c$. Hence, by the superposition theorem, series (12) satisfies conditions (9).

This example will be used in Sec. 31, where functions (10) are discovered and where it is shown how the results here can be used to complete the solution of a certain boundary value problem for temperatures in a slab.

EXAMPLE 2. It is easy to verify that if L is the linear operator

$$L = a^2\frac{\partial^2}{\partial x^2} - \frac{\partial^2}{\partial t^2}$$

and

$$y_n = \sin\frac{n\pi x}{c}\cos\frac{n\pi at}{c} \qquad (n = 1, 2, \ldots), \tag{13}$$

where a and c are positive constants, then $Ly_n = 0$ $(n = 1, 2, \ldots)$. Hence it follows from our theorem that $Ly = 0$ when y is the infinite series

$$y = \sum_{n=1}^{\infty} B_n\, y_n. \tag{14}$$

That is, the series

$$y(x, t) = \sum_{n=1}^{\infty} B_n \sin\frac{n\pi x}{c}\cos\frac{n\pi at}{c} \tag{15}$$

satisfies the wave equation (Sec. 25)

$$y_{tt}(x, t) = a^2 y_{xx}(x, t) \qquad (0 < x < c, t > 0). \tag{16}$$

Now write $L = 1$ and observe that Ly_n $(n = 1, 2, \ldots)$ has value zero when $x = 0$ and when $x = c$. In view of our superposition theorem, this shows that series (15) also satisfies the boundary conditions

$$y(0, t) = 0, \qquad y(c, t) = 0 \qquad (t > 0). \tag{17}$$

On the other hand, if $L = \partial/\partial t$, each Ly_n $(n = 1, 2, \ldots)$ has value zero when $t = 0$. So, by the superposition theorem, series (15) satisfies the condition

$$y_t(x, 0) = 0 \qquad (0 < x < c). \tag{18}$$

The differential equation (16) and boundary conditions (17) and (18) are part of a boundary value problem for a vibrating string that will be fully solved in Sec. 32, where it is shown how the functions (13) arise.

PROBLEMS

1. Show that if an operator L has the two properties

$$L(u_1 + u_2) = Lu_1 + Lu_2, \qquad L(c_1 u_1) = c_1 Lu_1$$

for all functions u_1, u_2 in some space and for every constant c_1, then L is linear; that is, show that it has property (1), Sec. 29.

2. Use the linear operators $L = x$ and $M = \partial/\partial x$ to illustrate the fact that products LM and ML are not always the same.

3. Verify that each of the functions

$$u_0 = y, \qquad u_n = \sinh ny \cos nx \qquad (n = 1, 2, \ldots)$$

satisfies Laplace's equation

$$u_{xx}(x, y) + u_{yy}(x, y) = 0 \qquad (0 < x < \pi,\ 0 < y < 2)$$

and the three boundary conditions

$$u_x(0, y) = u_x(\pi, y) = 0, \qquad u(x, 0) = 0.$$

Then use the superposition principle in Sec. 30 to show formally, without considering questions of convergence, differentiability, or continuity, that the series

$$u(x, y) = A_0\, y + \sum_{n=1}^{\infty} A_n \sinh ny \cos nx$$

satisfies the same differential equation and boundary conditions.

4. Show that each of the functions

$$y_1 = \frac{1}{x} \qquad \text{and} \qquad y_2 = \frac{1}{1+x}$$

satisfies the *nonlinear* differential equation

$$y' + y^2 = 0.$$

Then show that the sum $y_1 + y_2$ fails to satisfy that equation. Also show that if c is a constant, where $c \neq 0$ and $c \neq 1$, neither cy_1 nor cy_2 satisfies the equation.

5. Let u_1 and u_2 satisfy a linear *nonhomogeneous* differential equation $Lu = f$, where f is a function of the independent variables only. Prove that the linear combination $c_1u_1 + c_2u_2$ fails to satisfy that equation when $c_1 + c_2 \neq 1$.

6. Let L denote a linear differential operator, and suppose that f is a function of the independent variables. Show that the solutions u of the equation $Lu = f$ are of the form $u = u_1 + u_2$, where the u_1 are the solutions of the equation $Lu_1 = 0$ and u_2 is any particular solution of $Lu_2 = f$. (This is a principle of superposition of solutions for *nonhomogeneous* differential equations.)

7. Use mathematical induction on the integer N to verify property (3), Sec. 29, of a linear operator:

$$L\left(\sum_{n=1}^{N} c_n u_n\right) = \sum_{n=1}^{N} c_n L u_n.$$

Suggestion: Point out that the property is true when $N = 1$ and then show that if it is true when N is any positive integer M, it must be true for $N = M + 1$.

31. A TEMPERATURE PROBLEM

The linear boundary value problem

$$u_t(x, t) = k u_{xx}(x, t) \qquad (0 < x < c,\ t > 0), \tag{1}$$

$$u_x(0, t) = 0, \qquad u_x(c, t) = 0 \qquad (t > 0), \tag{2}$$

$$u(x, 0) = f(x) \qquad (0 < x < c) \tag{3}$$

is a problem for the temperatures $u(x, t)$ in an infinite slab of material, bounded by the planes $x = 0$ and $x = c$, if its faces are insulated and the initial temperature distribution is a prescribed function $f(x)$ of the distance from the face $x = 0$. (See Fig. 25.) We assume that the thermal conductivity k of the material is constant throughout the slab and that no heat is generated within it.

In this section, we illustrate the Fourier method for solving linear boundary value problems by solving the temperature problem just stated. A number of the

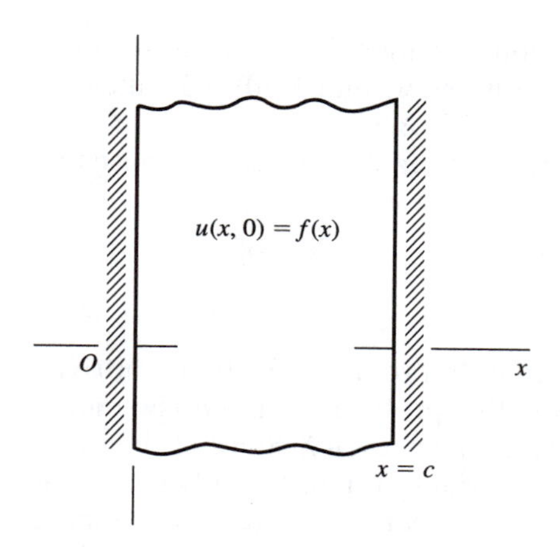

FIGURE 25

steps to be taken here are only *formal,* or manipulative. A verification of the final solution can be found in Chap. 11 (Sec. 95).

To determine nontrivial ($u \not\equiv 0$) functions that satisfy the homogeneous conditions (1) and (2), we seek *separated solutions* of those conditions,† or functions of the form

$$u = X(x)T(t) \tag{4}$$

that satisfy them. Note that X is a function of x alone and T is a function of t alone. Note, too, that X and T must be nontrivial ($X \not\equiv 0$, $T \not\equiv 0$).

If $u = XT$ satisfies equation (1), then

$$X(x)T'(t) = kX''(x)T(t);$$

and, for values of x and t such that the product $X(x)T(t)$ is nonzero, we can divide by $kX(x)T(t)$ to separate the variables:

$$\frac{T'(t)}{kT(t)} = \frac{X''(x)}{X(x)}.$$

Since the left-hand side here is a function of t alone, it does not vary with x. However, it is equal to a function of x alone, and so it cannot vary with t. Hence the two sides must have some constant value $-\lambda$ in common. That is,

$$\frac{T'(t)}{kT(t)} = -\lambda, \qquad \frac{X''(x)}{X(x)} = -\lambda.$$

Our choice of $-\lambda$, rather than λ, for the *separation constant* is, of course, a minor matter of notation. It is only for convenience later on (Chap. 8) that we have written $-\lambda$.

---

†This terminology is borrowed from the book by Pinsky (2003), which is listed in the Bibliography.

If $u = XT$ is to satisfy the first of conditions (2), then $X'(0)T(t)$ must vanish for all t ($t > 0$). With our requirement that $T \not\equiv 0$, it follows that $X'(0) = 0$. Likewise, the second of conditions (2) is satisfied by $u = XT$ if $X'(c) = 0$.

Thus $u = XT$ satisfies conditions (1) and (2) when X and T satisfy these two homogeneous problems:

$$X''(x) + \lambda X(x) = 0, \qquad X'(0) = 0, \qquad X'(c) = 0, \tag{5}$$

$$T'(t) + \lambda k T(t) = 0, \tag{6}$$

where the parameter λ has the *same* value in both problems. To find nontrivial solutions of this pair of problems, we first note that problem (6) has no boundary conditions. Hence it has nontrivial solutions for all values of λ. Since problem (5) has two boundary conditions, it may have nontrivial solutions for only particular values of λ. Problem (5) is called a *Sturm-Liouville problem*. The general theory of such problems is developed in Chap. 8, where it is shown that λ *must be real-valued* in order for there to be nontrivial solutions.

If $\lambda = 0$, the differential equation in problem (5) becomes $X''(x) = 0$. Its general solution is $X(x) = Ax + B$, where A and B are constants. Since $X'(x) = A$, the boundary conditions $X'(0) = 0$ and $X'(c) = 0$ require that $A = 0$. So $X(x) = B$; and, except for a constant factor, problem (5) has the solution $X(x) = 1$ if $\lambda = 0$. Note that any nonzero value of B might have been selected here.

If $\lambda > 0$, we can write $\lambda = \alpha^2$ ($\alpha > 0$). The differential equation in problem (5) then takes the form $X''(x) + \alpha^2 X(x) = 0$, its general solution being

$$X(x) = C_1 \cos \alpha x + C_2 \sin \alpha x.$$

Writing

$$X'(x) = -C_1 \alpha \sin \alpha x + C_2 \alpha \cos \alpha x$$

and keeping in mind that α is positive and, in particular, nonzero, we see that the condition $X'(0) = 0$ implies that $C_2 = 0$. Also, from the condition $X'(c) = 0$, it follows that $C_1 \alpha \sin \alpha c = 0$. Now if $X(x)$ is to be a nontrivial solution of problem (5), $C_1 \neq 0$. Hence α must be a positive root of the equation $\sin \alpha c = 0$. That is,

$$\alpha = \frac{n\pi}{c} \qquad (n = 1, 2, \ldots).$$

So, except for the constant factor C_1,

$$X(x) = \cos \frac{n\pi x}{c} \qquad (n = 1, 2, \ldots).$$

If $\lambda < 0$, we write $\lambda = -\alpha^2$ ($\alpha > 0$). This time, the differential equation in problem (5) has the general solution

$$X(x) = C_1 e^{\alpha x} + C_2 e^{-\alpha x}.$$

Since

$$X'(x) = C_1 \alpha e^{\alpha x} - C_2 \alpha e^{-\alpha x},$$

the condition $X'(0) = 0$ implies that $C_2 = C_1$. Hence

$$X(x) = C_1 (e^{\alpha x} + e^{-\alpha x}),$$

or

$$X(x) = 2C_1 \cosh \alpha x.$$

But the condition $X'(c) = 0$ requires that $C_1 \sinh \alpha c = 0$; and, since $\sinh \alpha c \neq 0$, it follows that $C_1 = 0$. So problem (5) has only the trivial solution $X(x) \equiv 0$ if $\lambda < 0$.

The values

$$\text{(7)} \qquad \lambda_0 = 0, \qquad \lambda_n = \left(\frac{n\pi}{c}\right)^2 \qquad (n = 1, 2, \ldots)$$

of λ for which problem (5) has nontrivial solutions are called *eigenvalues* of that problem, and the solutions

$$\text{(8)} \qquad X_0(x) = 1, \qquad X_n(x) = \cos\frac{n\pi x}{c} \qquad (n = 1, 2, \ldots)$$

are the corresponding *eigenfunctions*.

Turning to the differential equation (6), we need to determine its solutions $T_0(t)$ and $T_n(t)$ $(n = 1, 2, \ldots)$ corresponding to each of the eigenvalues λ_0 and λ_n $(n = 1, 2, \ldots)$. Those solutions are found to be constant multiples of

$$\text{(9)} \qquad T_0(t) = 1, \quad T_n(t) = \exp\left(-\frac{n^2\pi^2 k}{c^2}t\right) \qquad (n = 1, 2, \ldots).$$

Hence each of the products

$$\text{(10)} \qquad u_0 = X_0(x)T_0(t) = 1$$

and

$$\text{(11)} \qquad u_n = X_n(x)T_n(t) = \exp\left(-\frac{n^2\pi^2 k}{c^2}t\right)\cos\frac{n\pi x}{c} \qquad (n = 1, 2, \ldots)$$

satisfies the homogeneous conditions (1) and (2). The procedure just used to obtain them is called the *method of separation of variables*.

Now, as already shown in Example 1, Sec. 30, the superposition principle in that section tells us that the generalized linear combination

$$\text{(12)} \qquad u(x, t) = A_0 + \sum_{n=1}^{\infty} A_n \exp\left(-\frac{n^2\pi^2 k}{c^2}t\right)\cos\frac{n\pi x}{c}$$

of the functions (10) and (11) satisfies conditions (1) and (2). The constants A_n $(n = 0, 1, 2, \ldots)$ in expression (12) are readily obtained from the nonhomogeneous condition (3), namely $u(x, 0) = f(x)$. More precisely, by writing $t = 0$ in expression (12), we have

$$f(x) = \frac{2A_0}{2} + \sum_{n=1}^{\infty} A_n \cos\frac{n\pi x}{c} \qquad (0 < x < c).$$

Since this is a Fourier cosine series on $0 < x < c$ (see Sec. 14), it follows that

$$\text{(13)} \qquad A_0 = \frac{1}{c}\int_0^c f(x)\,dx$$

and

$$A_n = \frac{2}{c}\int_0^c f(x)\cos\frac{n\pi x}{c}\,dx \qquad (n = 1, 2, \ldots). \tag{14}$$

The formal solution of our temperature problem is now complete. It consists of expression (12) together with coefficients (13) and (14). Note that the steady-state temperatures, occurring when t tends to infinity, are A_0. That constant temperature is evidently the mean, or average, value of the initial temperatures $f(x)$ over the interval $0 < x < c$.

EXAMPLE. Suppose that the thickness c of the slab is unity and that the initial temperatures are $f(x) = x$ $(0 \le x \le 1)$. Here

$$A_0 = \int_0^1 x\,dx = \frac{1}{2}.$$

Using integration by parts, or Kronecker's method (Sec. 5), and observing that

$$\sin n\pi = 0 \qquad \text{and} \qquad \cos n\pi = (-1)^n$$

when n is an integer, we find that

$$A_n = 2\int_0^1 x\cos n\pi x\,dx = 2\left[\frac{x\sin n\pi x}{n\pi} + \frac{\cos n\pi x}{n^2\pi^2}\right]_0^1 = \frac{2}{\pi^2}\cdot\frac{(-1)^n - 1}{n^2}$$

$$(n = 1, 2, \ldots).$$

When $c = 1$ and these values for A_n $(n = 0, 1, 2, \ldots)$ are used, expression (12) becomes

$$u(x,t) = \frac{1}{2} + \frac{2}{\pi^2}\sum_{n=1}^{\infty}\frac{(-1)^n - 1}{n^2}\exp(-n^2\pi^2 kt)\cos n\pi x,$$

or[†]

$$u(x,t) = \frac{1}{2} - \frac{4}{\pi^2}\sum_{n=1}^{\infty}\frac{\exp[-(2n-1)^2\pi^2 kt]}{(2n-1)^2}\cos(2n-1)\pi x. \tag{15}$$

PROBLEMS

1. In Problem 3, Sec. 30, the functions

$$u_0 = y, \qquad u_n = \sinh ny\cos nx \qquad (n = 1, 2, \ldots)$$

were shown to satisfy Laplace's equation

$$u_{xx}(x,y) + u_{yy}(x,y) = 0 \qquad (0 < x < \pi,\ 0 < y < 2)$$

[†]Since the coefficients in this series are zero when n is even, the index n can be replaced by $2n - 1$ wherever it appears after the summation symbol. Compare with Example 1, Sec. 3.

and the homogeneous boundary conditions

$$u_x(0, y) = u_x(\pi, y) = 0, \qquad u(x, 0) = 0.$$

After writing $u = X(x)Y(y)$ and separating variables, use the solutions of the Sturm-Liouville problem in Sec. 31 to show how the functions u_0 and u_n $(n = 1, 2, \ldots)$ can be discovered. Then, by proceeding formally, derive the following solution of the boundary value problem resulting when the nonhomogeneous condition $u(x, 2) = f(x)$ is included:

$$u(x, y) = A_0 y + \sum_{n=1}^{\infty} A_n \sinh ny \cos nx,$$

where

$$A_0 = \frac{1}{2\pi}\int_0^\pi f(x)\,dx, \qquad A_n = \frac{2}{\pi \sinh 2n}\int_0^\pi f(x)\cos nx\,dx \quad (n = 1, 2, \ldots).$$

2. Suppose that in Sec. 31 we had written

$$\frac{T'(t)}{T(t)} = k\frac{X''(x)}{X(x)} \quad \text{instead of} \quad \frac{T'(t)}{kT(t)} = \frac{X''(x)}{X(x)}.$$

Continuing with

$$\frac{T'(t)}{T(t)} = k\frac{X''(x)}{X(x)} = -\lambda,$$

where λ is a separation constant, show how the functions (10) and (11) in Sec. 31 still follow. (This illustrates how it is generally simpler to keep the physical constants out of the Sturm-Liouville problem, as we did in Sec. 31.)

3. For each of the following partial differential equations in $u = u(x, t)$, determine if it is possible to write $u = X(x)T(t)$ and separate variables to obtain ordinary differential equations in X and T. If it can be done, find those ordinary differential equations.

(*a*) $u_{xx} - xtu_{tt} = 0$; (*b*) $(x + t)u_{xx} - u_t = 0$;
(*c*) $xu_{xx} + u_{xt} + tu_{tt} = 0$; (*d*) $u_{xx} - u_{tt} - u_t = 0$.

32. A VIBRATING STRING PROBLEM

To illustrate further the Fourier method, we now consider a boundary value problem for displacements in a vibrating string. This time, the nonhomogeneous condition will require us to expand a function $f(x)$ into a sine series, rather than a cosine series.

Let us find an expression for the transverse displacements $y(x, t)$ in a string, stretched between the points $x = 0$ and $x = c$ on the x axis and with no external forces acting along it, if the string is initially displaced into a position $y = f(x)$ and released at rest from that position. The function $y(x, t)$ must satisfy the wave equation (Sec. 25)

$$(1) \qquad y_{tt}(x, t) = a^2 y_{xx}(x, t) \qquad (0 < x < c, t > 0).$$

It must also satisfy the boundary conditions

$$(2) \qquad y(0, t) = 0, \qquad y(c, t) = 0, \qquad y_t(x, 0) = 0,$$

$$y(x, 0) = f(x) \qquad (0 \le x \le c), \tag{3}$$

where the prescribed displacement function f is continuous on the interval $0 \le x \le c$ and where $f(0) = f(c) = 0$.

We assume a product solution

$$y = X(x)T(t) \tag{4}$$

of the homogeneous conditions (1) and (2) and substitute it into those conditions. This leads to the two homogeneous problems

$$X''(x) + \lambda X(x) = 0, \qquad X(0) = 0, \qquad X(c) = 0, \tag{5}$$

$$T''(t) + \lambda a^2 T(t) = 0, \qquad T'(0) = 0. \tag{6}$$

Problem (5) is another instance of a Sturm-Liouville problem. The method of solution that was used to solve the one in Sec. 31 can be applied here. It turns out (Problem 2) that the eigenvalues are

$$\lambda_n = \left(\frac{n\pi}{c}\right)^2 \qquad (n = 1, 2, \ldots) \tag{7}$$

and that the corresponding eigenfunctions are

$$X_n(x) = \sin\frac{n\pi x}{c} \qquad (n = 1, 2, \ldots). \tag{8}$$

When $\lambda = \lambda_n$, problem (6) becomes

$$T''(t) + \left(\frac{n\pi a}{c}\right)^2 T(t) = 0, \qquad T'(0) = 0;$$

and it follows that except for a constant factor, the solution is

$$T_n(t) = \cos\frac{n\pi at}{c} \qquad (n = 1, 2, \ldots). \tag{9}$$

Consequently, each of the products

$$y_n = X_n(x)T_n(t) = \sin\frac{n\pi x}{c}\cos\frac{n\pi at}{c} \qquad (n = 1, 2, \ldots) \tag{10}$$

satisfies the homogeneous conditions (1) and (2).

According to Example 2 in Sec. 30, the generalized linear combination

$$y(x, t) = \sum_{n=1}^{\infty} B_n \sin\frac{n\pi x}{c}\cos\frac{n\pi at}{c} \tag{11}$$

also satisfies the homogeneous conditions (1) and (2), provided that the constants B_n can be restricted so that the infinite series is suitably convergent and differentiable. That series will satisfy the nonhomogeneous condition (3) if the B_n are such that

$$f(x) = \sum_{n=1}^{\infty} B_n \sin\frac{n\pi x}{c} \qquad (0 < x < c). \tag{12}$$

Because representation (12) is a Fourier sine series representation on the interval $0 < x < c$, we know from Sec. 14 that

$$B_n = \frac{2}{c} \int_0^c f(x) \sin \frac{n\pi x}{c}\, dx \qquad (n = 1, 2, \ldots). \tag{13}$$

The formal solution of our boundary value problem is, therefore, series (11) with coefficients (13). Note that it converges to zero at the endpoints $x = 0$ and $x = c$ of the interval $0 < x < c$.

EXAMPLE. Suppose that the string has length $c = 2$ and that its midpoint is initially raised to a height h above the horizontal axis. The rest position from which the string is released thus consists of two line segments (Fig. 26).

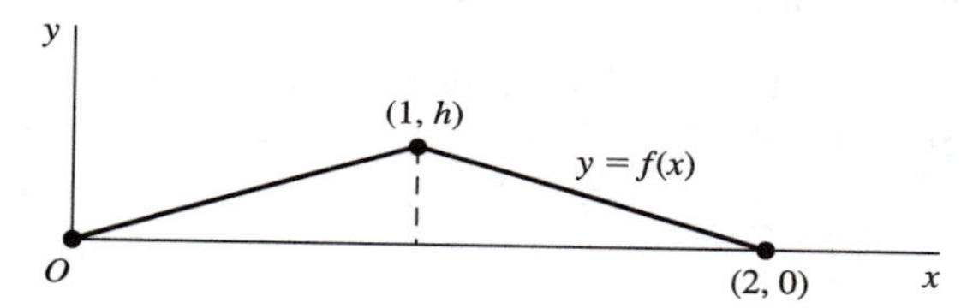

FIGURE 26

The function f, which describes the initial position of this plucked string, is given by the equations

$$f(x) = \begin{cases} hx & \text{when } 0 \le x \le 1, \\ -h(x-2) & \text{when } 1 < x \le 2; \end{cases} \tag{14}$$

and the coefficients B_n in the Fourier sine series for that function on the interval $0 < x < 2$ can be written

$$B_n = \int_0^2 f(x) \sin \frac{n\pi x}{2}\, dx = h \int_0^1 x \sin \frac{n\pi x}{2}\, dx - h \int_1^2 (x-2) \sin \frac{n\pi x}{2}\, dx.$$

After integrating by parts, or using Kronecker's method (Sec. 5), and simplifying, we find that

$$B_n = \frac{8h}{\pi^2} \cdot \frac{1}{n^2} \sin \frac{n\pi}{2} \qquad (n = 1, 2, \ldots).$$

Series (11) then becomes

$$y(x,t) = \frac{8h}{\pi^2} \sum_{n=1}^{\infty} \frac{1}{n^2} \sin \frac{n\pi}{2} \sin \frac{n\pi x}{2} \cos \frac{n\pi a t}{2}. \tag{15}$$

Since

$$\sin \frac{n\pi}{2} = 0$$

when n is even and since

$$\sin \frac{(2n-1)\pi}{2} = \sin\left(n\pi - \frac{\pi}{2}\right) = -\cos n\pi = (-1)^{n+1} \qquad (n = 1, 2, \ldots),$$

expression (15) for the displacements of points on the string in question can also be written†

$$(16) \qquad y(x,t) = \frac{8h}{\pi^2}\sum_{n=1}^{\infty}\frac{(-1)^{n+1}}{(2n-1)^2}\sin\frac{(2n-1)\pi x}{2}\cos\frac{(2n-1)\pi at}{2}.$$

PROBLEMS

1. By assuming a product solution $y = X(x)T(t)$, obtain conditions (5) and (6) on X and T in Sec. 32 from the homogeneous conditions (1) and (2) of the string problem there.
2. Derive the eigenvalues and eigenfunctions, stated in Sec. 32, of the Sturm-Liouville problem

$$X''(x) + \lambda X(x) = 0, \qquad X(0) = 0, \qquad X(c) = 0.$$

3. Point out how it follows from expression (11), Sec. 32, that, for each fixed x, the displacement function $y(x, t)$ is periodic in t with period

$$T_0 = \frac{2c}{a}.$$

33. HISTORICAL DEVELOPMENT

Mathematical sciences experienced a burst of activity following the invention of calculus by Newton (1642–1727) and Leibnitz (1646–1716). Among topics in mathematical physics that attracted the attention of great scientists during that period were boundary value problems in vibrations of strings, elastic bars, and columns of air, all associated with mathematical theories of musical vibrations. Early contributors to the theory of vibrating strings included the English mathematician Brook Taylor (1685–1731), the Swiss mathematicians Daniel Bernoulli (1700–1782) and Leonhard Euler (1707–1783), and Jean d'Alembert (1717–1783) in France.

By the 1750s d'Alembert, Bernoulli, and Euler had advanced the theory of vibrating strings to the stage where the wave equation $y_{tt} = a^2 y_{xx}$ was known and a solution of a boundary value problem for strings had been found from the general solution of that equation. Also, the concept of fundamental modes of vibration led those men to the notion of superposition of solutions, to a solution of the form (11), Sec. 32, where a series of trigonometric functions appears, and thus to the matter of representing arbitrary functions by trigonometric series. Euler later found expressions for the coefficients in those series. But the general concept of a function had not been clarified, and a lengthy controversy took place over the question of representing arbitrary functions on a bounded interval by such series. The question of representation was finally settled by the German mathematician Peter Gustav Lejeune Dirichlet (1805–1859) about 70 years later.

The French mathematical physicist Jean Baptiste Joseph Fourier (1768–1830) presented many instructive examples of expansions in trigonometric series

---

†See the footnote with the example in Sec. 31.

in connection with boundary value problems in the conduction of heat. His book *Théorie analytique de la chaleur,* published in 1822, is a classic in the theory of heat conduction. It was actually the third version of a monograph that he originally submitted to the Institut de France on December 21, 1807.[†] He effectively illustrated the basic procedures of separation of variables and superposition, and his work did much toward arousing interest in trigonometric series representations.

But Fourier's contributions to the representation problem did not include conditions of validity; he was interested in applications and methods. As noted above, Dirichlet was the first to give such conditions. In 1829 he firmly established general conditions on a function sufficient to ensure that it can be represented by a series of sine and cosine functions.[‡]

Representation theory has been refined and greatly extended since the time of Dirichlet. It is still growing.

[†] A. Freeman's early translation of Fourier's book into English was first reprinted by Dover, New York, in 1955. The original 1807 monograph itself remained unpublished until 1972, when the critical edition by Grattan-Guinness that is listed in the Bibliography appeared.

[‡] For supplementary reading on the history of these series, see the articles by Langer (1947) and Van Vleck (1914), listed in the Bibliography.

CHAPTER 5

BOUNDARY VALUE PROBLEMS

This chapter is devoted to the application of Fourier series in solving various boundary value problems that are mathematical formulations of problems in physics. The basic method has already been described in Chap. 4. Except for the final section of this chapter (Sec. 43), we shall limit our attention to problems whose solutions follow from the solutions of the two Sturm-Liouville problems encountered in Secs. 31 and 32 of Chap. 4. To be specific, we saw there that the Sturm-Liouville problem

$$X''(x) + \lambda X(x) = 0, \qquad X'(0) = 0, \qquad X'(c) = 0 \tag{1}$$

on the interval $0 \le x \le c$ has nontrivial solutions only when λ is one of the eigenvalues

$$\lambda_0 = 0, \qquad \lambda_n = \left(\frac{n\pi}{c}\right)^2 \qquad (n = 1, 2, \ldots)$$

and that the corresponding solutions, or eigenfunctions, are

$$X_0(x) = 1, \qquad X_n(x) = \cos\frac{n\pi x}{c} \qquad (n = 1, 2, \ldots).$$

For the Sturm-Liouville problem

$$X''(x) + \lambda X(x) = 0, \qquad X(0) = 0, \qquad X(c) = 0, \tag{2}$$

on the same interval $0 \le x \le c$,

$$\lambda_n = \left(\frac{n\pi}{c}\right)^2 \qquad (n = 1, 2, \ldots)$$

and

$$X_n(x) = \sin \frac{n\pi x}{c} \qquad (n = 1, 2, \ldots).$$

As illustrated in Chap. 4, the solutions of problems (1) and (2) lead to Fourier cosine and sine series representations, respectively. A third Sturm-Liouville problem, to be solved in Sec. 43, leads to Fourier series with both cosines and sines. Boundary value problems whose solutions involve terms other than

$$\cos \frac{n\pi x}{c} \qquad \text{and} \qquad \sin \frac{n\pi x}{c}$$

are taken up in Chap. 8, where the general theory of Sturm-Liouville problems is developed, and in subsequent chapters.

Once it is shown that a solution found for a given boundary value problem truly satisfies the partial differential equation and all the boundary conditions and continuity requirements, the solution is rigorously established. But, even for many of the simpler problems, the verification of solutions may be lengthy or difficult. The boundary value problems in this chapter will be solved only *formally* in the sense that we shall not always explicitly mention needed conditions on functions whose Fourier series are used and we shall not verify the solutions.

We shall also ignore questions of uniqueness, but the physics of a given boundary value problem that is well posed generally suggests that there should be only one solution of that problem. In Chap. 11 we shall give some attention to uniqueness of solutions, in addition to their verification.

34. A SLAB WITH FACES AT PRESCRIBED TEMPERATURES

We consider here the problem of finding temperatures in the same slab as in Sec. 31 when its faces, or boundary surfaces, are kept at certain specified temperatures. For convenience, however, we take the thickness of the slab to be π units, so that $c = \pi$. As illustrated in Example 1 of Sec. 35, temperature formulas for a slab of arbitrary thickness c follow readily once they are found when $c = \pi$. In each of the examples below, the temperature function $u = u(x, t)$ is to satisfy the one-dimensional heat equation

$$(1) \qquad u_t(x, t) = ku_{xx}(x, t) \qquad (0 < x < \pi, t > 0).$$

EXAMPLE 1. If both faces of the slab are kept at temperature zero and the initial temperatures are $f(x)$ (Fig. 27), then

$$(2) \qquad u(0, t) = 0, \qquad u(\pi, t) = 0, \qquad \text{and} \qquad u(x, 0) = f(x).$$

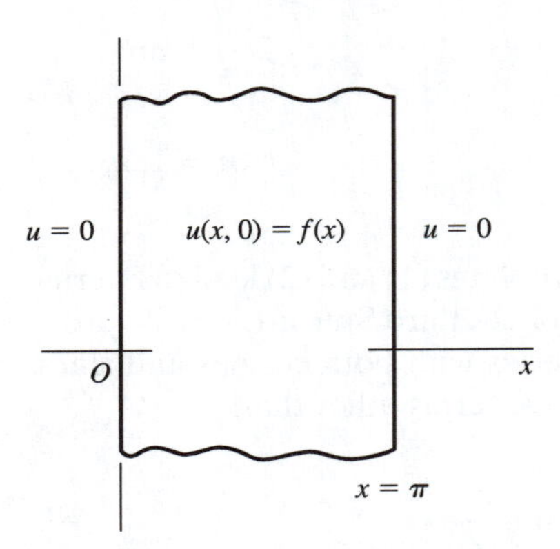

FIGURE 27

Conditions (1) and (2) make up the boundary value problem; and, by separation of variables, we find that a function $u = X(x)T(t)$ satisfies the homogeneous conditions if

$$X''(x) + \lambda X(x) = 0, \qquad X(0) = 0, \qquad X(\pi) = 0 \tag{3}$$

and

$$T'(t) + \lambda k T(t) = 0. \tag{4}$$

According to Sec. 32, the Sturm-Liouville problem (3) has eigenvalues and eigenfunctions

$$\lambda = n^2, \qquad X_n(x) = \sin nx \qquad (n = 1, 2, \ldots).$$

The corresponding functions of t arising from equation (4) are, except for constant factors,

$$T_n(t) = e^{-n^2 kt} \qquad (n = 1, 2, \ldots).$$

Formally, then, the function

$$u(x, t) = \sum_{n=1}^{\infty} B_n e^{-n^2 kt} \sin nx \tag{5}$$

satisfies all the conditions in the boundary value problem, including the nonhomogeneous condition $u(x, 0) = f(x)$, if

$$f(x) = \sum_{n=1}^{\infty} B_n \sin nx \qquad (0 < x < \pi). \tag{6}$$

Let us assume that f is piecewise smooth on the interval $0 < x < \pi$. Then $f(x)$ is represented by its Fourier sine series (6), where

$$B_n = \frac{2}{\pi} \int_0^{\pi} f(x) \sin nx \, dx \qquad (n = 1, 2, \ldots). \tag{7}$$

The function (5), with coefficients (7), is our formal solution of the boundary value problem (1)–(2). It can be expressed more compactly in the form

$$u(x,t) = \frac{2}{\pi}\sum_{n=1}^{\infty} e^{-n^2kt}\sin nx \int_0^{\pi} f(s)\sin ns\, ds,$$

where the variable of integration s is used in order to avoid confusion with the free variable x.

EXAMPLE 2. If the slab is initially at temperature zero throughout and the face $x = 0$ is kept at that temperature, while the face $x = \pi$ is kept at a constant temperature u_0 when $t > 0$, then

$$u(0,t) = 0, \qquad u(\pi,t) = u_0, \qquad u(x,0) = 0. \tag{8}$$

The boundary value problem consisting of equations (1) and (8) is not in proper form for the method of separation of variables to be applied because one of the two-point boundary conditions is nonhomogeneous. If we write

$$u(x,t) = U(x,t) + \Phi(x), \tag{9}$$

however, those equations become

$$U_t(x,t) = k[U_{xx}(x,t) + \Phi''(x)]$$

and

$$U(0,t) + \Phi(0) = 0, \qquad U(\pi,t) + \Phi(\pi) = u_0, \qquad U(x,0) + \Phi(x) = 0.$$

Suppose now that

$$\Phi''(x) = 0 \quad \text{and} \quad \Phi(0) = 0, \qquad \Phi(\pi) = u_0. \tag{10}$$

Then $U(x,t)$ satisfies the boundary value problem

$$U_t(x,t) = kU_{xx}(x,t), \quad U(0,t) = 0, \quad U(\pi,t) = 0, \quad U(x,0) = -\Phi(x). \tag{11}$$

Conditions (10) tell us that

$$\Phi(x) = \frac{u_0}{\pi}x. \tag{12}$$

Hence problem (11) is a special case of the one in Example 1, where

$$f(x) = -\frac{u_0}{\pi}x. \tag{13}$$

When $f(x)$ is this particular function, the coefficients B_n in solution (5) can be found by evaluating the integrals in expression (7). But since we already know from Example 1, Sec. 5, that

$$x = \sum_{n=1}^{\infty} 2\frac{(-1)^{n+1}}{n}\sin nx \qquad (0 < x < \pi) \tag{14}$$

and since the numbers B_n are the coefficients in the Fourier sine series for the function (13) on the interval $0 < x < \pi$, we can see at once that

$$B_n = -\frac{u_0}{\pi} 2 \frac{(-1)^{n+1}}{n} = \frac{u_0}{\pi} 2 \cdot \frac{(-1)^n}{n} \qquad (n = 1, 2, \ldots).$$

Consequently,

$$U(x,t) = \frac{u_0}{\pi} 2 \sum_{n=1}^{\infty} \frac{(-1)^n}{n} e^{-n^2kt} \sin nx;$$

and so, in view of expressions (9) and (12),

$$u(x,t) = \frac{u_0}{\pi}\left[x + 2\sum_{n=1}^{\infty} \frac{(-1)^n}{n} e^{-n^2kt} \sin nx\right]. \tag{15}$$

By letting t tend to infinity in solution (15), we find that the function (12) represents *steady-state* temperatures in the slab. In fact, conditions (10) consist of Laplace's equation in one dimension together with the conditions that the temperature be 0 and u_0 at $x = 0$ and $x = \pi$, respectively. Expression (9), in the form

$$U(x,t) = u(x,t) - \Phi(x),$$

reveals that $U(x,t)$ is merely the desired solution minus the steady-state temperatures.

Finally, note that one can replace the term x in solution (15) by its representation (14) and write that solution as

$$u(x,t) = \frac{2u_0}{\pi} \sum_{n=1}^{\infty} \frac{(-1)^{n+1}}{n} (1 - e^{-n^2kt}) \sin nx. \tag{16}$$

This alternative form can be more useful in approximating $u(x,t)$ by a few terms of the series when t is small, since the factors $1 - \exp(-n^2kt)$ are then small compared to the factors $\exp(-n^2kt)$ in expression (15). Hence the terms that are discarded are smaller. The terms in series (15) are, of course, smaller when t is large.

PROBLEMS†

1. Let the initial temperature distribution be uniform over the slab in Example 1, Sec. 34, so that $f(x) = u_0$. Find $u(x,t)$ and the flux $\Phi(x_0,t) = -Ku_x(x_0,t)$ (see Sec. 20) across a plane $x = x_0$ $(0 \le x_0 \le \pi)$ when $t > 0$. Show that no heat flows across the center plane $x = \pi/2$.

2. Suppose that $f(x) = \sin x$ in Example 1, Sec. 34. Find $u(x,t)$ and verify the result fully.
Suggestion: Use the integration formula obtained in Problem 9, Sec. 5.
Answer: $u(x,t) = e^{-kt} \sin x$.

---

†Only formal solutions of the boundary value problems here and in the sets of problems to follow are expected, unless the problem specifically states that the solution is to be fully verified. Partial verification is often easy and helpful.

3. Let $v(x, t)$ and $w(x, t)$ denote the solutions found in Examples 1 and 2 in Sec. 34. Assuming that those solutions are valid, show that the sum $u = v + w$ gives a temperature formula for a slab $0 \le x \le \pi$ whose faces $x = 0$ and $x = \pi$ are kept at temperatures 0 and u_0, respectively, and whose initial temperature distribution is $f(x)$.
4. Suppose that the face $x = \pi$ of the slab in Example 2, Sec. 34, is maintained at temperatures $u(\pi, t) = F(t)$, instead of $u(\pi, t) = u_0$, where $F(t)$ is continuous and differentiable when $t \ge 0$ and where $F(0) = 0$. Use solution (15) when $u_0 = 1$ in that example, together with the special case of Duhamel's theorem in Sec. 24, to solve this new temperature problem.

$$\textit{Answer: } u(x, t) = \frac{2k}{\pi}\sum_{n=1}^{\infty}(-1)^{n+1}\, n \sin nx \int_0^t e^{-n^2k(t-\tau)}F(\tau)\,d\tau.$$

35. RELATED PROBLEMS

As indicated in Example 2, Sec. 34, a given boundary value problem can sometimes be reduced to one already solved. The examples below illustrate this further.

EXAMPLE 1. Consider the boundary value problem consisting of the conditions

$$u_t(x, t) = ku_{xx}(x, t) \qquad (0 < x < c, t > 0), \tag{1}$$

$$u(0, t) = 0, \qquad u(c, t) = 0, \qquad u(x, 0) = f(x). \tag{2}$$

This is the problem for the temperatures in an infinite slab that was solved in Example 1, Sec. 34, when $c = \pi$. It is also the problem of determining temperatures in a bar of uniform cross section, such as one in the shape of a right circular cylinder (Fig. 28), when its bases in the planes $x = 0$ and $x = c$ are kept at temperature zero, its lateral surface is insulated and parallel to the x axis, and its initial temperatures are $f(x)$ $(0 < x < c)$.

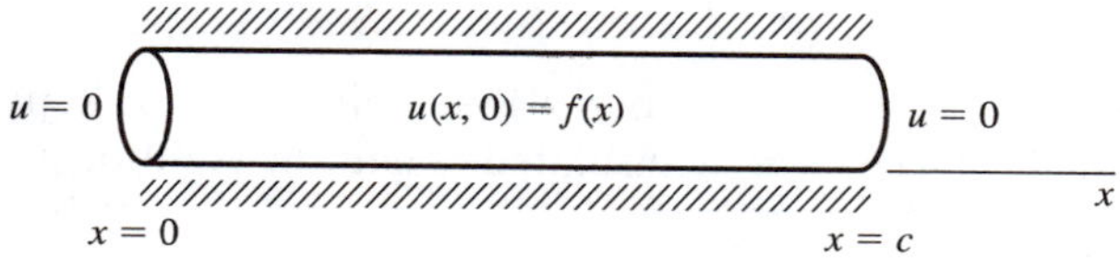

FIGURE 28

The problem in Example 1, Sec. 34, where $c = \pi$, suggests that we make the substitution

$$s = \frac{\pi x}{c} \tag{3}$$

in the problem here and then refer to the solution in that earlier example. Since

$$\frac{\partial u}{\partial x} = \frac{\partial u}{\partial s}\frac{ds}{dx} = \frac{\pi}{c}\frac{\partial u}{\partial s}$$

and

$$\frac{\partial^2 u}{\partial x^2} = \frac{\partial}{\partial x}\left(\frac{\partial u}{\partial x}\right) = \frac{\partial}{\partial s}\left(\frac{\pi}{c}\frac{\partial u}{\partial s}\right)\frac{ds}{dx} = \frac{\pi^2}{c^2}\frac{\partial^2 u}{\partial s^2},$$

equation (1) becomes

$$\frac{\partial u}{\partial t} = \left(\frac{\pi^2 k}{c^2}\right)\frac{\partial^2 u}{\partial s^2} \qquad (0 < s < \pi, t > 0). \tag{4}$$

Conditions (2) tell us that

$$u = 0 \quad \text{when} \quad s = 0 \qquad \text{and} \qquad u = 0 \quad \text{when} \quad s = \pi \tag{5}$$

and that

$$u = f\left(\frac{cs}{\pi}\right) \qquad \text{when} \quad t = 0. \tag{6}$$

Except for notation, conditions (4), (5), and (6) make up the earlier boundary value problem in Example 1, Sec. 34. From that example, we know that

$$u = \sum_{n=1}^{\infty} B_n \exp\left(-\frac{n^2\pi^2 k}{c^2}t\right)\sin ns,$$

where

$$B_n = \frac{2}{\pi}\int_0^{\pi} f\left(\frac{cs}{\pi}\right)\sin ns\, ds \qquad (n = 1, 2, \ldots).$$

Finally, then, substitution (3) gives us the solution of the boundary value problem (1)–(2):

$$u(x,t) = \sum_{n=1}^{\infty} B_n \exp\left(-\frac{n^2\pi^2 k}{c^2}t\right)\sin\frac{n\pi x}{c}, \tag{7}$$

where

$$B_n = \frac{2}{c}\int_0^{c} f(x)\sin\frac{n\pi x}{c}\, dx \qquad (n = 1, 2, \ldots). \tag{8}$$

EXAMPLE 2. Suppose that the face $x = 0$ of a slab of thickness π is kept at temperature zero and that the face $x = \pi$ is insulated. Then, in addition to satisfying the heat equation

$$u_t(x,t) = ku_{xx}(x,t) \qquad (0 < x < \pi, t > 0), \tag{9}$$

u satisfies the conditions

$$u(0,t) = 0 \qquad \text{and} \qquad u_x(\pi,t) = 0 \qquad (t > 0). \tag{10}$$

Also, let the initial temperatures be

$$u(x,0) = f(x) \qquad (0 < x < \pi), \tag{11}$$

where f is piecewise smooth. By writing $u = X(x)T(t)$ and separating variables, we find that

$$X''(x) + \lambda X(x) = 0, \qquad X(0) = 0, \qquad X'(\pi) = 0.$$

Although this problem in X can be treated by methods to be developed in Chap. 8, we are not fully prepared to handle it at this time. The stated temperature problem can, however, be solved here by considering a related problem in a larger slab $0 \le x \le 2\pi$ (Fig. 29).

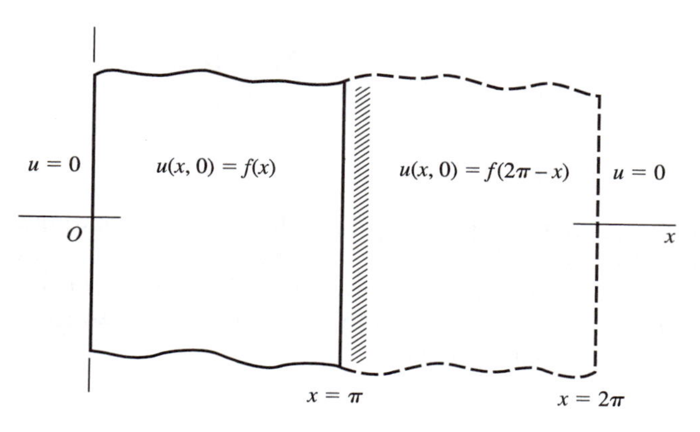

FIGURE 29

Let the two faces $x = 0$ and $x = 2\pi$ of that larger slab be kept at temperature zero, and let the initial temperatures be

$$u(x, 0) = F(x) \qquad (0 < x < 2\pi) \tag{12}$$

where

$$F(x) = \begin{cases} f(x) & \text{when} \quad 0 < x < \pi, \\ f(2\pi - x) & \text{when} \quad \pi < x < 2\pi. \end{cases} \tag{13}$$

The function F is a piecewise smooth extension of the function f on the interval $0 < x < 2\pi$, and the graph of $y = F(x)$ is symmetric with respect to the line $x = \pi$. This procedure is suggested by the fact that with the initial condition (12), no heat will flow across the midsection $x = \pi$ of the larger slab. So, when the variable x is restricted to the interval $0 < x < \pi$, the temperature function for the larger slab will be the desired one for the original slab.

According to Example 1, the temperature function for the larger slab is

$$u(x, t) = \sum_{n=1}^{\infty} B_n \exp\left(-\frac{n^2 k}{4} t\right) \sin \frac{nx}{2}, \tag{14}$$

where the B_n are the coefficients in the Fourier sine series for the function F on the interval $0 < x < 2\pi$:

$$B_n = \frac{1}{\pi} \int_0^{2\pi} F(x) \sin \frac{nx}{2} \, dx \qquad (n = 1, 2, \ldots).$$

This integral can be written in terms of the original function $f(x)$ by simply referring to Problem 6, Sec. 14, which tells us that

$$B_n = \frac{1 - (-1)^n}{\pi} \int_0^{\pi} f(x) \sin \frac{nx}{2} \, dx \qquad (n = 1, 2, \ldots);$$

that is, $B_{2n} = 0$ and

$$B_{2n-1} = \frac{2}{\pi}\int_0^{\pi} f(x)\sin\frac{(2n-1)x}{2}\,dx \qquad (n = 1, 2, \ldots). \tag{15}$$

Solution (14) then becomes

$$u(x,t) = \sum_{n=1}^{\infty} B_{2n-1}\exp\left[-\frac{(2n-1)^2 k}{4}t\right]\sin\frac{(2n-1)x}{2}, \tag{16}$$

with coefficients (15).

EXAMPLE 3. Let $u(r, t)$ denote temperatures in a solid sphere $r \le a$, where r is the spherical coordinate (Sec. 22), when that solid is initially at temperatures $f(r)$ and its surface $r = a$ is kept at temperature zero (Fig. 30). Because u is independent of the spherical coordinates ϕ and θ in this problem, it follows from expression (8), Sec. 22, for the laplacian $\nabla^2 u$ that

$$\nabla^2 u = \frac{1}{r}(ru)_{rr},$$

and hence that the heat equation $u_t = k\nabla^2 u$ here is

$$u_t = \frac{k}{r}(ru)_{rr} \qquad (0 < r < a, t > 0). \tag{17}$$

This differential equation and the boundary conditions

$$u(a,t) = 0, \qquad u(r,0) = f(r) \tag{18}$$

evidently make up the boundary value problem for the temperatures in the sphere.

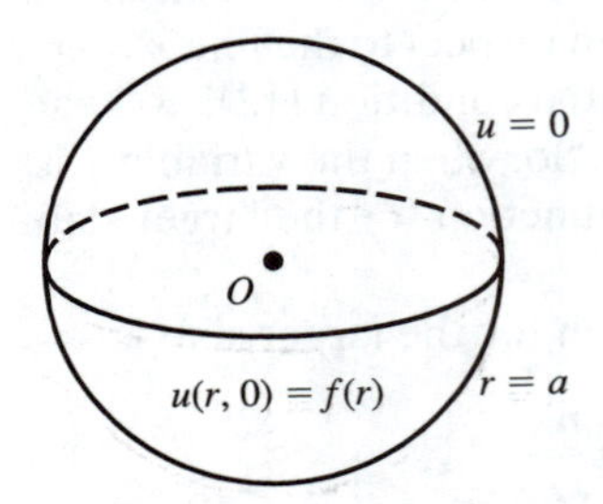

FIGURE 30

To solve this problem, we introduce the function

$$v(r,t) = ru(r,t) \tag{19}$$

to obtain the new boundary value problem

$$v_t(r,t) = kv_{rr}(r,t) \qquad (0 < r < a, t > 0), \tag{20}$$

$$v(0,t) = 0, \qquad v(a,t) = 0, \qquad v(r,0) = rf(r) \tag{21}$$

in $v(r, t)$. The first of conditions (21) follows from relation (19) and the fact that $u(r, t)$ is expected to be continuous at the center $r = 0$ of the sphere.

The solution of the problem consisting of conditions (20) and (21) is immediate from Example 1 since only minor changes in notation give us an expression for $v(r, t)$. Then, in view of relation (19), we arrive at the following expression for the temperatures in our sphere:

$$u(r, t) = \frac{1}{r}\sum_{n=1}^{\infty} B_n \exp\left(-\frac{n^2\pi^2 k}{a^2}t\right)\sin\frac{n\pi r}{a}, \tag{22}$$

where

$$B_n = \frac{2}{a}\int_0^a rf(r)\sin\frac{n\pi r}{a}\,dr \qquad (n = 1, 2, \ldots). \tag{23}$$

PROBLEMS

1. The initial temperature of a slab $0 \le x \le \pi$ is zero throughout, and the face $x = 0$ is kept at that temperature. Heat is supplied through the face $x = \pi$ at a constant rate $A\,(A > 0)$ per unit area, so that $Ku_x(\pi, t) = A$ (see Sec. 24). Write

$$u(x, t) = U(x, t) + \Phi(x)$$

and use the solution of the problem in Example 2, Sec. 35, to derive the expression

$$u(x, t) = \frac{A}{K}\left\{x + \frac{8}{\pi}\sum_{n=1}^{\infty}\frac{(-1)^n}{(2n-1)^2}\exp\left[-\frac{(2n-1)^2 k}{4}t\right]\sin\frac{(2n-1)x}{2}\right\}$$

for the temperatures in this slab.

2. A solid spherical body 40 cm in diameter, initially at 100°C throughout, is cooled by keeping its surface at 0°C. Use the temperature formula for $u(r, t)$ in Example 3, Sec. 35, and the fact that $(\sin\theta)/\theta$ tends to unity as θ tends to zero to show formally that

$$u(0+, t) = 200\sum_{n=1}^{\infty}(-1)^{n+1}\exp\left(-\frac{n^2\pi^2 k}{400}t\right).$$

Thus find the approximate temperature at the center of the sphere 10 min after cooling begins when the material is (*a*) iron, for which $k = 0.15$ cgs unit; (*b*) concrete, for which $k = 0.005$ cgs unit.

Answers: (*a*) 22°C; (*b*) 100°C.

3. A solid sphere $r \le 1$ is initially at temperature zero, and its surface is kept at that temperature. Heat is generated at a constant uniform rate per unit volume throughout the interior of the sphere, so that the temperature function $u = u(r, t)$ satisfies the nonhomogeneous heat equation (see Sec. 21)

$$\frac{\partial u}{\partial t} = \frac{k}{r}\frac{\partial^2}{\partial r^2}(ru) + q_0 \qquad (0 < r < 1, t > 0),$$

where q_0 is a positive constant. Make the substitution

$$u(r, t) = U(r, t) + \Phi(r)$$

in the temperature problem for this sphere, where U and Φ are to be continuous when $r = 0$. [Note that this continuity condition implies that $r\Phi(r)$ tends to zero as r tends to zero.] Then refer to the solution derived in Example 3, Sec. 35, to write the solution of a new boundary value problem for $U(r, t)$ and thus show that

$$u(r,t) = \frac{q_0}{kr}\left[\frac{1}{6}r(1-r^2) + \frac{2}{\pi^3}\sum_{n=1}^{\infty}\frac{(-1)^n}{n^3}e^{-n^2\pi^2kt}\sin n\pi r\right].$$

Suggestion: It is useful to note that in view of the formula for the coefficients in a Fourier sine series, the values of certain integrals that arise are, except for a constant factor, the coefficients in the series [Problem 4(*a*), Sec. 8]

$$x(1-x^2) = \frac{12}{\pi^3}\sum_{n=1}^{\infty}\frac{(-1)^{n+1}}{n^3}\sin n\pi x \qquad (0 < x < 1).$$

4. Let $v(x, t)$ denote temperatures in a slender wire lying along the x axis. Variations of the temperature over each cross section are to be neglected. At the lateral surface, the linear law of surface heat transfer between the wire and its surroundings is assumed to apply (see Problem 6, Sec. 24). Let the surroundings be at temperature zero; then

$$v_t(x,t) = kv_{xx}(x,t) - bv(x,t),$$

where b is a positive constant. The ends $x = 0$ and $x = c$ of the wire are insulated (Fig. 31), and the initial temperature distribution is $f(x)$. Solve the boundary value problem for v by separation of variables. Then show that

$$v(x,t) = u(x,t)\,e^{-bt},$$

where u is the temperature function found in Sec. 31.

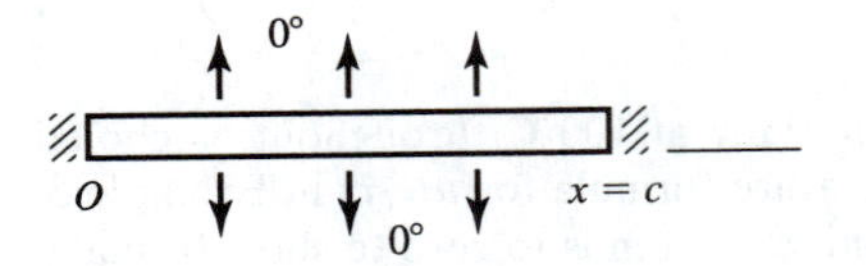

FIGURE 31

5. Solve the boundary value problem consisting of the differential equation

$$u_t(x,t) = u_{xx}(x,t) - b\,u(x,t) \qquad (0 < x < \pi, t > 0),$$

where b is a positive constant, and the boundary conditions

$$u(0,t) = 0, \qquad u(\pi,t) = 1, \qquad u(x,0) = 0.$$

Also, give a physical interpretation of this problem (see Problem 4).

Suggestion: The Fourier series for $\sinh ax$ in Example 2, Sec. 17, is useful here.

Answer: $u(x,t) = \dfrac{\sinh x\sqrt{b}}{\sinh \pi\sqrt{b}} + \dfrac{2}{\pi}e^{-bt}\displaystyle\sum_{n=1}^{\infty}(-1)^n\frac{n}{n^2+b}e^{-n^2t}\sin nx.$

36. A SLAB WITH INTERNALLY GENERATED HEAT

We consider here the same infinite slab $0 \le x \le \pi$ as in Sec. 34, but we assume that there is a source that generates heat at a rate per unit volume which depends on time. The slab is initially at temperatures $f(x)$, and both faces are maintained at temperature zero. According to Sec. 20, the temperatures $u(x, t)$ in the slab must satisfy the modified form

$$u_t(x, t) = ku_{xx}(x, t) + q(t) \qquad (0 < x < \pi, t > 0) \tag{1}$$

of the one-dimensional heat equation, where $q(t)$ is assumed to be a continuous function of t. The conditions

$$u(0, t) = 0, \qquad u(\pi, t) = 0, \qquad \text{and} \qquad u(x, 0) = f(x) \tag{2}$$

complete the statement of this boundary value problem.

Since the differential equation (1) is nonhomogeneous, the method of separation of variables cannot be applied directly. We shall use here, instead, the *method of variation of parameters*. Also called the method of eigenfunction expansions, it is often useful when the differential equation is nonhomogeneous, especially when the term making it so is time-dependent. To be specific, we seek a solution of the boundary value problem in the form

$$u(x, t) = \sum_{n=1}^{\infty} B_n(t) \sin nx \tag{3}$$

of a Fourier sine series whose coefficients $B_n(t)$ are differentiable functions of t. The form (3) is suggested by Example 1, Sec. 34, where the problem is the same as this one when $q(t) \equiv 0$ in equation (1). We anticipate that the function $q(t)$ in equation (1) will cause the coefficients B_n in the solution

$$u(x, t) = \sum_{n=1}^{\infty} B_n e^{-n^2kt} \sin nx$$

that we obtained for the homogeneous part of that earlier problem to depend on t. Instead of writing the coefficients of $\sin nx$ as

$$B_n(t) e^{-n^2kt},$$

we simply write $B_n(t)$ since it is only important that these coefficients depend on t and that they do not depend on x. So our approach here is to start with a generalized linear combination, with coefficients depending on t, of the eigenfunctions

$$X_n(x) = \sin nx \qquad (n = 1, 2, \ldots)$$

of the Sturm-Liouville problem arising in Example 1, Sec. 34. The reader will note that the method of finding a solution of the form (3) is similar in spirit to the method of variation of parameters used in solving linear *ordinary* differential equations that are nonhomogeneous.

We assume that series (3) can be differentiated term by term. Then, by substituting it into equation (1) and recalling [Problem 1(*b*), Sec. 5] that

$$1 = \sum_{n=1}^{\infty} \frac{2[1-(-1)^n]}{n\pi} \sin nx \qquad (0 < x < \pi),$$

we may write

$$\sum_{n=1}^{\infty} B_n'(t)\sin nx = k\sum_{n=1}^{\infty}[-n^2 B_n(t)]\sin nx + q(t)\sum_{n=1}^{\infty}\frac{2[1-(-1)^n]}{n\pi}\sin nx,$$

or

$$\sum_{n=1}^{\infty}[B_n'(t) + n^2kB_n(t)]\sin nx = \sum_{n=1}^{\infty}\frac{2[1-(-1)^n]}{n\pi}q(t)\sin nx.$$

By identifying the coefficients in the sine series on each side of this last equation, we now see that

$$B_n'(t) + n^2kB_n(t) = \frac{2[1-(-1)^n]}{n\pi}q(t) \qquad (n = 1, 2, \ldots). \tag{4}$$

Moreover, according to the third of conditions (2),

$$\sum_{n=1}^{\infty} B_n(0)\sin nx = f(x) \qquad (0 < x < \pi);$$

and this means that

$$B_n(0) = b_n \qquad (n = 1, 2, \ldots), \tag{5}$$

where b_n are the coefficients

$$b_n = \frac{2}{\pi}\int_0^{\pi} f(x)\sin nx\,dx \qquad (n = 1, 2, \ldots) \tag{6}$$

in the Fourier sine series for $f(x)$ on the interval $0 < x < \pi$.

For each value of n, equations (4) and (5) make up an initial value problem in ordinary differential equations. To solve the linear differential equation (4), we observe that an integrating factor is[†]

$$\exp\int n^2k\,dt = \exp n^2kt.$$

Multiplication through equation (4) by this integrating factor puts it in the form

$$\frac{d}{dt}[e^{n^2kt}B_n(t)] = \frac{2[1-(-1)^n]}{n\pi}e^{n^2kt}q(t),$$

where the left-hand side is an exact derivative. If we replace the variable t here by τ and integrate each side from $\tau = 0$ to $\tau = t$, we find that

$$\left[e^{n^2k\tau}B_n(\tau)\right]_0^t = \frac{2[1-(-1)^n]}{n\pi}\int_0^t e^{n^2k\tau}q(\tau)\,d\tau.$$

[†]The reader will recall that any linear first-order equation $y' + p(t)y = g(t)$ has an integrating factor of the form $\exp\int p(t)\,dt$. See, for instance, the book by Rainville, Bedient, and Bedient (1997, chap. 2), listed in the Bibliography.

In view of condition (5), then,

$$B_n(t) = b_n e^{-n^2kt} + \frac{2[1-(-1)^n]}{n\pi} \int_0^t e^{-n^2k(t-\tau)} q(\tau)\, d\tau. \tag{7}$$

Finally, by substituting this expression for $B_n(t)$ into series (3), we arrive at the formal solution of our boundary value problem:

$$\begin{aligned} u(x,t) &= \sum_{n=1}^{\infty} b_n e^{-n^2kt} \sin nx \\ &\quad + \frac{4}{\pi} \sum_{n=1}^{\infty} \frac{\sin(2n-1)x}{2n-1} \int_0^t e^{-(2n-1)^2k(t-\tau)} q(\tau)\, d\tau. \end{aligned} \tag{8}$$

Observe that the first of these series represents the solution of the boundary value problem in Example 1, Sec. 34, where $q(t) \equiv 0$.

To illustrate how interesting special cases of solution (8) are readily obtained, suppose now that $f(x) \equiv 0$ in the third of conditions (2) and that $q(t)$ is the constant function $q(t) = q_0$. Since $b_n = 0$ $(n = 1, 2, \ldots)$ and

$$\int_0^t e^{-(2n-1)^2k(t-\tau)} q_0\, d\tau = \frac{q_0}{k} \cdot \frac{1 - \exp[-(2n-1)^2kt]}{(2n-1)^2},$$

solution (8) reduces to[†]

$$u(x,t) = \frac{4q_0}{\pi k} \sum_{n=1}^{\infty} \frac{1 - \exp[-(2n-1)^2kt]}{(2n-1)^3} \sin(2n-1)x. \tag{9}$$

In view of the Fourier sine series representation (Problem 5, Sec. 5)

$$x(\pi - x) = \frac{8}{\pi} \sum_{n=1}^{\infty} \frac{\sin(2n-1)x}{(2n-1)^3} \qquad (0 < x < \pi),$$

solution (9) can also be written

$$u(x,t) = \frac{q_0}{2k} x(\pi - x) - \frac{4q_0}{\pi k} \sum_{n=1}^{\infty} \frac{\exp[-(2n-1)^2kt]}{(2n-1)^3} \sin(2n-1)x. \tag{10}$$

(See remarks at the end of Example 2, Sec. 34.)

PROBLEMS

1. The boundary value problem

$$u_t(x,t) = u_{xx}(x,t) + xp(t) \qquad (0 < x < 1, t > 0),$$
$$u(0,t) = 0, \qquad u(1,t) = 0, \qquad u(x,0) = 0$$

[†]This result occurs, for example, in the theory of gluing of wood with the aid of radio-frequency heating. See G. H. Brown, *Proc. Inst. Radio Engrs.*, vol. 31, no. 10, pp. 537–548, 1943, where operational methods are used.

describes temperatures in an internally heated slab, where the units for t are chosen so that the thermal conductivity k of the material can be taken as unity (compare with Problem 3, Sec. 21). Solve this problem with the aid of the expansion (see Problem 1, Sec. 8)

$$x = \frac{2}{\pi}\sum_{n=1}^{\infty}\frac{(-1)^{n+1}}{n}\sin n\pi x \qquad (0 < x < 1)$$

and using the method of variation of parameters.

Answer: $u(x,t) = \frac{2}{\pi}\sum_{n=1}^{\infty}\frac{(-1)^{n+1}}{n}\sin n\pi x \int_0^t e^{-n^2\pi^2(t-\tau)}p(\tau)\,d\tau.$

2. Let $u(x,t)$ denote temperatures in a slab $0 \le x \le 1$ that is initially at temperature zero throughout and whose faces are at temperatures

$$u(0,t) = 0 \qquad \text{and} \qquad u(1,t) = F(t),$$

where $F(t)$ and $F'(t)$ are continuous when $t \ge 0$ and where $F(0) = 0$. The unit of time is chosen so that the one-dimensional heat equation has the form $u_t(x,t) = u_{xx}(x,t)$. Write

$$u(x,t) = U(x,t) + xF(t),$$

and observe how it follows from the stated conditions on the faces of the slab that

$$U(0,t) = 0 \qquad \text{and} \qquad U(1,t) = 0.$$

Transform the remaining conditions on $u(x,t)$ into conditions on $U(x,t)$, and then refer to the solution found in Problem 1 to show that

$$u(x,t) = xF(t) + \frac{2}{\pi}\sum_{n=1}^{\infty}\frac{(-1)^n}{n}\sin n\pi x\int_0^t e^{-n^2\pi^2(t-\tau)}F'(\tau)\,d\tau.$$

(Compare with Problem 4, Sec. 34.)

3. Show that when $F(t) = At$, where A is a constant, the expression for $u(x,t)$ derived in Problem 2 becomes

$$u(x,t) = A\left[xt + \frac{2}{\pi^3}\sum_{n=1}^{\infty}(-1)^n\frac{1-\exp(-n^2\pi^2 t)}{n^3}\sin n\pi x\right].$$

4. A bar, with its lateral surface insulated, is initially at temperature zero, and its ends $x = 0$ and $x = c$ are kept at that temperature. Because of internally generated heat, the temperatures in the bar satisfy the differential equation

$$u_t(x,t) = ku_{xx}(x,t) + q(x,t) \qquad (0 < x < c,\ t > 0).$$

Use the method of variation of parameters to derive the temperature formula

$$u(x,t) = \frac{2}{c}\sum_{n=1}^{\infty} I_n(t)\sin\frac{n\pi x}{c},$$

where $I_n(t)$ denotes the iterated integrals

$$I_n(t) = \int_0^t \exp\left[-\frac{n^2\pi^2 k}{c^2}(t-\tau)\right]\int_0^c q(x,\tau)\sin\frac{n\pi x}{c}\,dx\,d\tau \qquad (n = 1, 2, \ldots).$$

Suggestion: Write

$$q(x,t) = \sum_{n=1}^{\infty} b_n(t) \sin \frac{n\pi x}{c} \qquad \text{where} \qquad b_n(t) = \frac{2}{c}\int_0^c q(x,t)\sin\frac{n\pi x}{c}\,dx.$$

5. By writing $c = 1$, $k = 1$, and $q(x,t) = xp(t)$ in the solution found in Problem 4, obtain the solution already found in Problem 1.

6. Solve the boundary value problem (1)–(2) in Sec. 36 when $q(t) = q_0$ by writing

$$u(x,t) = U(x,t) + \Phi(x)$$

and referring to the solution of the problem in Example 1, Sec. 34.

Answer: $u(x,t) = \dfrac{q_0}{2k}x(\pi - x) + \displaystyle\sum_{n=1}^{\infty} b_n e^{-n^2kt}\sin nx,$

where

$$b_n = \frac{2}{\pi}\int_0^{\pi}\left[f(x) - \frac{q_0}{2k}x(\pi - x)\right]\sin nx\,dx \qquad (n = 1, 2, \ldots).$$

7. Show that when $f(x) \equiv 0$, the solution obtained in Problem 6 can be put in the form (9), Sec. 36.

8. Using a series of the form

$$u(x,t) = A_0(t) + \sum_{n=1}^{\infty} A_n(t)\cos\frac{n\pi x}{c}$$

and the expansion (see the example in Sec. 8)

$$x^2 = \frac{c^2}{3} + \frac{4c^2}{\pi^2}\sum_{n=1}^{\infty}\frac{(-1)^n}{n^2}\cos\frac{n\pi x}{c} \qquad (0 < x < c),$$

solve the following temperature problem for a slab $0 \le x \le c$ with insulated faces:

$$u_t(x,t) = ku_{xx}(x,t) + ax^2 \qquad (0 < x < c,\ t > 0),$$

$$u_x(0,t) = 0, \qquad u_x(c,t) = 0, \qquad u(x,0) = 0,$$

where a is a constant. Thus show that

$$u(x,t) = ac^2\left\{\frac{t}{3} + \frac{4c^2}{\pi^4 k}\sum_{n=1}^{\infty}\frac{(-1)^n}{n^4}\left[1 - \exp\left(-\frac{n^2\pi^2 k}{c^2}t\right)\right]\cos\frac{n\pi x}{c}\right\}.$$

9. The boundary value problem

$$\frac{\partial u}{\partial t} = \frac{1}{r}\frac{\partial^2}{\partial r^2}(ru) + q(t) \qquad (0 < r < 1,\ t > 0),$$

$$u(1,t) = 0, \qquad u(r,0) = 0$$

for temperatures $u = u(r,t)$ in a solid sphere with heat generated internally reduces to Problem 3, Sec. 35, with $k = 1$, when $q(t) = q_0$.

(*a*) By writing $v(r,t) = ru(r,t)$ and transforming the problem here into a new one for $v(r,t)$ and then using the solution of Problem 1 above, show that

$$u(r,t) = \frac{2}{\pi r}\sum_{n=1}^{\infty}\frac{(-1)^{n+1}}{n}\sin n\pi r\int_0^t e^{-n^2\pi^2(t-\tau)}q(\tau)\,d\tau.$$

(*b*) Show that when $q(t) = q_0$, the solution obtained in part (*a*) becomes

$$u(r,t) = \frac{2q_0}{\pi^3 r}\sum_{n=1}^{\infty}\frac{(-1)^{n+1}}{n^3}\left(1 - e^{-n^2\pi^2 t}\right)\sin n\pi r.$$

(*c*) Use the Fourier sine series in the suggestion with Problem 3, Sec. 35, to put the solution in part (*b*) in the form

$$u(r,t) = \frac{q_0}{r}\left[\frac{1}{6}r(1-r^2) + \frac{2}{\pi^3}\sum_{n=1}^{\infty}\frac{(-1)^n}{n^3}e^{-n^2\pi^2 t}\sin n\pi r\right].$$

(This is the solution found in Problem 3, Sec. 35, when $k = 1$ there.)

10. Use the method of variation of parameters to solve the temperature problem

$$u_t(x,t) = u_{xx}(x,t) - b(t)\,u(x,t) + q_0 \qquad (0 < x < \pi, t > 0),$$

$$u(0,t) = 0, \qquad u(\pi,t) = 0, \qquad u(x,0) = 0,$$

where q_0 is a constant.† (See Problem 6, Sec. 24.)

Answer: $$u(x,t) = \frac{4q_0}{\pi a(t)}\sum_{n=1}^{\infty}\frac{\sin(2n-1)x}{2n-1}\int_0^t e^{-(2n-1)^2(t-\tau)}a(\tau)\,d\tau,$$

where

$$a(t) = \exp\int_0^t b(\sigma)\,d\sigma.$$

37. STEADY TEMPERATURES IN A RECTANGULAR PLATE

We now illustrate the use of the Fourier method in finding *steady temperatures* in rectangular plates whose faces are insulated. According to Sec. 21, these temperatures $u(x, y)$ must satisfy Laplace's equation $\nabla^2 u = 0$ in the interior of the regions occupied by the plates and are said to be harmonic there. The boundary value problem in the following example is, in fact, a Dirichlet problem (Sec. 28) since it has prescribed values of $u(x, y)$ along the entire boundary of the region.

EXAMPLE. Let $u(x, y)$ be harmonic in the interior of a rectangular region $0 \le x \le a, 0 \le y \le b$, so that

(1) $$u_{xx}(x,y) + u_{yy}(x,y) = 0 \qquad (0 < x < a, 0 < y < b).$$

The boundary values are (Fig. 32)

(2) $$u(0,y) = 0, \qquad u(a,y) = 0 \qquad (0 < y < b),$$

(3) $$u(x,0) = f(x), \qquad u(x,b) = 0 \qquad (0 < x < a).$$

---

†In finding an integrating factor for the ordinary differential equation that arises, it is useful to note that $\int_0^t b(\sigma)\,d\sigma$ is an antiderivative of $b(t)$.

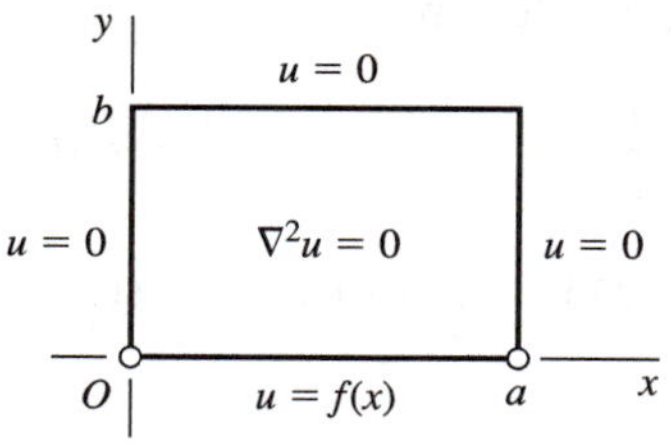

FIGURE 32

With these boundary conditions, the function $u(x, y)$ represents steady temperatures in a plate $0 < x < a, 0 < y < b$ when $u = f(x)$ on the edge $y = 0$ and $u = 0$ on the other three edges. The function $u(x, y)$ also represents the electrostatic potential in a space formed by the planes $x = 0$, $x = a$, $y = 0$, and $y = b$ when the space is free of charges and the planar surfaces are kept at potentials given by conditions (2) and (3).

Separation of variables, with $u = X(x)Y(y)$, leads to the Sturm-Liouville problem

$$X''(x) + \lambda X(x) = 0, \qquad X(0) = 0, \qquad X(a) = 0, \tag{4}$$

whose eigenvalues and eigenfunctions are (Sec. 32)

$$\lambda_n = \left(\frac{n\pi}{a}\right)^2, \qquad X_n(x) = \sin\frac{n\pi x}{a} \qquad (n = 1, 2, \ldots),$$

and to the conditions

$$Y''(y) - \lambda Y(y) = 0, \qquad Y(b) = 0. \tag{5}$$

When λ is a particular eigenvalue λ_n of the Sturm-Liouville problem (4), the function $Y_n(y)$ satisfying conditions (5) is found to be

$$Y_n(y) = C_1\left[\exp\frac{n\pi y}{a} - \exp\frac{n\pi(2b - y)}{a}\right],$$

where C_1 denotes an arbitrary nonzero constant. Instead of setting $C_1 = 1$, as we have always done in such cases, let us write

$$C_1 = -\frac{1}{2}\exp\left(-\frac{n\pi b}{a}\right).$$

Then $Y_n(y)$ takes the compact form

$$Y_n(y) = \sinh\frac{n\pi(b - y)}{a}.$$

Thus the function

$$u(x, y) = \sum_{n=1}^{\infty} B_n \sinh\frac{n\pi(b - y)}{a} \sin\frac{n\pi x}{a} \tag{6}$$

formally satisfies all the conditions (1) through (3), provided that

$$(7) \qquad f(x) = \sum_{n=1}^{\infty} B_n \sinh \frac{n\pi b}{a} \sin \frac{n\pi x}{a} \qquad (0 < x < a).$$

We assume that f is piecewise smooth. Then series (7) is the Fourier sine series representation of $f(x)$ on the interval $0 < x < a$ if

$$B_n \sinh \frac{n\pi b}{a} = \frac{2}{a} \int_0^a f(x) \sin \frac{n\pi x}{a}\, dx \qquad (n = 1, 2, \ldots).$$

The function defined by equation (6), with coefficients

$$(8) \qquad B_n = \frac{2}{a \sinh (n\pi b/a)} \int_0^a f(x) \sin \frac{n\pi x}{a}\, dx \qquad (n = 1, 2, \ldots)$$

is, therefore, our formal solution.

If y is replaced by the new variable $b - y$ in the example above, as well as in its solution, and if $f(x)$ is replaced by $g(x)$, the nonhomogeneous condition satisfied by u becomes $u(x, b) = g(x)$. An interchange of x and y then places non-homogeneous conditions on the edge $x = 0$ or $x = a$. Superposition of the four solutions thus gives the harmonic function whose values are prescribed as functions of position along the entire boundary of the rectangular domain, except for the corners.

PROBLEMS

1. The faces and edges $x = 0$ and $x = \pi$ $(0 < y < \pi)$ of a square plate $0 \le x \le \pi, 0 \le y \le \pi$ are insulated. The edges $y = 0$ and $y = \pi$ $(0 < x < \pi)$ are kept at temperatures 0 and $f(x)$, respectively. Let $u(x, y)$ denote steady temperatures in the plate and derive the expression

$$u(x, y) = A_0\, y + \sum_{n=1}^{\infty} A_n \sinh ny \cos nx,$$

where

$$A_0 = \frac{1}{\pi^2} \int_0^\pi f(x)\, dx \qquad \text{and} \qquad A_n = \frac{2}{\pi \sinh n\pi} \int_0^\pi f(x) \cos nx\, dx$$

$$(n = 1, 2, \ldots).$$

Find $u(x, y)$ when $f(x) = u_0$, where u_0 is a constant.

2. The faces and edge $y = 0$ $(0 < x < \pi)$ of a rectangular plate $0 \le x \le \pi, 0 \le y \le y_0$ are insulated. The other three edges are maintained at the temperatures indicated in Fig. 33. By making the substitution $u(x, y) = U(x, y) + \Phi(x)$ in the boundary value problem for the steady temperatures $u(x, y)$ in the plate and using the method described in Example 2, Sec. 34, derive the temperature formula

$$u(x, y) = \frac{1}{\pi} \left[x + 2 \sum_{n=1}^{\infty} \frac{(-1)^n}{n} \cdot \frac{\cosh ny}{\cosh ny_0} \sin nx \right].$$

Suggestion: The series representation (Example 1, Sec. 5)

$$x = 2\sum_{n=1}^{\infty} \frac{(-1)^{n+1}}{n} \sin nx \qquad (0 < x < \pi)$$

is useful in finding $U(x, y)$.

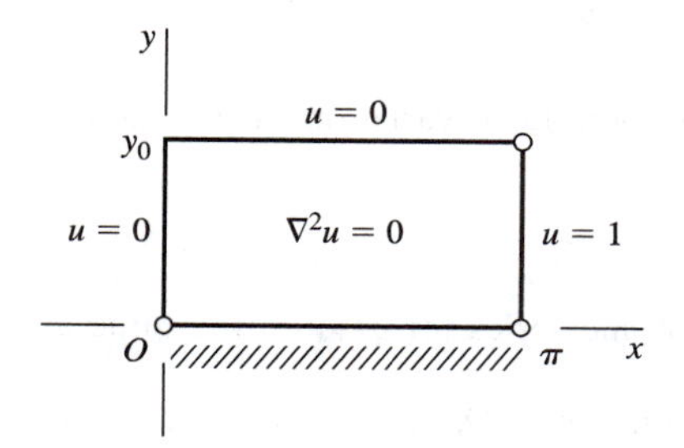

FIGURE 33

3. One edge of a square plate with insulated faces is kept at a uniform temperature u_0, and the other three edges are kept at temperature zero. Without solving a boundary value problem, but by superposition of solutions of like problems to obtain the trivial case in which all four edges are at temperature u_0, show why the steady temperature at the center of the given plate must be $u_0/4$.
4. Let $u(x, y)$ denote the bounded steady temperatures in the semi-infinite plate $x \geq 0$, $0 \leq y \leq \pi$, whose faces are insulated, when the edges are kept at the temperatures shown in Fig. 34. (The boundedness condition serves as a condition at the missing right-hand end of the plate.) Assuming that the function f is piecewise smooth, derive the temperature formula

$$u(x, y) = \sum_{n=1}^{\infty} B_n e^{-nx} \sin ny,$$

where

$$B_n = \frac{2}{\pi}\int_0^{\pi} f(y) \sin ny \, dy \qquad (n = 1, 2, \ldots).$$

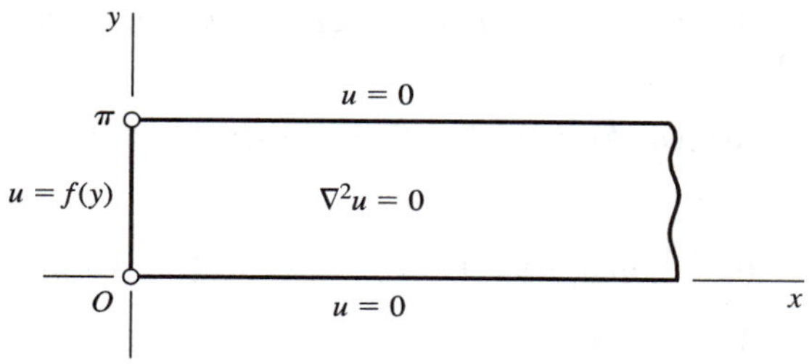

FIGURE 34

5. Suppose that in the plate described in Problem 4 there is a heat source depending on the variable y and that the entire boundary is kept at temperature zero. According to

Sec. 21, the steady temperatures $u(x, y)$ in the plate must now satisfy *Poisson's equation*

$$u_{xx}(x, y) + u_{yy}(x, y) + q(y) = 0 \qquad (x > 0, 0 < y < \pi).$$

(*a*) By assuming a (bounded) solution of the form

$$u(x, y) = \sum_{n=1}^{\infty} B_n(x) \sin ny$$

of this temperature problem and using the method of variation of parameters (Sec. 36), show formally that

$$B_n(x) = \frac{q_n}{n^2}(1 - e^{-nx}) \qquad (n = 1, 2, \ldots),$$

where q_n are the coefficients in the Fourier sine series for $q(y)$ on the interval $0 < y < \pi$.

(*b*) Show that when $q(y)$ is the constant function $q(y) = Q$, the solution in part (*a*) becomes

$$u(x, y) = \frac{4Q}{\pi} \sum_{n=1}^{\infty} \frac{1 - \exp[-(2n-1)x]}{(2n-1)^3} \sin(2n-1)y.$$

Suggestion: In part (*a*), recall that the general solution of a linear second-order equation $y'' + p(x)y = g(x)$ is of the form $y = y_c + y_p$, where y_p is any particular solution and y_c is the general solution of the complementary equation

$$y'' + p(x)y = 0.^\dagger$$

6. Derive an expression for the bounded steady temperatures $u(x, y)$ in a semi-infinite slab $0 \le x \le c$, $y \ge 0$ whose faces in the planes $x = 0$ and $x = c$ are insulated and where $u(x, 0) = f(x)$. Assume that f is piecewise smooth on the interval $0 < x < c$.

38. CYLINDRICAL COORDINATES

In this section, we solve two boundary value problems involving cylindrical coordinates. The first is a Dirichlet problem, and the second involves steady temperatures in a long rod part of whose surface is insulated.

EXAMPLE 1. Let $u(\rho, \phi)$ denote a function of the cylindrical, or polar, coordinates ρ and ϕ that is harmonic in the domain $1 < \rho < b, 0 < \phi < \pi$ of the plane $z = 0$ (Fig. 35). Thus (Sec. 22)

$$\text{(1)} \qquad \rho^2 u_{\rho\rho}(\rho, \phi) + \rho u_\rho(\rho, \phi) + u_{\phi\phi}(\rho, \phi) = 0 \qquad (1 < \rho < b, 0 < \phi < \pi).$$

Suppose further that

$$\text{(2)} \qquad u(\rho, 0) = 0, \qquad u(\rho, \pi) = 0 \qquad (1 < \rho < b),$$

$$\text{(3)} \qquad u(1, \phi) = 0, \qquad u(b, \phi) = u_0 \qquad (0 < \phi < \pi),$$

where u_0 is a constant.

---

†See, for instance, the book by Boyce and DiPrima (2005, Sec. 3.6), listed in the Bibliography.

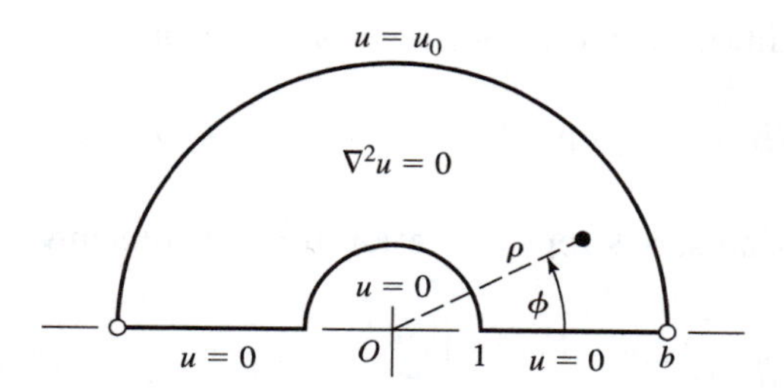

FIGURE 35

Substituting $u = R(\rho)\Phi(\phi)$ into the homogeneous conditions and separating variables, we find that

$$\rho^2 R''(\rho) + \rho R'(\rho) - \lambda R(\rho) = 0, \qquad R(1) = 0 \tag{4}$$

and

$$\Phi''(\phi) + \lambda\Phi(\phi) = 0, \qquad \Phi(0) = 0, \qquad \Phi(\pi) = 0. \tag{5}$$

Except for notation, the problem in Φ is the Sturm-Liouville problem in Sec. 32 with $c = \pi$. The eigenvalues and eigenfunctions are

$$\lambda_n = n^2, \qquad \Phi_n(\phi) = \sin n\phi \qquad (n = 1, 2, \ldots).$$

The corresponding functions $R_n(\rho)$ are determined by solving the differential equation

$$\rho^2 R''(\rho) + \rho R'(\rho) - n^2 R(\rho) = 0 \qquad (1 < \rho < b),$$

where $R(1) = 0$. This is a Cauchy-Euler equation (see Problem 1), and the substitution $\rho = \exp s$ transforms it into the differential equation

$$\frac{d^2R}{ds^2} - n^2 R = 0.$$

Hence

$$R = C_1 e^{ns} + C_2 e^{-ns};$$

and because $s = \ln \rho$,

$$R(\rho) = C_1 e^{n \ln \rho} + C_2 e^{-n \ln \rho} = C_1 \rho^n + C_2 \rho^{-n}.$$

In view of the condition $R(1) = 0$, it follows that except for constant factors, the desired functions of ρ are

$$R_n(\rho) = \rho^n - \rho^{-n} \qquad (n = 1, 2, \ldots).$$

Thus, formally,

$$u(\rho, \phi) = \sum_{n=1}^{\infty} B_n (\rho^n - \rho^{-n}) \sin n\phi$$

where, according to the second of conditions (3), the constants B_n are such that

$$u_0 = \sum_{n=1}^{\infty} B_n\,(b^n - b^{-n}) \sin n\phi \qquad (0 < \phi < \pi).$$

Since this is in the form of a Fourier sine series representation for the constant function u_0 on the interval $0 < \phi < \pi$,

$$B_n\,(b^n - b^{-n}) = \frac{2}{\pi}\int_0^{\pi} u_0 \sin n\phi\, d\phi = \frac{2u_0}{\pi}\cdot\frac{1-(-1)^n}{n} \qquad (n = 1, 2, \ldots).$$

The complete solution of our Dirichlet problem is, therefore,

$$u(\rho, \phi) = \frac{2u_0}{\pi}\sum_{n=1}^{\infty}\frac{\rho^n - \rho^{-n}}{b^n - b^{-n}}\cdot\frac{1-(-1)^n}{n}\sin n\phi,$$

or

$$u(\rho, \phi) = \frac{4u_0}{\pi}\sum_{n=1}^{\infty}\frac{\rho^{2n-1} - \rho^{-(2n-1)}}{b^{2n-1} - b^{-(2n-1)}}\cdot\frac{\sin(2n-1)\phi}{2n-1}. \tag{6}$$

EXAMPLE 2. Using cylindrical coordinates, let us derive an expression for the steady temperatures $u = u(\rho, \phi)$ in a long rod, with a uniform semicircular cross section and occupying the region $0 \le \rho \le a, 0 \le \phi \le \pi$, which is insulated on its planar surface and maintained at temperatures $f(\phi)$ on the semicircular part (Fig. 36).

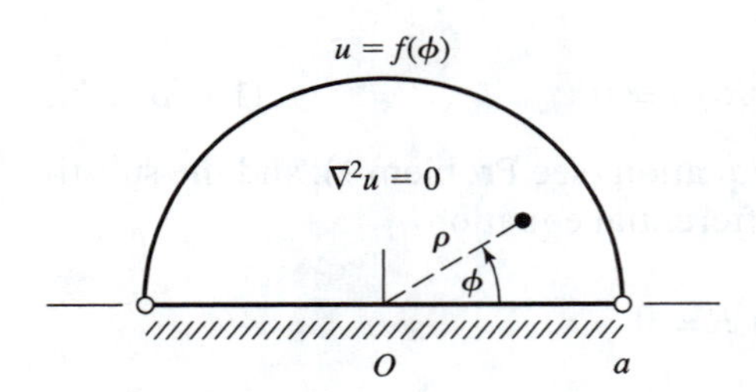

FIGURE 36

As in Example 1, $u(\rho, \phi)$ satisfies Laplace's equation

$$\rho^2 u_{\rho\rho}(\rho, \phi) + \rho u_\rho(\rho, \phi) + u_{\phi\phi}(\rho, \phi) = 0, \tag{7}$$

but now in the domain $0 < \rho < a, 0 < \phi < \pi$. It also satisfies the homogeneous conditions (see Example 2, Sec. 24)

$$u_\phi(\rho, 0) = 0, \qquad u_\phi(\rho, \pi) = 0 \qquad (0 < \rho < a), \tag{8}$$

as well as the nonhomogeneous one

$$u(a, \phi) = f(\phi) \qquad (0 < \phi < \pi). \tag{9}$$

The function f is understood to be piecewise smooth and therefore bounded. We assume further that $|u(\rho, \phi)| \le M$, where M denotes some positive constant.

The need for such a boundedness condition is physically evident and has been only tacitly assumed in most earlier problems. Here it serves as a condition at the origin, which may be thought of as the limiting case of a smaller semicircle (compare with Fig. 35 in Example 1) as its radius tends to zero.

The substitution $u = R(\rho)\Phi(\phi)$ in the homogeneous conditions (7) and (8) leads to the condition

$$\rho^2 R''(\rho) + \rho R'(\rho) - \lambda R(\rho) = 0 \qquad (0 < \rho < a) \tag{10}$$

on $R(\rho)$ and to the Sturm-Liouville problem

$$\Phi''(\phi) + \lambda\Phi(\phi) = 0, \qquad \Phi'(0) = 0, \qquad \Phi'(\pi) = 0, \tag{11}$$

whose eigenvalues and eigenfunctions are

$$\lambda_0 = 0, \qquad \lambda_n = n^2 \qquad (n = 1, 2, \ldots)$$

and

$$\Phi_0(\phi) = 1, \qquad \Phi_n(\phi) = \cos n\phi \qquad (n = 1, 2, \ldots),$$

according to Sec. 31.

Equation (10) is a Cauchy-Euler equation; and, with the substitution $\rho = \exp s$, Problem 1 below tells us that it becomes

$$\frac{d^2R}{ds^2} = 0$$

when $\lambda = \lambda_0$. So, for the eigenvalue λ_0,

$$R = As + B = A \ln\rho + B \qquad (0 < \rho < a),$$

where A and B are constants. But since the product $R(\rho)\Phi(\phi)$ is expected to be bounded in the domain $0 < \rho < a$, $0 < \phi < \pi$ and since $\ln \rho$ tends to $-\infty$ as ρ tends to 0 through positive values, the constant A must be zero. Hence, except for a constant factor,

$$R_0(\rho) = 1.$$

Similarly, when $\lambda = \lambda_n$ $(n = 1, 2, \ldots)$, our boundedness condition requires that the constant C_2 in the general solution

$$R(\rho) = C_1\,\rho^n + C_2\,\rho^{-n} = C_1\,\rho^n + \frac{C_2}{\rho^n} \qquad (0 < \rho < a)$$

of equation (10) be zero. Hence we may write

$$R_n(\rho) = \rho^n \qquad (n = 1, 2, \ldots),$$

and the homogeneous conditions (7) and (8) are formally satisfied by the function

$$u(\rho, \phi) = A_0 + \sum_{n=1}^{\infty} A_n\,\rho^n \cos n\phi, \tag{12}$$

where the constants A_n $(n = 0, 1, 2, \ldots)$ are yet to be determined.

In view of the nonhomogeneous condition (9),

$$f(\phi) = \frac{2A_0}{2} + \sum_{n=1}^{\infty} (A_n a^n) \cos n\phi \qquad (0 < \phi < \pi).$$

Consequently,

$$(13) \qquad A_0 = \frac{1}{\pi} \int_0^{\pi} f(\phi)\, d\phi \quad \text{and} \quad A_n = \frac{2}{\pi a^n} \int_0^{\pi} f(\phi) \cos n\phi \, d\phi \qquad (n = 1, 2, \ldots).$$

The complete solution of our boundary value problem is, then, given by series (12) with coefficients (13). This solution can, of course, be alternatively written as

$$(14) \quad u(\rho, \phi) = \frac{1}{\pi} \int_0^{\pi} f(\psi)\, d\psi + \frac{2}{\pi} \sum_{n=1}^{\infty} \left(\frac{\rho}{a}\right)^n \cos n\phi \int_0^{\pi} f(\psi) \cos n\psi \, d\psi,$$

where the variable of integration ψ is to be distinguished from the free variable ϕ.

PROBLEMS

1. If A, B, and C, are constants, the differential equation

$$Ax^2 y'' + Bxy' + C y = 0$$

is called a *Cauchy-Euler equation*. Show that with the substitution $x = e^s$ $(s = \ln x)$, it can be transformed into the constant-coefficient differential equation

$$A \frac{d^2y}{ds^2} + (B - A) \frac{dy}{ds} + C y = 0.$$

Suggestion: Use the chain rule to show that

$$y' = \frac{dy}{dx} = \frac{dy}{ds}\frac{ds}{dx} = e^{-s} \frac{dy}{ds}$$

and then

$$y'' = (y')' = e^{-s} \frac{dy'}{ds} = e^{-s} \frac{d}{ds}\left(e^{-s} \frac{dy}{ds}\right) = e^{-2s}\left(\frac{d^2y}{ds^2} - \frac{dy}{ds}\right).$$

2. Let the faces of a plate in the shape of a wedge $0 \le \rho \le a$, $0 \le \phi \le \alpha$ (Fig. 37) be insulated. Find the steady temperatures $u(\rho, \phi)$ in the plate when $u = 0$ on the two

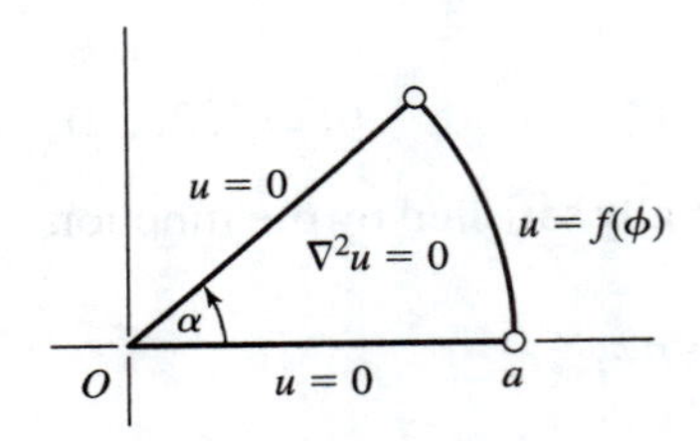

FIGURE 37

rays $\phi = 0, \phi = \alpha$ $(0 < \rho < a)$ and $u = f(\phi)$ on the arc $\rho = a$ $(0 < \phi < \alpha)$. Assume that f is piecewise smooth and that u is bounded.

$$\textit{Answer: } u(\rho, \phi) = \frac{2}{\alpha}\sum_{n=1}^{\infty}\left(\frac{\rho}{a}\right)^{n\pi/\alpha} \sin\frac{n\pi\phi}{\alpha}\int_0^{\alpha} f(\psi)\sin\frac{n\pi\psi}{\alpha}\,d\psi.$$

3. Let ρ, ϕ, z be cylindrical coordinates. Find the harmonic function $u(\rho, \phi)$ in the domain $1 < \rho < b, 0 < \phi < \pi/2$ of the plane $z = 0$ when

$$u(1, \phi) = 0, \qquad u(b, \phi) = f(\phi) \qquad (0 < \phi < \pi/2)$$

and

$$u_\phi(\rho, 0) = 0, \qquad u_\phi(\rho, \pi/2) = 0 \qquad (1 < \rho < b).$$

Give a physical interpretation of this problem.

$$\textit{Answer: } u(\rho, \phi) = A_0 \ln\rho + \sum_{n=1}^{\infty} A_n\,(\rho^{2n} - \rho^{-2n})\cos 2n\phi,$$

where

$$A_0 = \frac{2}{\pi \ln b}\int_0^{\pi/2} f(\phi)\,d\phi$$

and

$$A_n = \frac{4}{\pi(b^{2n} - b^{-2n})}\int_0^{\pi/2} f(\phi)\cos 2n\phi\,d\phi \qquad (n = 1, 2, \ldots).$$

39. A STRING WITH PRESCRIBED INITIAL CONDITIONS

Section 32 was devoted to solving the boundary value problem

$$y_{tt}(x, t) = a^2 y_{xx}(x, t) \qquad (0 < x < c,\ t > 0), \tag{1}$$

$$y(0, t) = 0, \qquad y(c, t) = 0, \tag{2}$$

$$y(x, 0) = f(x), \qquad y_t(x, 0) = 0 \tag{3}$$

for the transverse displacements $y(x, t)$ in a finite string that starts with displacements $y = f(x)$ and is initially at rest. We recall that f was continuous on $0 \le x \le c$ and that

$$f(0) = f(c) = 0.$$

We obtained the solution

$$y(x, t) = \sum_{n=1}^{\infty} B_n \sin\frac{n\pi x}{c}\cos\frac{n\pi a t}{c}, \tag{4}$$

where

$$B_n = \frac{2}{c}\int_0^c f(x)\sin\frac{n\pi x}{c}\,dx \qquad (n = 1, 2, \ldots). \tag{5}$$

Solution (4) is easily written in a closed form that does not involve infinite series. We can do this with the aid of the trigonometric identity

$$2\sin A\cos B = \sin(A + B) + \sin(A - B)$$

by first noting that

$$\sin\frac{n\pi x}{c}\cos\frac{n\pi at}{c} = \frac{1}{2}\left[\sin\frac{n\pi(x+at)}{c} + \sin\frac{n\pi(x-at)}{c}\right]. \tag{6}$$

Series (4) then becomes

$$y(x,t) = \frac{1}{2}\left[\sum_{n=1}^{\infty} B_n \sin\frac{n\pi(x+at)}{c} + \sum_{n=1}^{\infty} B_n \sin\frac{n\pi(x-at)}{c}\right]. \tag{7}$$

Next, we let F denote the odd periodic extension, with period $2c$, of the function f. That is (see Fig. 38),

$$F(x) = f(x) \qquad \text{when } 0 \le x \le c$$

and

$$F(-x) = -F(x), \qquad F(x+2c) = F(x) \qquad \text{for all } x.$$

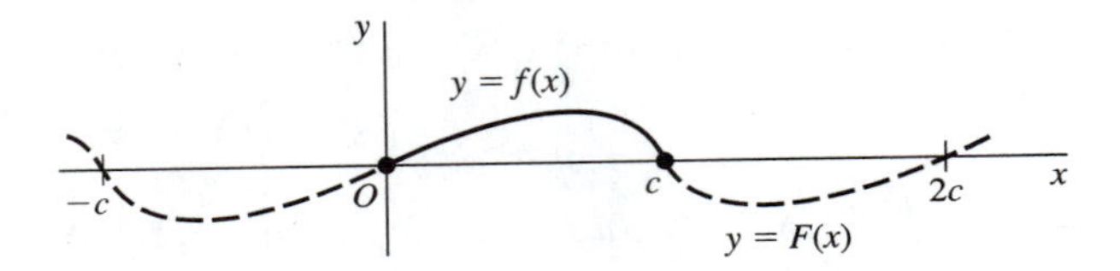

FIGURE 38

Under the assumption that the Fourier sine series representation

$$f(x) = \sum_{n=1}^{\infty} B_n \sin\frac{n\pi x}{c} \qquad (0 < x < c)$$

is valid when coefficients (5) are used, it follows from the fact that each of the functions $\sin(n\pi x/c)$ is odd and periodic with period $2c$ that $F(x)$ is represented for all x by the same series:

$$F(x) = \sum_{n=1}^{\infty} B_n \sin\frac{n\pi x}{c} \qquad (-\infty < x < \infty). \tag{8}$$

Consequently, expression (7) can be written

$$y(x,t) = \frac{1}{2}[F(x+at) + F(x-at)]. \tag{9}$$

(Compare with d'Alembert's solution in the example in Sec. 27.) The form (9) of our solution will be verified in Chap. 11 (Sec. 97).

When, initially, the string is in its position of equilibrium $y = 0$ and has a prescribed distribution of velocities $g(x)$ parallel to the y axis, the boundary value problem for the displacements $y(x,t)$ becomes

$$y_{tt}(x,t) = a^2 y_{xx}(x,t) \qquad (0 < x < c,\ t > 0), \tag{10}$$

$$y(0,t) = 0, \qquad y(c,t) = 0, \tag{11}$$

$$y(x,0) = 0, \qquad y_t(x,0) = g(x). \tag{12}$$

If the xy plane, with the string lying on the x axis, is moving parallel to the y axis and is brought to rest at the instant $t = 0$, the function $g(x)$ is a constant. The hammer action in a piano may produce approximately a uniform initial velocity over a short span of a piano wire, in which case $g(x)$ may be considered to be a step function.

As in Sec. 32, we seek functions of the type $y = X(x)T(t)$ that satisfy all the homogeneous conditions in the boundary value problem. The Sturm-Liouville problem that arises is the same as the one in Sec. 32:

$$X''(x) + \lambda X(x) = 0, \qquad X(0) = 0, \qquad X(c) = 0.$$

We recall that the eigenvalues and corresponding eigenfunctions are

$$\lambda_n = \left(\frac{n\pi}{c}\right)^2, \qquad X_n(x) = \sin\frac{n\pi x}{c} \qquad (n = 1, 2, \ldots).$$

Since the conditions on $T(t)$ are

$$T''(t) + \lambda a^2 T(t) = 0, \qquad T(0) = 0,$$

the corresponding functions of t are, except for constant factors,

$$T_n(t) = \sin\frac{n\pi at}{c} \qquad (n = 1, 2, \ldots).$$

Thus the homogeneous conditions in the boundary value problem are formally satisfied by the function

$$y(x, t) = \sum_{n=1}^{\infty} C_n \sin\frac{n\pi x}{c} \sin\frac{n\pi at}{c}, \tag{13}$$

where the constants C_n must be determined. To find those constants, we use expression (13) to write

$$y_t(x, t) = \sum_{n=1}^{\infty} \frac{n\pi a}{c} C_n \sin\frac{n\pi x}{c} \cos\frac{n\pi at}{c}; \tag{14}$$

and, from this and the second of conditions (12), we see that

$$g(x) = \sum_{n=1}^{\infty} \frac{n\pi a}{c} C_n \sin\frac{n\pi x}{c} \qquad (0 < x < c). \tag{15}$$

Because this is a Fourier sine series for $g(x)$ on $0 < x < c$, then,

$$\frac{n\pi a}{c} C_n = \frac{2}{c}\int_0^c g(x)\sin\frac{n\pi x}{c}\,dx \qquad (n = 1, 2, \ldots).$$

That is,

$$C_n = \frac{2}{n\pi a}\int_0^c g(x)\sin\frac{n\pi x}{c}\,dx \qquad (n = 1, 2, \ldots). \tag{16}$$

We can sum the series (13), with the aid of expressions (14) and (6), by writing

$$y_t(x, t) = \frac{1}{2}\left[\sum_{n=1}^{\infty} \frac{n\pi a}{c} C_n \sin \frac{n\pi(x+at)}{c} + \sum_{n=1}^{\infty} \frac{n\pi a}{c} C_n \sin \frac{n\pi(x-at)}{c}\right]$$

and noting how it follows from representation (15) that

$$y_t(x, t) = \frac{1}{2}[G(x+at) + G(x-at)]$$

where G is the odd periodic extension, with period $2c$, of the function g. Then, since $y(x, 0) = 0$, we have†

$$y(x, t) = \frac{1}{2}\left[\int_0^t G(x+a\tau)\,d\tau + \int_0^t G(x-a\tau)\,d\tau\right].$$

After substituting $s = x + a\tau$ in the first of these integrals and $s = x - a\tau$ in the second, we arrive at the expression

$$y(x, t) = \frac{1}{2a}\left[\int_x^{x+at} G(s)\,ds - \int_x^{x-at} G(s)\,ds\right],$$

or

$$y(x, t) = \frac{1}{2a}\int_{x-at}^{x+at} G(s)\,ds. \tag{17}$$

If points on the string are given both nonzero initial displacements and nonzero initial velocities, so that

$$y(x, 0) = f(x) \quad \text{and} \quad y_t(x, 0) = g(x), \tag{18}$$

the displacements $y(x, t)$ can be written as the sum

$$y(x, t) = \frac{1}{2}[F(x+at) + F(x-at)] + \frac{1}{2a}\int_{x-at}^{x+at} G(s)\,ds \tag{19}$$

of the solutions (9) and (17) above. (Compare with Problem 2, Sec. 27.) To see that this is so, let $Y(x, t)$ and $Z(x, t)$ denote those two solutions, respectively. The principle of superposition in Sec. 30, applied to just two functions, ensures that the sum

$$y(x, t) = Y(x, t) + Z(x, t)$$

satisfies the linear homogeneous conditions (1) and (2), which are the same as conditions (10) and (11). Furthermore, in view of conditions (3) and (12),

$$y(x, 0) = Y(x, 0) + Z(x, 0) = f(x) + 0 = f(x)$$

and

$$y_t(x, 0) = Y_t(x, 0) + Z_t(x, 0) = 0 + g(x) = g(x).$$

†See also the footnote with Problem 10, Sec. 36, regarding antiderivatives.

In general, the solution of a linear problem containing more than one nonhomogeneous condition can be written as a sum of solutions of problems each of which contains only one nonhomogeneous condition. The resolution of the original problem in this way, although not an essential step, often simplifies the process of solving it.

PROBLEMS

1. A string is stretched between the fixed points 0 and 1 on the x axis and released at rest from the position $y = A \sin \pi x$, where A is a constant. Obtain from expression (9), Sec. 39, the subsequent displacements $y(x, t)$, and verify the result fully. Sketch the position of the string at several instants of time.

Answer: $y(x, t) = A \sin \pi x \cos \pi a t$.

2. Solve Problem 1 when the initial displacement there is changed to $y = B \sin 2\pi x$, where B is a constant.

Answer: $y(x, t) = B \sin 2\pi x \cos 2\pi a t$.

3. Show why the sum of the two functions $y(x, t)$ found in Problems 1 and 2 represents the displacements after the string is released at rest from the position

$$y = A \sin \pi x + B \sin 2\pi x.$$

4. A string, stretched between the points 0 and π on the x axis and initially at rest, is released from the position $y = f(x)$. Its motion is opposed by air resistance, which is proportional to the velocity at each point (Sec. 25). Let the unit of time be chosen so that the equation of motion becomes

$$y_{tt}(x, t) = y_{xx}(x, t) - 2\beta y_t(x, t) \qquad (0 < x < \pi, t > 0),$$

where β is a positive constant. Assuming that $0 < \beta < 1$, derive the expression

$$y(x, t) = e^{-\beta t} \sum_{n=1}^{\infty} B_n \left(\cos \alpha_n t + \frac{\beta}{\alpha_n} \sin \alpha_n t \right) \sin nx,$$

where

$$\alpha_n = \sqrt{n^2 - \beta^2}, \qquad B_n = \frac{2}{\pi} \int_0^{\pi} f(x) \sin nx \, dx \qquad (n = 1, 2, \ldots),$$

for the transverse displacements.

5. Suppose that the string in Problem 4 is initially straight with a uniform velocity in the direction of the y axis, as if a moving frame supporting the endpoints is brought to rest at the instant $t = 0$. The transverse displacements $y(x, t)$ thus satisfy the same differential equation, where $0 < \beta < 1$, and the boundary conditions

$$y(0, t) = y(\pi, t) = 0, \qquad y(x, 0) = 0, \qquad y_t(x, 0) = v_0.$$

Derive this expression for those displacements:

$$y(x, t) = \frac{4v_0}{\pi} e^{-\beta t} \sum_{n=1}^{\infty} \frac{\sin(2n-1)x}{(2n-1)\alpha_n} \sin \alpha_n t,$$

where $\alpha_n = \sqrt{(2n-1)^2 - \beta^2}$.

6. The ends of a stretched string are fixed at the origin and at the point $x = \pi$ on the horizontal x axis. The string is initially at rest along the x axis and then drops under its own weight. The vertical displacements $y(x,t)$ thus satisfy the differential equation (Sec. 25)

$$y_{tt}(x,t) = a^2 y_{xx}(x,t) - g \qquad (0 < x < \pi, t > 0),$$

where g is acceleration due to gravity.

(*a*) Use the method of variation of parameters (Sec. 36) to derive the expression

$$y(x,t) = \frac{4g}{\pi a^2}\left[\sum_{n=1}^{\infty} \frac{\sin(2n-1)x}{(2n-1)^3}\cos(2n-1)at - \frac{\pi}{8}x(\pi - x)\right]$$

for those displacements.

(*b*) With the aid of the trigonometric identity

$$2\sin A\cos B = \sin(A+B) + \sin(A-B),$$

show that the expression found in part (*a*) can be put in the closed form

$$y(x,t) = \frac{g}{2a^2}\left[\frac{P(x+at) + P(x-at)}{2} - x(\pi - x)\right]$$

where the function $P(x)$ is the odd periodic extension, with period 2π, of the function $x(\pi - x)$ $(0 \le x \le \pi)$.

Suggestion: In both parts (*a*) and (*b*), the Fourier sine series representation (Problem 5, Sec. 5)

$$x(\pi - x) = \frac{8}{\pi}\sum_{n=1}^{\infty} \frac{\sin(2n-1)x}{(2n-1)^3} \qquad (0 \le x \le \pi)$$

is needed. Also, for part (*a*), see the suggestion with Problem 5, Sec. 37.

40. RESONANCE

A stretched string, of length unity and with fixed ends, is initially at rest in its position of equilibrium. A simple periodic transverse force acts uniformly on all elements of the string, so that the transverse displacements $y(x,t)$ satisfy this modified form (see Sec. 25) of the wave equation:

$$(1) \qquad y_{tt}(x,t) = y_{xx}(x,t) + A\sin\omega t \qquad (0 < x < 1, t > 0),$$

where A is a constant. Equation (1) and the boundary conditions

$$(2) \qquad y(0,t) = 0, \qquad y(1,t) = 0,$$

$$(3) \qquad y(x,0) = 0, \qquad y_t(x,0) = 0,$$

just described, make up a boundary value problem to which the method of variation of parameters (Sec. 36) can be applied.

We note from Sec. 32 that if the constant A were actually zero, the Sturm-Liouville problem arising would have eigenfunctions $\sin n\pi x$ $(n = 1, 2, \ldots)$. Hence we seek a solution of our boundary value problem having the form

$$(4) \qquad y(x,t) = \sum_{n=1}^{\infty} B_n(t)\sin n\pi x,$$

where the coefficients $B_n(t)$ are to be determined. Substituting series (4) into equation (1) and using the Fourier sine series representation

$$1 = \sum_{n=1}^{\infty} \frac{2[1-(-1)^n]}{n\pi} \sin n\pi x \qquad (0 < x < 1),$$

which is easily obtained by first replacing x by πx in the known [Problem 1(*b*), Sec. 5] representation

$$1 = \sum_{n=1}^{\infty} \frac{2[1-(-1)^n]}{n\pi} \sin nx \qquad (0 < x < \pi),$$

we write

$$\sum_{n=1}^{\infty} B_n''(t) \sin n\pi x = \sum_{n=1}^{\infty} [-(n\pi)^2 B_n(t)] \sin n\pi x + \sum_{n=1}^{\infty} \frac{2A[1-(-1)^n]}{n\pi} \sin \omega t \sin n\pi x.$$

Thus

$$B_n''(t) + (n\pi)^2 B_n(t) = \frac{2A[1-(-1)^n]}{n\pi} \sin \omega t \qquad (n = 1, 2, \ldots), \tag{5}$$

and conditions (3) yield the initial conditions

$$B_n(0) = 0 \quad \text{and} \quad B_n'(0) = 0 \qquad (n = 1, 2, \ldots) \tag{6}$$

on $B_n(t)$.

When n is replaced by $2n$ in equations (5) and (6), we have

$$B_{2n}''(t) + (2n\pi)^2 B_{2n}(t) = 0, \tag{7}$$

$$B_{2n}(0) = 0, \qquad B_{2n}'(0) = 0. \tag{8}$$

Solving this initial value problem, we find that $B_{2n}(t) \equiv 0$ $(n = 1, 2, \ldots)$. That is, $B_n(t)$ is identically equal to zero when n is even.

To find $B_n(t)$ when n is odd, we replace n by $2n - 1$ in equations (5) and (6) and write

$$\omega_n = (2n-1)\pi \qquad (n = 1, 2, \ldots). \tag{9}$$

The initial value problem for $B_{2n-1}(t)$ is then

$$B_{2n-1}''(t) + \omega_n^2 B_{2n-1}(t) = \frac{4A}{\omega_n} \sin \omega t, \tag{10}$$

$$B_{2n-1}(0) = 0, \qquad B_{2n-1}'(0) = 0. \tag{11}$$

We may now refer to Problem 3 below. In that problem, methods learned in an introductory course in ordinary differential equations are used to solve the initial value problem

$$y''(t) + a^2 y(t) = b \sin \omega t, \tag{12}$$

$$y(0) = 0, \qquad y'(0) = 0, \tag{13}$$

where a and b are constants.

To be specific, if $\omega \neq a$,

$$y(t) = \frac{b}{\omega^2 - a^2}\left(\frac{\omega}{a}\sin at - \sin \omega t\right). \tag{14}$$

Thus we see that if $\omega \neq \omega_n$ for any value of n ($n = 1, 2, \ldots$), the solution of problem (10)–(11) is

$$B_{2n-1}(t) = \frac{4A}{\omega_n(\omega^2 - \omega_n^2)}\left(\frac{\omega}{\omega_n}\sin \omega_n t - \sin \omega t\right) \qquad (n = 1, 2, \ldots);$$

and it follows from equation (4) that

$$y(x,t) = 4A\sum_{n=1}^{\infty}\frac{\sin \omega_n x}{\omega_n(\omega^2 - \omega_n^2)}\left(\frac{\omega}{\omega_n}\sin \omega_n t - \sin \omega t\right). \tag{15}$$

It is also shown in Problem 3 that if $\omega = a$, the solution of differential equation (12), with conditions (13), is

$$y(t) = \frac{b}{2a}\left(\frac{1}{a}\sin at - t\cos at\right). \tag{16}$$

Hence, when there is a value N of n ($n = 1, 2, \ldots$) such that $\omega = \omega_N$,

$$B_{2N-1}(t) = \frac{2A}{\omega_N^2}\left(\frac{1}{\omega_N}\sin \omega_N t - t\cos \omega_N t\right). \tag{17}$$

Because of the factor t with the cosine function here, this means that series (4) contains an *unstable* component. Such a phenomenon is called *resonance*. The periodic external force is evidently in resonance with the string when the frequency ω of that force coincides with any one of the resonant frequencies (9). Those frequencies depend, in general, on the physical properties of the string and the manner in which it is supported.

PROBLEMS

1. The boundary value problem

$$y_{tt}(x,t) = a^2 y_{xx}(x,t) + Ax\sin \omega t \qquad (0 < x < c, t > 0),$$

$$y(0,t) = y(c,t) = 0, \qquad y(x,0) = y_t(x,0) = 0$$

describes transverse displacements in a vibrating string. [Compare with equation (1), Sec. 40, where the term that was $A \sin \omega t$ is now $Ax \sin \omega t$.] Show that resonance occurs when ω has one of the values

$$\omega_n = \frac{n\pi a}{c} \qquad (n = 1, 2, \ldots).$$

2. Let a, b, and ω denote nonzero constants. The general solution of the ordinary differential equation

$$y''(t) + a^2 y(t) = b\sin \omega t$$

is of the form $y = y_c + y_p$, where y_c is the general solution of the complementary equation $y''(t) + a^2 y(t) = 0$ and y_p is any particular solution of the original nonhomogeneous equation.[†]

(*a*) Suppose that $\omega \neq a$. After substituting

$$y_p = A\cos\omega t + B\sin\omega t,$$

where A and B are constants, into the given differential equation, determine values of A and B such that y_p is a solution. Thus derive the general solution

$$y(t) = C_1\cos at + C_2\sin at + \frac{b}{a^2 - \omega^2}\sin\omega t$$

of that equation.

(*b*) Suppose that $\omega = a$ and find constants A and B such that

$$y_p = At\cos\omega t + Bt\sin\omega t$$

is a particular solution of the given differential equation. Thus obtain the general solution

$$y(t) = C_1\cos at + C_2\sin at - \frac{b}{2a}t\cos at.$$

3. Use the general solutions derived in Problem 2 to obtain the following solutions of the initial value problem

$$y''(t) + a^2 y(t) = b\sin\omega t, \qquad y(0) = 0, \qquad y'(0) = 0:$$

$$y(t) = \begin{cases} \dfrac{b}{\omega^2 - a^2}\left(\dfrac{\omega}{a}\sin at - \sin\omega t\right) & \text{when } \omega \neq a: \\[2ex] \dfrac{b}{2a}\left(\dfrac{1}{a}\sin at - t\cos at\right) & \text{when } \omega = a. \end{cases}$$

41. AN ELASTIC BAR

Boundary value problems for longitudinal displacements in elastic bars are similar to problems for transverse displacements in a stretched string.

EXAMPLE. A cylindrical bar of natural length c is initially stretched by an amount bc (Fig. 39) and starts its motion from rest. The initial longitudinal

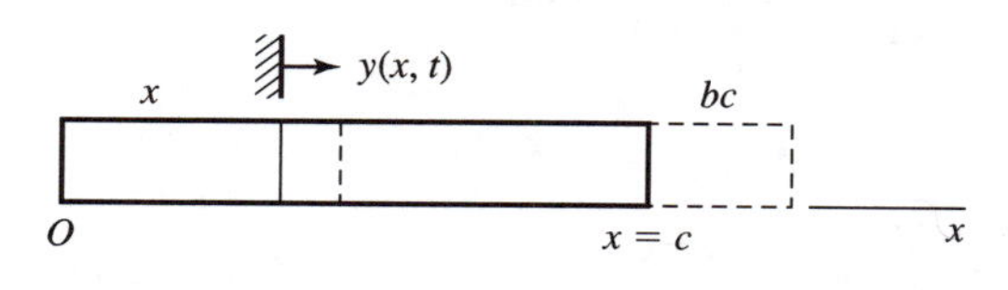

FIGURE 39

[†]For the method of solution to be used here, which is known as the *method of undetermined coefficients*, see, for instance, the book by Boyce and DiPrima (2005) or the one by Rainville, Bedient, and Bedient (1997). Both books are listed in the Bibliography.

displacements of its sections are then proportional to the distance from the fixed end $x = 0$. At the instant $t = 0$, both ends are released and left free. The longitudinal displacements $y(x, t)$ satisfy the following boundary value problem, where $a^2 = E/\delta$ (Sec. 26):

$$y_{tt}(x, t) = a^2 y_{xx}(x, t) \qquad (0 < x < c, t > 0), \tag{1}$$

$$y_x(0, t) = 0, \qquad y_x(c, t) = 0, \tag{2}$$

$$y(x, 0) = bx, \qquad y_t(x, 0) = 0. \tag{3}$$

The homogeneous two-point boundary conditions (2) state that the force per unit area on the end sections is zero.

The function $y(x, t)$ can also be interpreted as representing transverse displacements in a stretched string, released at rest from the position $y(x, 0) = bx$, when the ends are looped around perfectly smooth rods lying along the lines $x = 0$ and $x = c$. In that case, $a^2 = H/\delta$; and the boundary conditions (2) state that no forces act in the y direction at the ends of the string (see Sec. 25).

The products $y = X(x)T(t)$ satisfy all the homogeneous conditions above when $X(x)$ is an eigenfunction of the problem

$$X''(x) + \lambda X(x) = 0, \qquad X'(0) = 0, \qquad X'(c) = 0 \tag{4}$$

and when, for the same eigenvalue λ,

$$T''(t) + \lambda a^2 T(t) = 0, \qquad T'(0) = 0. \tag{5}$$

The eigenvalues are (Sec. 31)

$$\lambda_0 = 0 \qquad \text{and} \qquad \lambda_n = \left(\frac{n\pi}{c}\right)^2 \qquad (n = 1, 2, \ldots),$$

with eigenfunctions

$$X_0(x) = 1 \qquad \text{and} \qquad X_n(x) = \cos\frac{n\pi x}{c} \qquad (n = 1, 2, \ldots).$$

The corresponding functions of t are

$$T_0(t) = 1 \qquad \text{and} \qquad T_n(t) = \cos\frac{n\pi at}{c} \qquad (n = 1, 2, \ldots).$$

Formally, then, the generalized linear combination

$$y(x, t) = A_0 + \sum_{n=1}^{\infty} A_n \cos\frac{n\pi x}{c} \cos\frac{n\pi at}{c} \tag{6}$$

satisfies conditions (1) through (3), provided that

$$bx = A_0 + \sum_{n=1}^{\infty} A_n \cos\frac{n\pi x}{c} \qquad (0 < x < c). \tag{7}$$

The coefficients in this Fourier cosine series, which is actually valid on the closed interval $0 \le x \le c$, are

$$A_0 = \frac{b}{c}\int_0^c x\,dx, \qquad A_n = \frac{2b}{c}\int_0^c x\cos\frac{n\pi x}{c}\,dx \qquad (n = 1, 2, \ldots).$$

Consequently,

$$(8) \qquad A_0 = \frac{bc}{2}, \qquad A_n = -\frac{2bc}{\pi^2} \cdot \frac{1-(-1)^n}{n^2} \qquad (n = 1, 2, \ldots);$$

and we arrive at the solution

$$(9) \qquad y(x,t) = \frac{bc}{2} - \frac{4bc}{\pi^2} \sum_{n=1}^{\infty} \frac{1}{(2n-1)^2} \cos\frac{(2n-1)\pi x}{c} \cos\frac{(2n-1)\pi a t}{c}.$$

By a method already used in Sec. 39, we can put this series solution in closed form, involving the *even* periodic extension $P(x)$, with period $2c$, of the function bx $(0 \le x \le c)$. To be specific, we know from the trigonometric identity

$$2\cos A \cos B = \cos(A+B) + \cos(A-B)$$

that

$$2\cos\frac{n\pi x}{c}\cos\frac{n\pi a t}{c} = \cos\frac{n\pi(x+at)}{c} + \cos\frac{n\pi(x-at)}{c}.$$

Hence expression (6) can be written

$$(10) \quad y(x,t) = \frac{1}{2}\left[A_0 + \sum_{n=1}^{\infty} A_n \cos\frac{n\pi(x+at)}{c} + A_0 + \sum_{n=1}^{\infty} A_n \cos\frac{n\pi(x-at)}{c}\right].$$

But series (7) represents $P(x)$ for all values of x when the values (8) of the coefficients A_n $(n = 0, 1, 2, \ldots)$ are used. Hence expression (10), with those values of A_n, reduces to

$$(11) \qquad y(x,t) = \frac{1}{2}[P(x+at) + P(x-at)].$$

This is the desired closed form of solution (9).

PROBLEMS

1. Show that the motion of each cross section of the elastic bar treated in Sec. 41 is periodic in t, with period $2c/a$.
2. From expression (11), Sec. 41, show that $y(0,t) = P(at)$ and hence that the end $x = 0$ of the bar moves with the constant velocity ab during the half period $0 < t < c/a$ (see Problem 1) and with velocity $-ab$ during the next half-period.
3. The end $x = 0$ of an elastic bar is free, and a constant longitudinal force F_0 per unit area is applied at the end $x = c$ (Fig. 40). The bar is initially unstrained and at rest. Set up the boundary value problem for the longitudinal displacements $y(x,t)$, the conditions at the ends of the bar being $y_x(0,t) = 0$ and $y_x(c,t) = F_0/E$ (Sec. 26). After noting that the method of separation of variables cannot be applied directly, follow the steps below to find $y(x,t)$.

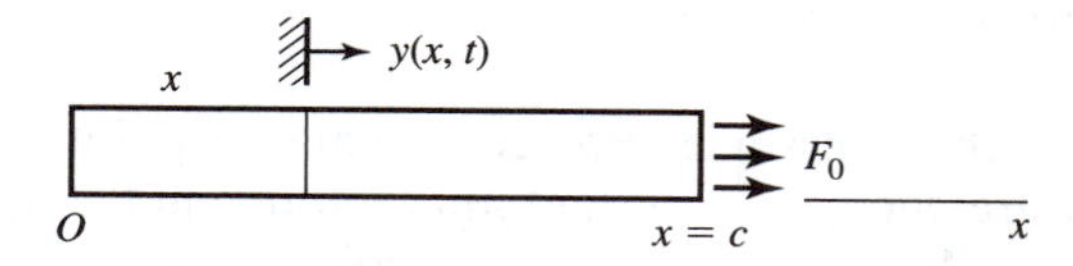

FIGURE 40

(*a*) By writing $y(x,t) = Y(x,t) + Ax^2$, determine a value of A that leads to the new boundary value problem

$$Y_{tt}(x,t) = a^2 Y_{xx}(x,t) + \frac{F_0 a^2}{cE} \qquad (0 < x < c, t > 0),$$

$$Y_x(0,t) = 0, \qquad Y_x(c,t) = 0,$$

$$Y(x,0) = -\frac{F_0}{2cE}x^2, \qquad Y_t(x,0) = 0.$$

(*b*) Point out why it is reasonable to expect that the new boundary value problem in part (*a*) has a solution of the form

$$Y(x,\ t) = A_0(t) + \sum_{n=1}^{\infty} A_n(t)\cos\frac{n\pi x}{c}.$$

Then use the method of variation of parameters (Sec. 36) to find $Y(x,t)$ and thereby derive the solution

$$y(x,t) = \frac{F_0}{6cE}\left[3(x^2 + a^2t^2) - c^2 + \frac{12c^2}{\pi^2}\sum_{n=1}^{\infty}\frac{(-1)^{n+1}}{n^2}\cos\frac{n\pi at}{c}\cos\frac{n\pi x}{c}\right]$$

of the original problem.

(*c*) Use the trigonometric identity

$$2\cos A\cos B = \cos(A+B) + \cos(A-B)$$

and the series representation

$$x^2 = \frac{c^2}{3} + \frac{4c^2}{\pi^2}\sum_{n=1}^{\infty}\frac{(-1)^n}{n^2}\cos\frac{n\pi x}{c} \qquad (-c \le x \le c),$$

which follows from the example in Sec. 8, to write the expression for $y(x,\ t)$ in part (*b*) as

$$y(x,t) = \frac{F_0}{2cE}\left[x^2 + a^2t^2 - \frac{P(x+at) + P(x-at)}{2}\right]$$

where $P(x)$ is the periodic extension, with period $2c$, of the function x^2 defined on the interval $-c \le x \le c$.

4. Show how it follows from the expression for $y(x,t)$ in Problem 3(*c*) that the end $x = 0$ of the bar remains at rest until time $t = c/a$ and then moves with velocity $v_0 = 2aF_0/E$ when $c/a < t < 3c/a$, with velocity $2v_0$ when $3c/a < t < 5c/a$, etc.

42. DOUBLE FOURIER SERIES

The Fourier method is readily adapted to boundary value problems that give rise to so-called double Fourier series. This is illustrated below.

EXAMPLE. Let $z(x,\ y,t)$ denote the transverse displacement at each point $(x,\ y)$ at time t in a membrane that is stretched across a rigid square frame in the xy plane. To simplify the notation, we select the origin and the point $(\pi,\ \pi)$ as ends of a diagonal of the frame. If the membrane is released at rest with a given initial displacement $f(x, y)$ that is continuous and vanishes on the boundary of the square, then (Sec. 26)

$$z_{tt} = a^2(z_{xx} + z_{yy}) \tag{1}$$

in the three-dimensional domain $0 < x < \pi, 0 < y < \pi, t > 0$ and

$$z(0, y, t) = z(\pi, y, t) = z(x, 0, t) = z(x, \pi, t) = 0, \tag{2}$$

$$z(x, y, 0) = f(x, y), \qquad z_t(x, y, 0) = 0, \tag{3}$$

where $0 \le x \le \pi, 0 \le y \le \pi$. We assume that the partial derivatives $f_x(x, y)$ and $f_y(x, y)$ are also continuous.

Functions of the type $z = X(x)Y(y)T(t)$ satisfy equation (1) if

$$\frac{T''(t)}{a^2T(t)} = \frac{X''(x)}{X(x)} + \frac{Y''(y)}{Y(y)} = -\lambda. \tag{4}$$

Separating variables again, this time in the second of equations (4), we have

$$\frac{Y''(y)}{Y(y)} = -\lambda - \frac{X''(x)}{X(x)} = -\mu,$$

where μ is another separation constant. So we are led to two Sturm-Liouville problems,

$$X''(x) + (\lambda - \mu)X(x) = 0, \qquad X(0) = 0, \qquad X(\pi) = 0$$

and

$$Y''(y) + \mu Y(y) = 0, \qquad Y(0) = 0, \qquad Y(\pi) = 0,$$

and to the conditions

$$T''(t) + \lambda a^2 T(t) = 0, \qquad T'(0) = 0$$

on T.

We turn to the Sturm-Liouville problem in Y first since it involves only one of the separation constants. According to Sec. 32, that problem has eigenvalues $\mu = m^2$ $(m = 1, 2, \ldots)$ and corresponding eigenfunctions

$$Y_m(y) = \sin my \qquad (m = 1, 2, \ldots).$$

When $\lambda - \mu = n^2$ $(n = 1, 2, \ldots)$, the eigenfunctions

$$X_n(x) = \sin nx \qquad (n = 1, 2, \ldots)$$

of the problem in X are also obtained. The conditions on T thus become

$$T''(t) + a^2(m^2 + n^2)T(t) = 0, \qquad T'(0) = 0,$$

where $m = 1, 2, \ldots$ and $n = 1, 2, \ldots$. For any fixed positive integers m and n, the solution of this problem in T is, except for a constant factor,

$$T_{mn}(t) = \cos\left(at\sqrt{m^2 + n^2}\right).$$

The formal solution of our boundary value problem is, therefore,

$$z(x, y, t) = \sum_{n=1}^{\infty} \sum_{m=1}^{\infty} B_{mn} \sin nx \sin my \cos\left(at\sqrt{m^2 + n^2}\right), \tag{5}$$

where the coefficients B_{mn} need to be determined so that

$$f(x, y) = \sum_{n=1}^{\infty} \sum_{m=1}^{\infty} B_{mn} \sin nx \sin my \tag{6}$$

when $0 \le x \le \pi$ and $0 \le y \le \pi$. By grouping terms in this double sine series so as to display the total coefficient of $\sin nx$ for each n, one can write, formally,

$$f(x, y) = \sum_{n=1}^{\infty} \left(\sum_{m=1}^{\infty} B_{mn} \sin my \right) \sin nx. \tag{7}$$

For each fixed y $(0 \le y \le \pi)$, equation (7) is a Fourier sine series representation of the function $f(x, y)$, with variable x $(0 \le x \le \pi)$, provided that

$$\sum_{m=1}^{\infty} B_{mn} \sin my = \frac{2}{\pi} \int_0^{\pi} f(x, y) \sin nx \, dx \qquad (n = 1, 2, \ldots). \tag{8}$$

The right-hand side here is a sequence of functions

$$F_n(y) = \frac{2}{\pi} \int_0^{\pi} f(x, y) \sin nx \, dx \qquad (n = 1, 2, \ldots) \tag{9}$$

of y, each represented by its Fourier sine series

$$F_n(y) = \sum_{m=1}^{\infty} B_{mn} \sin my$$

on the interval $0 \le y \le \pi$ when

$$B_{mn} = \frac{2}{\pi} \int_0^{\pi} F_n(y) \sin my \, dy \qquad (m = 1, 2, \ldots). \tag{10}$$

Combining expressions (9) and (10), we find that

$$B_{mn} = \frac{4}{\pi^2} \int_0^{\pi} \sin my \int_0^{\pi} f(x, y) \sin nx \, dx \, dy. \tag{11}$$

The solution of our membrane problem is now given by equation (5) with coefficients (11).

Since the numbers $\sqrt{m^2 + n^2}$ do not change by integral multiples of some fixed number as m and n vary through integral values, the cosine functions in equation (5) have no common period in the variable t; so the displacement z is not generally a periodic function of t. Consequently, the vibrating membrane, in contrast to the vibrating string, generally does not produce a musical note. It can be made to do so, however, by giving it the proper initial displacement. If, for example,

$$z(x, y, 0) = A \sin x \sin y$$

where A is a constant, the displacements (5) are given by a single term:

$$z(x, y, t) = A \sin x \sin y \cos\left(a\sqrt{2}t\right).$$

Then z is periodic in t, with period $\pi\sqrt{2}/a$.

PROBLEMS

1. All four faces of an infinitely long rectangular prism, formed by the planes $x=0$, $x = a$, $y = 0$, and $y = b$, are kept at temperature zero. Let the initial temperature distribution be $f(x, y)$, and derive this expression for the temperatures $u(x, y, t)$ in the prism:

$$u(x, y, t) = \sum_{n=1}^{\infty}\sum_{m=1}^{\infty} B_{mn} \exp\left[-\pi^2 kt\left(\frac{m^2}{a^2} + \frac{n^2}{b^2}\right)\right] \sin\frac{n\pi x}{a} \sin\frac{m\pi y}{b},$$

where

$$B_{mn} = \frac{4}{ab}\int_0^b \sin\frac{m\pi y}{b}\int_0^a f(x, y)\sin\frac{n\pi x}{a}\,dx\,dy.$$

2. Write $f(x, y) = g(x)h(y)$ in Problem 1 and show that the double series obtained there for u reduces to the product

$$u(x, y, t) = v(x, t)w(y, t)$$

of two single series, where v and w represent temperatures in the slabs $0 \le x \le a$ and $0 \le y \le b$ with faces at temperature zero and with initial temperatures $g(x)$ and $h(y)$, respectively.

3. Let the functions $v(x, t)$ and $w(y, t)$ satisfy the heat equation for one-dimensional flow:

$$v_t = kv_{xx}, \qquad w_t = kw_{yy}.$$

Show by differentiation that their product $u = vw$ satisfies the two-dimensional heat equation

$$u_t = k(u_{xx} + u_{yy}).$$

Use this result to arrive at the expression for $u(x, y, t)$ in Problem 2.

43. PERIODIC BOUNDARY CONDITIONS

The solutions of the boundary value problems in this chapter have been based on the solutions of just two Sturm-Liouville problems, which lead to Fourier cosine and sine series representations of prescribed functions. Although Chap. 8 is devoted to the theory and application of many other Sturm-Liouville problems, as well as to the precise definition of such a problem, we conclude this chapter by considering a third problem that arises in certain boundary value problems for regions with circular boundaries:

$$(1) \qquad X''(x) + \lambda X(x) = 0, \qquad X(-\pi) = X(\pi), \qquad X'(-\pi) = X'(\pi).$$

We include it here since its solutions also lead to Fourier series representations, but now involving *both* cosines and sines on the interval $-\pi < x < \pi$, and since the general theory of Sturm-Liouville problems is not actually required. We need accept only the fact, to be verified in Chap. 8 (Sec. 61), that each eigenvalue, or value of λ for which problem (1) has a nontrivial solution, is a real number. In anticipation of Chap. 8, we continue to refer to such values of λ as eigenvalues and to the nontrivial solutions as eigenfunctions.

To solve problem (1), we consider first the case in which $\lambda = 0$. It is easy to see that $X(x) = Ax + B$, where A and B are constants; and the boundary conditions are satisfied if $A = 0$. Since the conditions in problem (1) are all homogeneous, we thus find that, except for a constant factor, $X(x) = 1$.

When $\lambda > 0$, we write $\lambda = \alpha^2$ ($\alpha > 0$) and note that the general solution of the differential equation in problem (1) is

$$X(x) = C_1 \cos \alpha x + C_2 \sin \alpha x.$$

It is straightforward to show that in order for the boundary conditions to be satisfied,

$$C_2 \sin \alpha\pi = 0 \qquad \text{and} \qquad C_1 \sin \alpha\pi = 0.$$

Since the constants C_1 and C_2 cannot both vanish if $X(x)$ is to be nontrivial, it follows that the positive number α must, in fact, be a positive integer n. That is, $\lambda = n^2$ ($n = 1, 2, \ldots$); and the corresponding general solution of problem (1) is an arbitrary linear combination of the two linearly independent eigenfunctions $\cos nx$ and $\sin nx$.

It is left to the reader (Problem 1) to show that there are no negative eigenvalues. The eigenvalues and corresponding eigenfunctions are then

$$\lambda_0 = 0 \qquad \text{and} \qquad \lambda_n = n^2 \qquad (n = 1, 2, \ldots)$$

and

$$X_0(x) = 1 \qquad \text{and} \qquad X_n(x) = A_n \cos nx + B_n \sin nx \qquad (n = 1, 2, \ldots),$$

where A_n and B_n are arbitrary constants.

We now illustrate the use of this Sturm-Liouville problem, involving *periodic boundary conditions.*

EXAMPLE. Let $u(\rho, \phi)$ denote the steady temperatures in a thin disk $\rho \le 1$, with insulated surfaces, when its edge $\rho = 1$ is kept at temperatures $f(\phi)$. The variables ρ and ϕ are, of course, polar coordinates, and u satisfies Laplace's equation $\nabla^2 u = 0$. That is,

$$\rho^2 u_{\rho\rho}(\rho, \phi) + \rho u_\rho(\rho, \phi) + u_{\phi\phi}(\rho, \phi) = 0 \qquad (0 < \rho < 1, -\pi < \phi < \pi), \tag{2}$$

where

$$u(1, \phi) = f(\phi) \qquad (-\pi < \phi < \pi). \tag{3}$$

Also, u and its partial derivatives of the first and second orders are continuous and bounded in the interior of the disk. In particular, u and its first-order partial derivatives are continuous on the ray $\phi = \pi$ (Fig. 41).

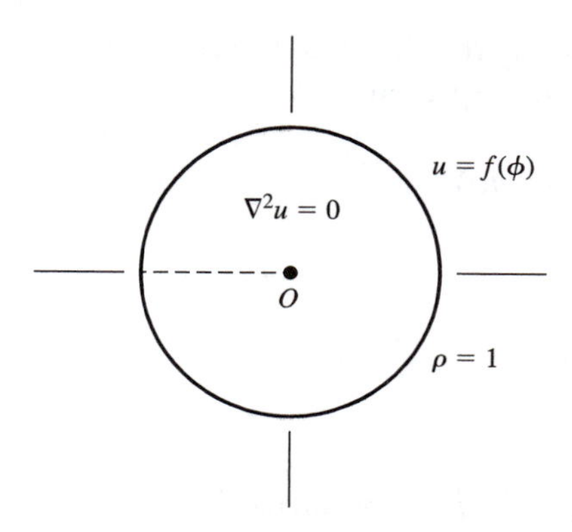

FIGURE 41

If functions of the type $u = R(\rho)\Phi(\phi)$ are to satisfy condition (2) and the continuity requirements, then

$$\rho^2 R''(\rho) + \rho R'(\rho) - \lambda R(\rho) = 0 \qquad (0 < \rho < 1) \tag{4}$$

and

$$\Phi''(\phi) + \lambda\Phi(\phi) = 0, \qquad \Phi(-\pi) = \Phi(\pi), \qquad \Phi'(-\pi) = \Phi'(\pi), \tag{5}$$

where λ is a separation constant. We now recognize that conditions (5) constitute a Sturm-Liouville problem in Φ, with eigenvalues $\lambda_0 = 0$ and $\lambda_n = n^2$ $(n = 1, 2, \ldots)$. The corresponding eigenfunctions are

$$\Phi_0(\phi) = 1 \qquad \text{and} \qquad \Phi_n(\phi) = A_n \cos n\phi + B_n \sin n\phi \qquad (n = 1, 2, \ldots).$$

Equation (4) is a Cauchy-Euler equation, and we know from Example 2, Sec. 38, that its bounded solutions are

$$R_0(\rho) = 1 \qquad \text{when } \lambda = 0$$

and

$$R_n(\rho) = \rho^n \qquad \text{when } \lambda = n^2 \qquad (n = 1, 2, \ldots).$$

By superposition, the generalized linear combination of our continuous functions

$$R_0(\rho)\Phi_0(\phi) = 1 \qquad \text{and} \qquad R_n(\rho)\Phi_n(\phi) = \rho^n(A_n \cos n\phi + B_n \sin n\phi) \qquad (n = 1, 2, \ldots)$$

is

$$u(\rho, \phi) = A_0 + \sum_{n=1}^{\infty} \rho^n (A_n \cos n\phi + B_n \sin n\phi). \tag{6}$$

Note that if we had multiplied the products $R_n(\rho)\Phi_n(\phi)$ $(n = 1, 2, \ldots)$ by arbitrary constants, those constants would have been absorbed into the arbitrary constants A_n and B_n.

The nonhomogeneous condition (3) evidently requires that the constants in expression (6) be the same as in the Fourier series representation

$$f(\phi) = \frac{2A_0}{2} + \sum_{n=1}^{\infty}(A_n \cos n\phi + B_n \sin n\phi) \qquad (-\pi < x < \pi).$$

Thus

$$A_0 = \frac{1}{2\pi}\int_{-\pi}^{\pi} f(\phi)\,d\phi \tag{7}$$

and

$$A_n = \frac{1}{\pi}\int_{-\pi}^{\pi} f(\phi)\cos n\phi\,d\phi, \qquad B_n = \frac{1}{\pi}\int_{-\pi}^{\pi} f(\phi)\sin n\phi\,d\phi \qquad (n = 1, 2, \ldots). \tag{8}$$

We assume that f is piecewise smooth.

PROBLEMS

1. Write $\lambda = -\alpha^2$ $(\alpha > 0)$, and show that the Sturm-Liouville problem

$$X''(x) + \lambda X(x) = 0, \qquad X(-\pi) = X(\pi), \qquad X'(-\pi) = X'(\pi)$$

in Sec. 43 has no negative eigenvalues.

Suggestion: Show that when $X(x) = C_1 e^{\alpha x} + C_2 e^{-\alpha x}$, an application of the boundary conditions leads to the equations

$$(C_1 - C_2)\sinh \alpha\pi = 0 \qquad \text{and} \qquad (C_1 + C_2)\sinh \alpha\pi = 0.$$

Then note that in each equation, the factor $\sinh \alpha\pi$ is nonzero and can, therefore, be canceled out.

2. Using the cylindrical coordinates ρ, ϕ, and z, let $u(\rho, \phi)$ denote steady temperatures in a long hollow cylinder $a \le \rho \le b$, $-\infty < z < \infty$ when the temperatures on the inner surface $\rho = a$ are $f(\phi)$ and the temperature of the outer surface $\rho = b$ is zero.

(*a*) Derive the temperature formula

$$u(\rho, \phi) = A_0 \frac{\ln b - \ln \rho}{\ln b - \ln a} + \sum_{n=1}^{\infty}\left(\frac{a}{\rho}\right)^n \frac{b^{2n} - \rho^{2n}}{b^{2n} - a^{2n}} (A_n \cos n\phi + B_n \sin n\phi),$$

where the coefficients A_n and B_n, including A_0, are given by equations (7) and (8) in Sec. 43.

(*b*) Use the result in part (*a*) to show that if $f(\phi) = \alpha + \beta \sin \phi$, where α and β are constants, then

$$u(\rho, \phi) = \alpha \frac{\ln b - \ln \rho}{\ln b - \ln a} + \beta \frac{a}{\rho} \cdot \frac{b^2 - \rho^2}{b^2 - a^2} \sin \phi.$$

3. Solve the boundary value problem

$$u_t(x, t) = k u_{xx}(x, t) \qquad (-\pi < x < \pi, t > 0),$$

$$u(-\pi, t) = u(\pi, t), \qquad u_x(-\pi, t) = u_x(\pi, t), \qquad u(x, 0) = f(x).$$

The solution $u(x, t)$ represents, for example, temperatures in an insulated wire of length 2π that is bent into a unit circle and has a given temperature distribution along it. For convenience, the wire is thought of as being cut at one point and laid on the x axis between $x = -\pi$ and $x = \pi$. The variable x then measures the distance along the wire, starting at the point $x = -\pi$; and the points $x = -\pi$ and $x = \pi$ denote the same point on the circle. The first two boundary conditions in the problem state that the temperatures and the flux must be the same for each of those values of x. This problem was of considerable interest to Fourier himself, and the wire has come to be known as *Fourier's ring*.

$$\textit{Answer: } u(x, t) = A_0 + \sum_{n=1}^{\infty} e^{-n^2 kt}(A_n \cos nx + B_n \sin nx),$$

where

$$A_0 = \frac{1}{2\pi}\int_{-\pi}^{\pi} f(x)\,dx$$

and

$$A_n = \frac{1}{\pi}\int_{-\pi}^{\pi} f(x)\cos nx\,dx, \qquad B_n = \frac{1}{\pi}\int_{-\pi}^{\pi} f(x)\sin nx\,dx \qquad (n = 1, 2, \ldots).$$

4. (*a*) By writing $A = n\theta$ and $B = \theta$ in the trigonometric identity

$$2\cos A\cos B = \cos(A + B) + \cos(A - B),$$

multiplying through the resulting equation by a^n $(-1 < a < 1)$, and then summing each side from $n = 1$ to $n = \infty$, derive the summation formula

$$\sum_{n=1}^{\infty} a^n \cos n\theta = \frac{a\cos\theta - a^2}{1 - 2a\cos\theta + a^2} \qquad (-1 < a < 1).$$

[One can readily see that this series is absolutely convergent by comparing it with the geometric series whose terms are a^n $(n = 1, 2, \ldots)$.]

(*b*) Write expression (6), with coefficients (7) and (8), in Sec. 43 for steady temperatures in a disk as

$$u(\rho, \phi) = \frac{1}{2\pi}\int_{-\pi}^{\pi} f(\psi)\left[1 + 2\sum_{n=1}^{\infty}\rho^n \cos n(\phi - \psi)\right]d\psi.$$

Then, with the aid of the summation formula in part (*a*), derive *Poisson's integral formula* for those temperatures:†

$$u(\rho, \phi) = \frac{1}{2\pi}\int_{-\pi}^{\pi} f(\psi)\frac{1 - \rho^2}{1 - 2\rho\cos(\phi - \psi) + \rho^2}\,d\psi \qquad (\rho < 1).$$

†This and related formulas are obtained by complex-variable methods in the authors' book (2004, chap. 12) that is listed in the Bibliography.

CHAPTER 6

FOURIER INTEGRALS AND APPLICATIONS

In Chap. 2 (Sec. 14), we saw that a periodic function, with period $2c$, has a Fourier series representation that is valid for all x when it satisfies certain conditions on the fundamental interval $-c < x < c$. In this chapter, we develop the theory of trigonometric representations for functions, defined for all x, that are *not* periodic. Such representations, which are analogous to Fourier series representations, involve improper integrals instead of infinite series.

44. THE FOURIER INTEGRAL FORMULA

From Problem 9, Sec. 8, we know that the Fourier series corresponding to a function $f(x)$ on an interval $-c < x < c$ can be written

$$(1) \qquad \frac{1}{2c}\int_{-c}^{c} f(s)ds + \frac{1}{c}\sum_{n=1}^{\infty}\int_{-c}^{c} f(s)\cos\left[\frac{n\pi}{c}(s-x)\right]ds;$$

and, from Theorem 1 in Sec. 14, we know conditions under which this series converges to $f(x)$ everywhere in the interval $-c < x < c$. Namely, it is sufficient that f be piecewise smooth on the interval and that the value of f at each of its points of discontinuity x be the mean value of the one-sided limits $f(x+)$ and $f(x-)$.

Suppose now that f satisfies such conditions on *every* bounded interval $-c<x<c$. Here c may be given any fixed positive value, arbitrarily large but finite, and series (1) will represent $f(x)$ over the large segment $-c<x<c$ of the x axis. But that series representation cannot apply over the rest of the x axis unless f is periodic, with period $2c$, because the sum of the series has that periodicity.

In seeking a representation that is valid for all real x when f is not periodic, it is natural to try to modify series (1) by letting c tend to infinity. The first term in the series will then vanish, provided that the improper integral

$$\int_{-\infty}^{\infty} f(s)\,ds$$

exists. If we write $\Delta\alpha = \pi/c$, the remaining terms take the form

$$\frac{1}{\pi}\sum_{n=1}^{\infty} \Delta\alpha \int_{-c}^{c} f(s)\cos\left[n\Delta\alpha\,(s-x)\right] ds, \tag{2}$$

which is the same as

$$\frac{1}{\pi}\sum_{n=1}^{\infty} F_c(n\Delta\alpha, x)\,\Delta\alpha \qquad \left(\Delta\alpha = \frac{\pi}{c}\right), \tag{3}$$

where

$$F_c(\alpha, x) = \int_{-c}^{c} f(s)\cos\alpha(s-x)\,ds. \tag{4}$$

Let the value of x be fixed and c be large, so that $\Delta\alpha$ is a small positive number. The points $n\Delta\alpha$ $(n = 1, 2, \ldots)$ are equally spaced along the entire positive α axis; and, because of the resemblance of the series in expression (3) to a sum of areas of rectangles used in defining definite integrals (see Fig. 42), one might expect the partial sums of that series to approach

$$\int_{0}^{\infty} F_c(\alpha, x)\,d\alpha, \tag{5}$$

or possibly

$$\int_{0}^{\infty} F_\infty(\alpha, x)\,d\alpha, \tag{6}$$

as $\Delta\alpha$ tends to zero. As the subscript in integral (6) indicates, however, the function $F_c(\alpha, x)$ changes with $\Delta\alpha$ because $c = \pi/\Delta\alpha$. Also, the limit of the series in expression (3) as $\Delta\alpha$ tends to zero is not, in fact, the definition of the improper integral (5) even if c could be kept fixed.

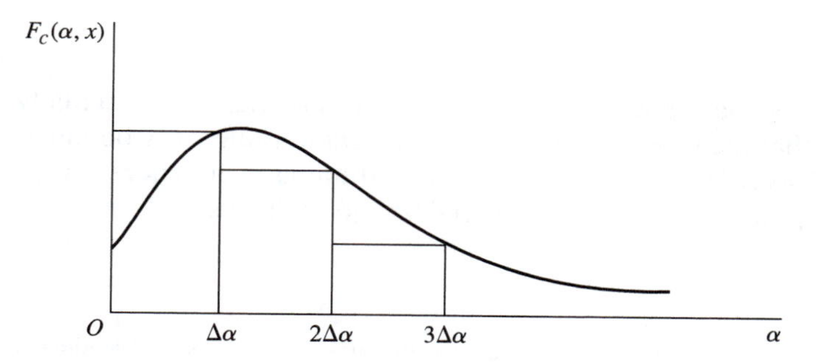

FIGURE 42

The above manipulations merely *suggest* that under appropriate conditions on f, the function may have the representation

$$f(x) = \frac{1}{\pi}\int_0^\infty \int_{-\infty}^\infty f(s)\cos\alpha(s-x)\,ds\,d\alpha \qquad (-\infty < x < \infty). \tag{7}$$

This is the *Fourier integral formula* for the function f, to be established in Sec. 47.

Observe how it follows from the trigonometric identity

$$\cos\alpha(s-x) = \cos\alpha s\cos\alpha x + \sin\alpha s\sin\alpha x$$

that

$$\int_{-\infty}^\infty f(s)\cos\alpha(s-x)\,ds = \int_{-\infty}^\infty f(s)\cos\alpha s\,ds\cos\alpha x + \int_{-\infty}^\infty f(s)\sin\alpha s\,ds\sin\alpha x.$$

Hence formula (7) can be written in terms of separate cosine and sine functions as follows:

$$f(x) = \int_0^\infty [A(\alpha)\cos\alpha x + B(\alpha)\sin\alpha x]\,d\alpha \qquad (-\infty < x < \infty), \tag{8}$$

where

$$A(\alpha) = \frac{1}{\pi}\int_{-\infty}^\infty f(x)\cos\alpha x\,dx, \qquad B(\alpha) = \frac{1}{\pi}\int_{-\infty}^\infty f(x)\sin\alpha x\,dx. \tag{9}$$

Expression (8), with coefficients (9), bears a resemblance to a Fourier series representation on $-\pi < x < \pi$.

A reader who wishes to accept the validity of Fourier integral representations in order to proceed more quickly to physical applications can at this time skip to Sec. 48 without serious disruption.

45. DIRICHLET'S INTEGRAL

Just as we prefaced the Fourier theorem in Sec. 12 with some preliminary theory, we include here and in Sec. 46 background that is essential to our proof of a theorem that gives conditions under which representation (7) in Sec. 44 is valid.

This section is devoted to the evaluation of an improper integral, known as *Dirichlet's integral,* that is prominent in applied mathematics. We show here that†

$$\int_0^\infty \frac{\sin x}{x}\,dx = \frac{\pi}{2}. \tag{1}$$

Our method of evaluation requires us to first show that the integral actually converges. We note that the integrand is piecewise continuous on every bounded interval $0 \le x \le c$. This is because that quotient is continuous everywhere except at $x = 0$, where l'Hôpital's rule shows that its right-hand limit exists.

---

†For another approach that is fairly standard, see, for instance, the book by Buck (2004). A method of evaluation involving complex variables is given in the authors' book (2004, pp. 269–270). Both books are listed in the Bibliography.

Since

$$\int_0^\infty \frac{\sin x}{x}\,dx = \lim_{c\to\infty}\int_0^c \frac{\sin x}{x}\,dx$$
$$= \lim_{c\to\infty}\left(\int_0^1 \frac{\sin x}{x}\,dx + \int_1^c \frac{\sin x}{x}\,dx\right)$$
$$= \int_0^1 \frac{\sin x}{x}\,dx + \lim_{c\to\infty}\int_1^c \frac{\sin x}{x}\,dx,$$

where c is any positive number, it suffices to show that the last limit here exists. To accomplish this, we use the method of integration by parts and write

$$\int_1^c \frac{\sin x}{x}\,dx = \cos 1 - \frac{\cos c}{c} - \int_1^c \frac{\cos x}{x^2}\,dx. \tag{2}$$

Because

$$\left|\frac{\cos c}{c}\right| \le \frac{1}{c} \quad \text{and} \quad \left|\frac{\cos x}{x^2}\right| \le \frac{1}{x^2},$$

the second term on the right in equation (2) tends to zero as c tends to infinity, and the improper integral

$$\int_1^\infty \frac{\cos x}{x^2}\,dx = \lim_{c\to\infty}\int_1^c \frac{\cos x}{x^2}\,dx$$

is (absolutely) convergent. The limit of the left-hand side of equation (2) as c tends to infinity therefore exists; that is, integral (1) converges.

Now that we have established that integral (1) converges to some number L, or that

$$\lim_{c\to\infty}\int_0^c \frac{\sin x}{x}\,dx = L,$$

we note that, in particular,

$$\lim_{N\to\infty}\int_0^{(\frac{1}{2}+N)\pi} \frac{\sin x}{x}\,dx = L \tag{3}$$

as N passes through positive integers. That is,

$$\lim_{N\to\infty}\int_0^\pi \frac{\sin\left(\frac{1}{2}+N\right)u}{u}\,du = L, \tag{4}$$

where the substitution

$$x = \left(\frac{1}{2}+N\right)u$$

has been made for the variable of integration. Observe that equation (4) can be written

$$\lim_{N\to\infty}\int_0^\pi g(u)\,D_N(u)\,du = L, \tag{5}$$

where

$$g(u) = \frac{2\sin\frac{u}{2}}{u} \tag{6}$$

and where $D_N(u)$ is the Dirichlet kernel (Sec. 11)

$$D_N(u) = \frac{\sin\left(\frac{u}{2} + Nu\right)}{2\sin\frac{u}{2}}. \tag{7}$$

The function $g(u)$, moreover, satisfies the conditions in Lemma 2, Sec. 11 (see Problem 1, Sec. 47); and $g(0+) = 1$. So, by that lemma, limit (5) has the value $\pi/2$; and, by uniqueness of limits, $L = \pi/2$. Integration formula (1) is now established.

46. TWO LEMMAS

The two lemmas in this section are analogues of the ones in Sec. 11, leading up to a convergence theorem for Fourier series. The statement and proof of the corresponding theorem for Fourier integrals appear in Sec. 47, where these lemmas are needed.

Lemma 1. *If a function $G(u)$ is piecewise continuous on an interval $0 < x < c$, then*

$$\lim_{r\to\infty} \int_0^c G(u)\sin ru\, du = 0. \qquad (r > 0). \tag{1}$$

This is the general statement of the *Riemann-Lebesgue lemma* involving a sine function. Lemma 1 in Sec. 11 is a special case of it, where $c = \pi$ and r tends to infinity through the half-integers

$$r = \frac{1}{2} + N \qquad (N = 1, 2, \ldots),$$

rather than continuously as it does here. This lemma also holds when $\sin ru$ is replaced by $\cos ru$; and the proof is similar to the one below involving $\sin ru$.

To verify limit (1), it is sufficient to show that if $G(u)$ is *continuous* at each point of an interval $a \le u \le b$, then

$$\lim_{r\to\infty} \int_a^b G(u)\sin ru\, du = 0. \tag{2}$$

For, in view of the discussion of integrals of piecewise continuous functions in Sec. 1, the integral in limit (1) can be expressed as the sum of a finite number of integrals of the type appearing in limit (2).

Assuming, then, that $G(u)$ is continuous on the closed bounded interval $a \le u \le b$, we note that it must also be *uniformly* continuous there. That is, for each positive number ε, there exists a positive number δ such that

$$|G(u) - G(v)| < \varepsilon$$

whenever u and v lie in the interval and satisfy the inequality $|u-v|<\delta$.† Writing

$$\varepsilon = \frac{\varepsilon_0}{2(b-a)},$$

where ε_0 is an arbitrary positive number, we are thus assured that there is a positive number δ such that

$$|G(u)-G(v)| < \frac{\varepsilon_0}{2(b-a)} \qquad \text{whenever } |u-v|<\delta. \tag{3}$$

To obtain the limit (2), divide the interval $a \le u \le b$ into N subintervals of equal length $(b-a)/N$ by means of the points $a = u_0, u_1, u_2, \ldots, u_N = b$, where $u_0 < u_1 < u_2 < \cdots < u_N$, and let N be so large that the length of each subinterval is less than the number δ in condition (3). Then write

$$\int_a^b G(u)\sin ru\,du = \sum_{n=1}^N \int_{u_{n-1}}^{u_n} G(u)\sin ru\,du,$$

or

$$\int_a^b G(u)\sin ru\,du$$
$$= \sum_{n=1}^N \int_{u_{n-1}}^{u_n} [G(u)-G(u_n)]\sin ru\,du + \sum_{n=1}^N G(u_n)\int_{u_{n-1}}^{u_n} \sin ru\,du,$$

from which it follows that

$$\left|\int_a^b G(u)\sin ru\,du\right| \tag{4}$$
$$\le \sum_{n=1}^N \int_{u_{n-1}}^{u_n} |G(u)-G(u_n)|\,|\sin ru|\,du + \sum_{n=1}^N |G(u_n)|\left|\int_{u_{n-1}}^{u_n} \sin ru\,du\right|.$$

In view of condition (3) and the fact that $|\sin ru| \le 1$, it is easy to see that

$$\int_{u_{n-1}}^{u_n} |G(u)-G(u_n)|\,|\sin ru|\,du < \frac{\varepsilon_0}{2(b-a)}\cdot\frac{b-a}{N} = \frac{\varepsilon_0}{2N} \qquad (n=1,2,\ldots,N).$$

Also, since $G(u)$ is continuous on the closed interval $a \le u \le b$, it is bounded there; that is, there is a positive number M such that $|G(u)| \le M$ for all u in that interval. Furthermore,

$$\left|\int_{u_{n-1}}^{u_n} \sin ru\,du\right| \le \frac{|\cos ru_n| + |\cos ru_{n-1}|}{r} \le \frac{2}{r} \qquad (n=1,2,\ldots,N).$$

With these observations, we find that inequality (4) yields the statement

$$\left|\int_a^b G(u)\sin ru\,du\right| < \frac{\varepsilon_0}{2} + \frac{2MN}{r}.$$

†See, for instance, the book by Taylor and Mann (1983, pp. 529–531), listed in the Bibliography.

Now write

$$R = \frac{4MN}{\varepsilon_0}$$

and observe that

$$\frac{2MN}{r} < \frac{\varepsilon_0}{2} \qquad \text{whenever } r > R.$$

Consequently,

$$\left| \int_a^b G(u) \sin ru \, du \right| < \frac{\varepsilon_0}{2} + \frac{\varepsilon_0}{2} = \varepsilon_0 \qquad \text{whenever } r > R;$$

and limit (2) is established.

Our second lemma makes direct use of the first one.

Lemma 2. *Suppose that a function $g(u)$ is piecewise continuous on every bounded interval of the positive u axis and that the right-hand derivative $g_R'(0)$ exists. If the improper integral*

$$\int_0^\infty |g(u)| \, du \tag{5}$$

converges, then

$$\lim_{r \to \infty} \int_0^\infty g(u) \frac{\sin ru}{u} \, du = \frac{\pi}{2} g(0+). \tag{6}$$

Observe that the integrand appearing in equation (6) is piecewise continuous on the same intervals as $g(u)$ and that when $u \geq 1$,

$$\left| g(u) \frac{\sin ru}{u} \right| \leq |g(u)|.$$

Thus the convergence of integral (5) ensures the existence of the integral in equation (6).

We begin the proof of the lemma by demonstrating its validity when the interval of integration is replaced by any *bounded* interval $0 < x < c$. That is, we first show that if a function $g(u)$ is piecewise continuous on a bounded interval $0 < x < c$ and $g_R'(0)$ exists, then

$$\lim_{r \to \infty} \int_0^c g(u) \frac{\sin ru}{u} \, du = \frac{\pi}{2} g(0+). \tag{7}$$

To prove this, we write

$$\int_0^c g(u) \frac{\sin ru}{u} \, du = I(r) + J(r),$$

where

$$I(r) = \int_0^c \frac{g(u) - g(0+)}{u} \sin ru \, du \qquad \text{and} \qquad J(r) = \int_0^c g(0+) \frac{\sin ru}{u} \, du.$$

Since the function

$$G(u) = \frac{g(u) - g(0+)}{u}$$

is piecewise continuous on the interval $0 < x < c$, where $G(0+) = g_R'(0)$, we need only refer to Lemma 1 to see that

$$\lim_{r\to\infty} I(r) = 0. \tag{8}$$

On the other hand, if we substitute $x = ru$ in the integral representing $J(r)$, the integration formula in Sec. 45 tells us that

$$\lim_{r\to\infty} J(r) = g(0+) \lim_{r\to\infty} \int_0^{cr} \frac{\sin x}{x}\, dx = \frac{\pi}{2}\, g(0+). \tag{9}$$

Limit (7) is evidently now a consequence of limits (8) and (9).

To actually obtain limit (6), we note that

$$\left| \int_c^\infty g(u) \frac{\sin ru}{u}\, du \right| \le \int_0^\infty |g(u)|\, du,$$

where we assume that $c \ge 1$. We then write

$$\begin{aligned} &\left| \int_0^\infty g(u) \frac{\sin ru}{u}\, du - \frac{\pi}{2} g(0+) \right| \\ &\quad \le \left| \int_0^c g(u) \frac{\sin ru}{u}\, du - \frac{\pi}{2} g(0+) \right| + \int_c^\infty |g(u)|\, du, \end{aligned} \tag{10}$$

choosing c to be so large that the value of the last integral on the right, which is the remainder of integral (5), is less than $\varepsilon/2$, where ε is an arbitrary positive number independent of the value of r. In view of limit (7), there exists a positive number R such that whenever $r > R$, the first absolute value on the right-hand side of inequality (10) is also less than $\varepsilon/2$. It then follows that

$$\left| \int_0^\infty g(u) \frac{\sin ru}{u}\, du - \frac{\pi}{2}\, g(0+) \right| < \frac{\varepsilon}{2} + \frac{\varepsilon}{2} = \varepsilon$$

whenever $r > R$, and this is the same as statement (6).

47. A FOURIER INTEGRAL THEOREM

The following theorem gives conditions under which the Fourier integral representation (7), Sec. 44, is valid.[†]

[†] For other conditions, see the books by Carslaw (1952, pp. 315ff) and Titchmarsh (1986, pp. 13ff), both listed in the Bibliography.

Theorem. *Let f denote a function that is piecewise continuous on every bounded interval of the x axis, and suppose that it is absolutely integrable over the x axis; that is, the improper integral*

$$\int_{-\infty}^{\infty} |f(x)|\, dx$$

converges. Then the Fourier integral

$$\frac{1}{\pi}\int_0^{\infty}\int_{-\infty}^{\infty} f(s)\cos\alpha(s-x)\, ds\, d\alpha \tag{1}$$

converges to the mean value

$$\frac{f(x+)+f(x-)}{2} \tag{2}$$

of the one-sided limits of f at each point x $(-\infty < x < \infty)$ where both of the one-sided derivatives $f_R'(x)$ and $f_L'(x)$ exist.

We begin our proof with the observation that integral (1) represents the limit as r tends to infinity of the integral

$$\frac{1}{\pi}\int_0^{r}\int_{-\infty}^{\infty} f(s)\cos\alpha(s-x)\, ds\, d\alpha = \frac{1}{\pi}[I(r,x)+J(r,x)], \tag{3}$$

where

$$I(r,x) = \int_0^{r}\int_{x}^{\infty} f(s)\cos\alpha(s-x)\, ds\, d\alpha,$$

$$J(r,x) = \int_0^{r}\int_{-\infty}^{x} f(s)\cos\alpha(s-x)\, ds\, d\alpha.$$

We now show that the individual integrals $I(r,x)$ and $J(r,x)$ exist; and assuming that $f_R'(x)$ and $f_L'(x)$ exist, we examine the behavior of these integrals as r tends to infinity.

Turning to $I(r,x)$ first, we introduce the new variable of integration $u = s - x$ and write that integral in the form

$$I(r,x) = \int_0^{r}\int_0^{\infty} f(x+u)\cos\alpha u\, du\, d\alpha. \tag{4}$$

Since

$$|f(x+u)\cos\alpha u| \le |f(x+u)|$$

and because

$$\int_0^{\infty} |f(x+u)|\, du = \int_x^{\infty} |f(s)|\, ds \le \int_{-\infty}^{\infty} |f(s)|\, ds,$$

the Weierstrass M-test for improper integrals applies to show that the integral

$$\int_0^{\infty} f(x+u)\cos\alpha u\, du$$

converges uniformly with respect to the variable α. Consequently, not only does the iterated integral (4) exist, but also the order of integration there can be reversed:[†]

$$I(r, x) = \int_0^\infty \int_0^r f(x+u)\cos \alpha u \, d\alpha \, du = \int_0^\infty f(x+u)\,\frac{\sin ru}{u}\, du.$$

Now the function $g(u) = f(x+u)$ satisfies the conditions in Lemma 2, Sec. 46 (compare with Sec. 12). So, applying that lemma to this last integral, we find that

$$\lim_{r\to\infty} I(r, x) = \frac{\pi}{2} f(x+). \tag{5}$$

The limit of $J(r, x)$ as r tends to infinity is treated similarly. Here we make the substitution $u = x - s$ and write

$$J(r, x) = \int_0^r \int_0^\infty f(x-u)\cos \alpha u \, du \, d\alpha = \int_0^\infty f(x-u)\frac{\sin ru}{u}\, du.$$

When $g(u) = f(x-u)$, the limit

$$\lim_{r\to\infty} J(r, x) = \frac{\pi}{2} f(x-) \tag{6}$$

also follows from Lemma 2 in Sec. 46.

Finally, in view of limits (5) and (6), we see that the limit of the left-hand side of equation (3) as r tends to infinity has the value (2), which is, then, the value of integral (1) at any point where the one-sided derivatives of f exist.

Note that since the integrals in expressions (9), Sec. 44, for the coefficients $A(\alpha)$ and $B(\alpha)$ exist when f satisfies the conditions stated in the theorem, the form (8), Sec. 44, of the Fourier integral formula is also justified.

PROBLEMS

1. Show that the function

$$g(u) = \frac{2\sin\dfrac{u}{2}}{u},$$

used in equation (5), Sec. 45, satisfies the conditions in Lemma 2, Sec. 11. To be precise, show that g is piecewise continuous on the interval $0 < x < \pi$ and that $g_R'(0)$ exists.

Suggestion: To obtain $g_R'(0)$, show that

$$g_R'(0) = \lim_{\substack{u\to 0\\ u>0}} \frac{2\sin\dfrac{u}{2} - u}{u^2}.$$

Then apply l'Hôpital's rule twice.

2. Verify that all the conditions in the theorem in Sec. 47 are satisfied by the function f defined by the equations

$$f(x) = \begin{cases} 1 & \text{when } |x| < 1, \\ 0 & \text{when } |x| > 1, \end{cases}$$

[†]Theorems on improper integrals used here are developed in the book by Buck (2004), listed in the Bibliography, as well as in most other texts on advanced calculus. The theorems are usually given for integrals with continuous integrands, but they are also valid when the integrands are piecewise continuous.

and $f(\pm 1) = 1/2$. Thus show that for every x $(-\infty < x < \infty)$,

$$f(x) = \frac{1}{\pi}\int_0^\infty \frac{\sin\alpha(1+x) + \sin\alpha(1-x)}{\alpha}\,d\alpha = \frac{2}{\pi}\int_0^\infty \frac{\sin\alpha\cos\alpha x}{\alpha}\,d\alpha.$$

3. Show that the function defined by the equations

$$f(x) = \begin{cases} 0 & \text{when } x < 0, \\ \exp(-x) & \text{when } x > 0, \end{cases}$$

and $f(0) = 1/2$ satisfies the conditions in the theorem in Sec. 47 and hence that

$$f(x) = \frac{1}{\pi}\int_0^\infty \frac{\cos\alpha x + \alpha\sin\alpha x}{1+\alpha^2}\,d\alpha \qquad (-\infty < x < \infty).$$

Verify this representation directly at the point $x = 0$.

4. Show how it follows from the result in Problem 3 that

$$\exp(-|x|) = \frac{2}{\pi}\int_0^\infty \frac{\cos\alpha x}{1+\alpha^2}\,d\alpha \qquad (-\infty < x < \infty).$$

5. Use the theorem in Sec. 47 to show that if

$$f(x) = \begin{cases} 0 & \text{when } x < 0 \text{ or } x > \pi, \\ \sin x & \text{when } 0 \le x \le \pi, \end{cases}$$

then

$$f(x) = \frac{1}{\pi}\int_0^\infty \frac{\cos\alpha x + \cos\alpha(\pi - x)}{1-\alpha^2}\,d\alpha \qquad (-\infty < x < \infty).$$

In particular, write $x = \pi/2$ to show that

$$\int_0^\infty \frac{\cos(\alpha\pi/2)}{1-\alpha^2}\,d\alpha = \frac{\pi}{2}.$$

6. Show why the Fourier integral formula fails to represent the function

$$f(x) = 1 \qquad (-\infty < x < \infty).$$

Also, point out which condition in the theorem in Sec. 47 is not satisfied by that function.

7. Give details showing that the integral $J(r, x)$ in Sec. 47 actually exists and that limit (6) in that section holds.

8. Let f be a nonzero function that is periodic, with period $2c$. Point out why the integrals

$$\int_{-\infty}^\infty f(x)\,dx \qquad \text{and} \qquad \int_{-\infty}^\infty |f(x)|\,dx$$

fail to exist.

9. Prove Lemma 1 in Sec. 46 when $\sin ru$ is replaced by $\cos ru$ in integral (1) there.

10. Assume that a function $f(x)$ has the Fourier integral representation (8), Sec. 44, which can be written

$$f(x) = \lim_{c\to\infty}\int_0^c [A(\alpha)\cos\alpha x + B(\alpha)\sin\alpha x]\,d\alpha.$$

Use the exponential forms (compare with Problem 8, Sec. 14)

$$\cos\theta = \frac{e^{i\theta} + e^{-i\theta}}{2}, \qquad \sin\theta = \frac{e^{i\theta} - e^{-i\theta}}{2i}$$

of the cosine and sine functions to show formally that

$$f(x) = \lim_{c\to\infty} \int_{-c}^{c} C(\alpha)\, e^{i\alpha x}\, d\alpha,$$

where

$$C(\alpha) = \frac{A(\alpha) - i\,B(\alpha)}{2}, \qquad C(-\alpha) = \frac{A(\alpha) + i\,B(\alpha)}{2} \qquad (\alpha > 0).$$

Then use expressions (9), Sec. 44, for $A(\alpha)$ and $B(\alpha)$ to obtain the single formula†

$$C(\alpha) = \frac{1}{2\pi} \int_{-\infty}^{\infty} f(x)\, e^{-i\alpha x}\, dx \qquad (-\infty < \alpha < \infty).$$

11. Let $A(\alpha)$ and $B(\alpha)$ denote the coefficients (9), Sec. 44, in the Fourier integral representation (8) in that section for a function $f(x)$ $(-\infty < x < \infty)$ that satisfies the conditions in the theorem in Sec. 47.

(*a*) By considering even and odd functions of α, point out why

$$\int_{-\infty}^{\infty} [A(\alpha)\cos\alpha x + B(\alpha)\sin\alpha x]\, d\alpha = 2f(x)$$

and

$$\int_{-\infty}^{\infty} [B(\alpha)\cos\alpha x + A(\alpha)\sin\alpha x]\, d\alpha = 0.$$

(*b*) By adding corresponding sides of the equations in part (*a*), obtain the following symmetric form of the Fourier integral formula:‡

$$f(x) = \frac{1}{\sqrt{2\pi}} \int_{-\infty}^{\infty} g(\alpha)(\cos\alpha x + \sin\alpha x)\, d\alpha \qquad (-\infty < x < \infty),$$

where

$$g(\alpha) = \frac{1}{\sqrt{2\pi}} \int_{-\infty}^{\infty} f(x)(\cos\alpha x + \sin\alpha x)\, dx.$$

48. THE COSINE AND SINE INTEGRALS

Let f denote a function satisfying the conditions stated in the theorem in Sec. 47. As pointed out in the final paragraph of that section, the Fourier integral representation of $f(x)$ remains valid when written in the form

$$f(x) = \int_0^{\infty} [A(\alpha)\cos\alpha x + B(\alpha)\sin\alpha x]\, d\alpha, \tag{1}$$

---

†The function $C(\alpha)$ is known as the *exponential Fourier transform* of $f(x)$ and is of particular importance in electrical engineering. For a development of this and other types of Fourier transforms, see, for example, the book by Churchill (1972), listed in the Bibliography.

‡This form is useful in certain types of transmission problems. See R. V. L. Hartley, *Proc. Inst. Radio Engrs.*, vol. 30, no. 3, pp. 144–150, 1942.

where

$$(2)\qquad A(\alpha) = \frac{1}{\pi}\int_{-\infty}^{\infty} f(x)\cos\alpha x\,dx, \qquad B(\alpha) = \frac{1}{\pi}\int_{-\infty}^{\infty} f(x)\sin\alpha x\,dx.$$

Also, in view of the theorem in Sec. 9, representation (1) is valid for any function f that is absolutely integrable over the entire x axis and *piecewise smooth* on every bounded interval of it.

Suppose now that $f(x)$ is defined only when $x > 0$ and that

(*i*) f is absolutely integrable over the positive x axis and piecewise smooth on every bounded interval of it;

(*ii*) $f(x)$ at each point of discontinuity of f is the mean value of the one-sided limits $f(x+)$ and $f(x-)$.

The following theorem regarding *Fourier cosine and sine integral formulas* is analogous to Theorem 2 in Sec. 14, which ensures the convergence to $f(x)$ of Fourier cosine and sine series on an interval $0 < x < c$. It is an immediate consequence of the Fourier integral theorem in Sec. 47 and will be especially useful in our applications.

Theorem. *Let f denote a function that is defined on the positive x axis and satisfies conditions (i) and (ii). The Fourier cosine integral representation*

$$(3)\qquad f(x) = \int_0^{\infty} A(\alpha)\cos\alpha x\,d\alpha,$$

where

$$(4)\qquad A(\alpha) = \frac{2}{\pi}\int_0^{\infty} f(x)\cos\alpha x\,dx,$$

is valid for each x ($x > 0$); and the same is true of the Fourier sine integral representation

$$(5)\qquad f(x) = \int_0^{\infty} B(\alpha)\sin\alpha x\,d\alpha,$$

where

$$(6)\qquad B(\alpha) = \frac{2}{\pi}\int_0^{\infty} f(x)\sin\alpha x\,dx.$$

Note how representations (3) and (5) can also be written

$$(7)\qquad f(x) = \frac{2}{\pi}\int_0^{\infty} \cos\alpha x \int_0^{\infty} f(s)\cos\alpha s\,ds\,d\alpha$$

and

$$(8)\qquad f(x) = \frac{2}{\pi}\int_0^{\infty} \sin\alpha x \int_0^{\infty} f(s)\sin\alpha s\,ds\,d\alpha,$$

respectively.

To start the proof of the theorem, we note that if f is *even*, $f(x)\sin\alpha x$ is odd in the variable x. The graph of $y = f(x)\sin\alpha x$ is therefore symmetric with respect to the origin. Hence $B(\alpha) = 0$ in representation (1), which reduces to equation (3). The function $f(x)\cos\alpha x$ is, moreover, even in x, and so the graph of $y = f(x)\cos\alpha x$ is symmetric with respect to the y axis. Consequently, the coefficient $A(\alpha)$ in representation (1) takes the form (4). If, on the other hand, f is *odd*, $A(\alpha) = 0$; and similar considerations lead to representation (5).

Suppose now that f is defined only when $x > 0$, as in the statement of the theorem. When the even extension is made, so that f is defined on the entire x axis except at $x = 0$, integral (3) represents that extension for every nonzero x and converges to $f(0+)$ when $x = 0$. Likewise, integral (5) represents the odd extension of $f(x)$ for every nonzero x and converges to zero when $x = 0$. So the theorem here follows from the Fourier integral theorem in Sec. 47.

Representation (3) is needed in various applications involving the solutions of the eigenvalue problem

$$(9) \qquad X''(x) + \lambda X(x) = 0, \qquad X'(0) = 0, \qquad |X(x)| < M \qquad (x > 0),$$

where M is a positive constant. This is another kind of Sturm-Liouville problem, which is basically different from the ones that first appeared in Secs. 31 and 32. The difference, to be described further in Chap. 8 (Sec. 60), is due to the fact that the fundamental interval $x > 0$ here is unbounded. We accept the fact, which can be verified using complex-variable methods, that the eigenvalues λ in this and the other eigenvalue problems in this chapter must be *real* numbers.

To solve problem (9), we observe first that if $\lambda = 0$, the function $X(x)$ must be a constant multiple of unity. If $\lambda > 0$ and we write $\lambda = \alpha^2$ $(\alpha > 0)$, we readily find that except for constant factors, the eigenfunctions are $X(x) = \cos\alpha x$, where α takes on all positive values. The eigenvalues $\lambda = \alpha^2$ are continuous rather than discrete. If $\lambda < 0$, or $\lambda = -\alpha^2$ $(\alpha > 0)$, the solution of the differential equation is $X(x) = C_1 e^{\alpha x} + C_2 e^{-\alpha x}$; and the boundary condition at $x = 0$ requires that

$$X(x) = C_1(e^{\alpha x} + e^{-\alpha x}) = 2C_1 \cosh \alpha x.$$

This is, however, unbounded on the half-line $x > 0$ unless $C_1 = 0$. So the case $\lambda < 0$ yields no eigenfunctions. The nontrivial solutions of problem (9) are, therefore,

$$(10) \qquad \lambda = \alpha^2, \qquad X(x) = \cos\alpha x \qquad (\alpha \geq 0).$$

Although the eigenfunctions $X(x) = \cos\alpha x$ $(\alpha \geq 0)$ have no orthogonality property, the Fourier cosine integral formula (3) gives representations of functions $f(x)$ on the interval $x > 0$ that are generalized linear combinations of those eigenfunctions.

Likewise, it is straightforward to show (Problem 7) that

$$(11) \qquad \lambda = \alpha^2, \qquad X(x) = \sin\alpha x \qquad (\alpha > 0)$$

are the eigenvalues and eigenfunctions of the problem

$$(12) \qquad X''(x) + \lambda X(x) = 0, \qquad X(0) = 0, \qquad |X(x)| < M \qquad (x > 0);$$

and formula (5) represents functions $f(x)$ in terms of $\sin\alpha x$.

PROBLEMS

1. By applying the Fourier sine integral formula and the theorem in Sec. 48 to the function defined by the equations

$$f(x) = \begin{cases} 1 & \text{when } 0 < x < b, \\ 0 & \text{when } \quad x > b, \end{cases}$$

and $f(b) = 1/2$, obtain the representation

$$f(x) = \frac{2}{\pi}\int_0^\infty \frac{1 - \cos b\alpha}{\alpha} \sin \alpha x \, d\alpha \qquad (x > 0).$$

2. Verify that the function $\exp(-bx)$, where b is a positive constant, satisfies the conditions in the theorem in Sec. 48, and show that the coefficient $B(\alpha)$ in the Fourier sine integral representation of that function is

$$B(\alpha) = \frac{2}{\pi}\int_0^\infty e^{-bx} \sin \alpha x \, dx = \frac{2}{\pi} \cdot \frac{\alpha}{\alpha^2 + b^2}.$$

Thus prove that

$$e^{-bx} = \frac{2}{\pi}\int_0^\infty \frac{\alpha \sin \alpha x}{\alpha^2 + b^2} \, d\alpha \qquad (b > 0, x > 0).$$

3. Verify the Fourier sine integral representation

$$\frac{x}{x^2 + b^2} = \frac{2}{\pi}\int_0^\infty \sin \alpha x \int_0^\infty \frac{s \sin \alpha s}{s^2 + b^2} \, ds \, d\alpha \qquad (b > 0, x \geq 0)$$

by first observing that according to the final result in Problem 2,

$$\int_0^\infty \frac{s \sin \alpha s}{s^2 + b^2} \, ds = \frac{\pi}{2} e^{-b\alpha} \qquad (b > 0, \alpha > 0).$$

Then, by referring to the expression for $B(\alpha)$ in Problem 2, complete the verification. Show that the function $x/(x^2 + b^2)$ is *not,* however, absolutely integrable over the positive x axis.

4. As already verified in Problem 2, the function $\exp(-bx)$, where b is a positive constant, satisfies the conditions in the theorem in Sec. 48. Show that the coefficient $A(\alpha)$ in the Fourier cosine integral representation of that function is

$$A(\alpha) = \frac{2}{\pi}\int_0^\infty e^{-bx} \cos \alpha x \, dx = \frac{2}{\pi} \cdot \frac{b}{\alpha^2 + b^2}.$$

Thus prove that

$$e^{-bx} = \frac{2b}{\pi}\int_0^\infty \frac{\cos \alpha x}{\alpha^2 + b^2} \, d\alpha \qquad (b > 0, x \geq 0).$$

5. By regarding the positive constant b in the final equation obtained in Problem 4 as a variable and then differentiating each side of that equation with respect to b, show *formally* that

$$(1 + x)\, e^{-x} = \frac{4}{\pi}\int_0^\infty \frac{\cos \alpha x}{(\alpha^2 + 1)^2} \, d\alpha \qquad (x \geq 0).$$

6. Verify that the function $e^{-x} \cos x$ satisfies the conditions in the theorem in Sec. 48, and show that the coefficient $A(\alpha)$ in the Fourier cosine integral representation of that function can be written

$$A(\alpha) = \frac{1}{\pi}\int_0^\infty e^{-x}\cos(\alpha+1)x\,dx + \frac{1}{\pi}\int_0^\infty e^{-x}\cos(\alpha-1)x\,dx.$$

Then use the expression for the corresponding coefficients in Problem 4 to prove that

$$e^{-x}\cos x = \frac{2}{\pi}\int_0^\infty \frac{\alpha^2+2}{\alpha^4+4}\cos\alpha x\,d\alpha \qquad (x \ge 0).$$

7. In Sec. 48, show that the solutions of the eigenvalue problem (12) are as stated there.

8. Show that the eigenvalues of the eigenvalue problem

$$X''(x) + \lambda X(x) = 0, \qquad |X(x)| < M \qquad (-\infty < x < \infty),$$

where M is a positive constant, are $\lambda = \alpha^2$ $(\alpha \ge 0)$ and that the corresponding eigenfunctions are constant multiples of unity when $\alpha = 0$ and arbitrary linear combinations of $\cos\alpha x$ and $\sin\alpha x$ when $\alpha > 0$.

49. MORE ON SUPERPOSITION OF SOLUTIONS

In Sec. 30 we showed that if $u_1, u_2, \ldots$ are solutions of a given linear homogeneous differential equation or boundary condition, then so is any generalized linear combination

$$u = \sum_{n=1}^{\infty} c_n u_n$$

of those functions, provided that needed differentiability and continuity conditions are satisfied. We thus had the basis of the technique for solving boundary value problems in Chap. 5. Another important version of that principle of superposition is illustrated in the following example, where superposition consists of integration with respect to a parameter α instead of summation with respect to an index n. It will enable us to solve certain boundary value problems in which Fourier integrals, rather than Fourier series, are required.

EXAMPLE. Consider the set of functions $e^{-\alpha y}\sin\alpha x$, where each function corresponds to a value of the parameter α $(\alpha > 0)$ and where α is independent of x and y. Each function satisfies Laplace's equation

$$u_{xx}(x,y) + u_{yy}(x,y) = 0 \qquad (x > 0,\ y > 0) \tag{1}$$

and the boundary condition

$$u(0,y) = 0 \qquad (y > 0). \tag{2}$$

These functions are bounded in the domain $x > 0$, $y > 0$ (Fig. 43) and are obtained from conditions (1) and (2) by the method of separation of variables when that boundedness condition is included (Problem 1, Sec. 50).

We now show that their superposition of the type

$$u(x,y) = \int_0^\infty B(\alpha)\,e^{-\alpha y}\sin\alpha x\,d\alpha \qquad (x > 0,\ y > 0) \tag{3}$$

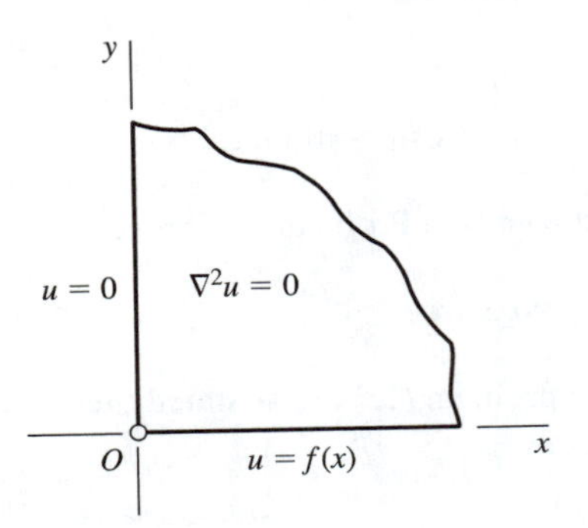

FIGURE 43

also represents a solution of the homogeneous conditions (1) and (2) which is bounded in the domain $x > 0$, $y > 0$ for each function $B(\alpha)$ that is bounded and continuous on the half-line $\alpha > 0$ and absolutely integrable over it.

To accomplish this, we use tests for improper integrals that are analogous to those for infinite series.† The integral in equation (3) converges absolutely and uniformly with respect to x and y because

$$|B(\alpha)\, e^{-\alpha y} \sin \alpha x| \le |B(\alpha)| \qquad (x \ge 0,\ y \ge 0) \tag{4}$$

and because $B(\alpha)$ is independent of x and y and absolutely integrable from zero to infinity with respect to α. Moreover, since

$$|u(x, y)| \le \int_0^\infty |B(\alpha)\, e^{-\alpha y} \sin \alpha x|\, d\alpha \le \int_0^\infty |B(\alpha)|\, d\alpha, \tag{5}$$

u is bounded. It is also a continuous function of x and y ($x \ge 0$, $y \ge 0$) because of the uniform convergence of the integral in equation (3) and the continuity of the integrand. Clearly, $u = 0$ when $x = 0$.

When $y > 0$,

$$\frac{\partial u}{\partial x} = \frac{\partial}{\partial x} \int_0^\infty B(\alpha)\, e^{-\alpha y} \sin \alpha x\, d\alpha = \int_0^\infty \frac{\partial}{\partial x} [B(\alpha)\, e^{-\alpha y} \sin \alpha x]\, d\alpha; \tag{6}$$

for if $|B(\alpha)| \le B_0$ and $y \ge y_0$, where y_0 is some small positive number, then the absolute value of the integrand of the integral on the far right does not exceed $B_0\, \alpha \exp(-\alpha y_0)$, which is independent of x and y and integrable over the semi-infinite interval $0 \le \alpha < \infty$. Hence that integral is uniformly convergent. Integral (3) is then differentiable with respect to x, and similarly for the other derivatives involved in the laplacian operator $\nabla^2 = \partial^2/\partial x^2 + \partial^2/\partial y^2$. Therefore,

$$\nabla^2 u = \int_0^\infty B(\alpha)\, \nabla^2 (e^{-\alpha y} \sin \alpha x)\, d\alpha = 0 \qquad (x > 0,\ y > 0). \tag{7}$$

†See the book by Kaplan (2003, pp. 447ff) or by Taylor and Mann (1983, pp. 682ff), listed in the Bibliography.

Suppose now that the function (3) is also required to satisfy the nonhomogeneous boundary condition

$$u(x, 0) = f(x) \qquad (x > 0), \tag{8}$$

where f is a given function satisfying conditions (*i*) and (*ii*) in Sec. 48. We need to determine the function $B(\alpha)$ in equation (3) so that

$$f(x) = \int_0^\infty B(\alpha) \sin \alpha x \, d\alpha \qquad (x > 0). \tag{9}$$

This is easily done since representation (9) is the Fourier sine integral formula in the theorem in Sec. 48 when

$$B(\alpha) = \frac{2}{\pi} \int_0^\infty f(x) \sin \alpha x \, dx \qquad (\alpha > 0). \tag{10}$$

We have shown here that the function (3), with $B(\alpha)$ given by equation (10), is a solution of the boundary value problem consisting of equations (1), (2), and (8), together with the requirement that u be bounded.

50. TEMPERATURES IN A SEMI-INFINITE SOLID

The face $x = 0$ of a semi-infinite solid $x \geq 0$ is kept at temperature zero (Fig. 44). Let us find the temperatures $u(x, t)$ in the solid when the initial temperature distribution is $f(x)$, assuming for now that f is piecewise smooth on each bounded interval of the positive x axis and that f is bounded and absolutely integrable from $x = 0$ to $x = \infty$.

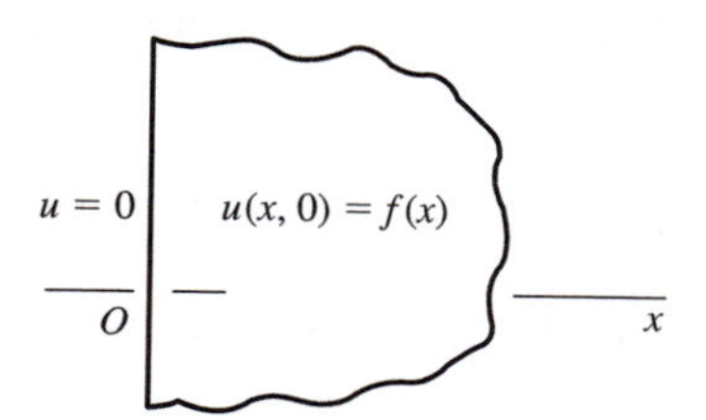

FIGURE 44

If the solid is considered as a limiting case of a slab $0 \leq x \leq c$ as c increases, some condition corresponding to a thermal condition on the face $x = c$ seems to be needed. Otherwise, the temperatures on that face may be increased in any manner as c increases. We require that our function u be bounded; that condition also implies that there is no instantaneous source of heat on the face $x = 0$ at the instant $t = 0$. Then

$$u_t(x, t) = k u_{xx}(x, t) \qquad (x > 0, t > 0), \tag{1}$$

$$u(0, t) = 0 \qquad (t > 0), \tag{2}$$

$$u(x, 0) = f(x) \qquad (x > 0), \tag{3}$$

and $|u(x, t)| < M$, where M is some positive constant.

Linear combinations of functions $u = X(x)T(t)$ will not ordinarily be bounded unless X and T are themselves bounded. Upon separating variables, we thus have the conditions

$$X''(x) + \lambda X(x) = 0, \qquad X(0) = 0, \qquad |X(x)| < M_1 \qquad (x > 0) \tag{4}$$

and

$$T'(t) + \lambda k T(t) = 0, \qquad |T(t)| < M_2 \qquad (t > 0), \tag{5}$$

where M_1 and M_2 are positive constants. As pointed out at the end of Sec. 48, the eigenvalue problem (4) has continuous eigenvalues $\lambda = \alpha^2$, where α represents *all positive real numbers;* $X(x) = \sin \alpha x$ are the eigenfunctions. In this case, the corresponding functions

$$T(t) = e^{-\alpha^2 kt} \qquad (a > 0)$$

are bounded. The generalized linear combination (see Sec. 49)

$$u(x,t) = \int_0^\infty B(\alpha)\, e^{-\alpha^2 kt} \sin \alpha x \, d\alpha \tag{6}$$

of the functions $X(x)T(t)$ will formally satisfy all the conditions in the boundary value problem, including condition (3), if $B(\alpha)$ can be determined so that

$$f(x) = \int_0^\infty B(\alpha) \sin \alpha x \, d\alpha \qquad (x > 0). \tag{7}$$

As in Sec. 49, we note that representation (7) is the Fourier sine integral formula (5), Sec. 48, for $f(x)$ if

$$B(\alpha) = \frac{2}{\pi} \int_0^\infty f(x) \sin \alpha x \, dx \qquad (\alpha > 0). \tag{8}$$

Our formal solution (6), with $B(\alpha)$ defined by equation (8), can also be written

$$u(x,t) = \frac{2}{\pi} \int_0^\infty e^{-\alpha^2 kt} \sin \alpha x \int_0^\infty f(s) \sin \alpha s \, ds \, d\alpha. \tag{9}$$

One can simplify this result by formally reversing the order of integration, replacing $2 \sin \alpha s \sin \alpha x$ by $\cos \alpha(s - x) - \cos \alpha(s + x)$, and then applying the integration formula (Problem 17)

$$\int_0^\infty e^{-\alpha^2 a} \cos \alpha b \, d\alpha = \frac{1}{2}\sqrt{\frac{\pi}{a}} \exp\left(-\frac{b^2}{4a}\right) \qquad (a > 0). \tag{10}$$

Equation (9) then becomes

$$u(x,t) = \frac{1}{2\sqrt{\pi k t}} \int_0^\infty f(s) \left\{ \exp\left[-\frac{(s-x)^2}{4kt}\right] - \exp\left[-\frac{(s+x)^2}{4kt}\right] \right\} ds, \tag{11}$$

or

$$u(x,t) = \frac{1}{2\sqrt{\pi kt}} \int_0^\infty f(s) \exp\left[-\frac{(s-x)^2}{4kt}\right] ds$$
$$- \frac{1}{2\sqrt{\pi kt}} \int_0^\infty f(s) \exp\left[-\frac{(s+x)^2}{4kt}\right] ds,$$

when $t > 0$. By writing

$$s = x + 2\sigma\sqrt{kt} \qquad \text{and} \qquad s = -x + 2\sigma\sqrt{kt}$$

in these last two integrals, respectively, we have this alternative form of expression (11):

$$u(x,t) = \frac{1}{\sqrt{\pi}} \int_{-x/(2\sqrt{kt})}^\infty f(x + 2\sigma\sqrt{kt}\,) e^{-\sigma^2} d\sigma$$
$$- \frac{1}{\sqrt{\pi}} \int_{x/(2\sqrt{kt})}^\infty f(-x + 2\sigma\sqrt{kt}\,) e^{-\sigma^2} d\sigma. \tag{12}$$

Our use of the Fourier sine integral formula in obtaining solution (9) suggests that we apply the theorem in Sec. 48 in verifying that solution. The forms (11) and (12) suggest, however, that the condition in the theorem that $|f(x)|$ be integrable from zero to infinity can be relaxed in the verification. More precisely, when s is kept fixed and $t > 0$, the functions

$$\frac{1}{\sqrt{t}} \exp\left[-\frac{(s \pm x)^2}{4kt}\right]$$

satisfy the heat equation (1). Then, under the assumption that $f(x)$ is continuous and bounded when $x \geq 0$, it is possible to show that the function (11) is bounded and satisfies the heat equation when $x_0 < x < x_1$ and $t_0 < t < t_1$, where x_0, x_1, t_0, and t_1 are any positive numbers. Conditions (2) and (3) can be verified by using expression (12). By adding step functions to f (see Problem 4), we can allow f to have a finite number of jumps on the half-line $x > 0$. Except for special cases, details in the verification of formal solutions of this problem are, however, tedious.

If $f(x) = u_0$, where u_0 is a constant, it follows from equation (12) that

$$u(x,t) = \frac{u_0}{\sqrt{\pi}} \left(\int_{-x/(2\sqrt{kt})}^\infty e^{-\sigma^2} d\sigma - \int_{x/(2\sqrt{kt})}^\infty e^{-\sigma^2} d\sigma \right). \tag{13}$$

In terms of the *error function*

$$\operatorname{erf}(x) = \frac{2}{\sqrt{\pi}} \int_0^x e^{-\sigma^2} d\sigma, \tag{14}$$

expression (13) can be written

$$u(x,t) = u_0 \operatorname{erf}\left(\frac{x}{2\sqrt{kt}}\right). \tag{15}$$

The verification of this is not difficult, since $\exp(-\sigma^2)$ is an even function of σ. Note that because (see Problem 16)

$$\int_0^\infty e^{-\sigma^2}\,d\sigma = \frac{\sqrt{\pi}}{2}, \tag{16}$$

$\mathrm{erf}(x)$ tends to unity as x tends to infinity. Thus, if $f(x) = u_0$, the temperatures $u(x,t)$ in the solid tend to u_0 when x tends to infinity, as would be expected.

PROBLEMS

1. Give details showing how the functions $e^{-\alpha y}\sin\alpha x$ $(\alpha > 0)$ arise by means of separation of variables from conditions (1) and (2), Sec. 49, and the condition that the function $u(x, y)$ there be bounded when $x > 0$, $y > 0$.

2. (*a*) Substitute expression (10), Sec. 49, for the function $B(\alpha)$ into equation (3) of that section. Then, by formally reversing the order of integration, show that the solution of the boundary value problem treated in Sec. 49 can be written

$$u(x, y) = \frac{y}{\pi}\int_0^\infty f(s)\left[\frac{1}{(s-x)^2+y^2} - \frac{1}{(s+x)^2+y^2}\right]ds.$$

(*b*) Show that when $f(x) = 1$, the form of the solution obtained in part (*a*) can be written in terms of the inverse tangent function as

$$u(x, y) = \frac{2}{\pi}\tan^{-1}\left(\frac{x}{y}\right).$$

3. Verify that when $x > 0$ and $t > 0$, the function (see Sec. 50)

$$u(x, t) = \mathrm{erf}\left(\frac{x}{2\sqrt{kt}}\right)$$

satisfies the heat equation $u_t(x, t) = ku_{xx}(x, t)$ as well as the conditions

$$u(0+, t) = 0, \qquad u(x, 0+) = 1, \qquad \text{and} \qquad |u(x, t)| < 1.$$

4. Show that if

$$f(x) = \begin{cases} 0 & \text{when } 0 < x < c, \\ 1 & \text{when } \quad x > c, \end{cases}$$

expression (12), Sec. 50, reduces to

$$u(x, t) = \frac{1}{2}\,\mathrm{erf}\left(\frac{c+x}{2\sqrt{kt}}\right) - \frac{1}{2}\,\mathrm{erf}\left(\frac{c-x}{2\sqrt{kt}}\right).$$

Verify this solution of the boundary value problem in Sec. 50 when f is this function.

5. (*a*) The face $x = 0$ of a semi-infinite solid $x \geq 0$ is kept at a constant temperature u_0 after the interior $x > 0$ is initially at temperature zero throughout. By writing $u(x, t) = U(x, t) + \Phi(x)$ (compare with Example 2, Sec. 34), where $\Phi(x)$ is to be bounded for $x \geq 0$, and referring to solution (15), Sec. 50, of the special case of the boundary value problem noted there, derive the temperature formula

$$u(x, t) = u_0\left[1 - \mathrm{erf}\left(\frac{x}{2\sqrt{kt}}\right)\right].$$

(*b*) Use equations (14) and (16) in Sec. 50 to rewrite the solution in part (*a*) as

$$u(x,t) = \frac{2u_0}{\sqrt{\pi}} \int_{x/(2\sqrt{kt})}^{\infty} e^{-\sigma^2} d\sigma.$$

6. Replace the constant temperature u_0 in Problem 5 by a time-dependent temperature $F(t)$, where $F(t)$ is continuous and differentiable when $t \geq 0$ and $F(0) = 0$. Then use the solution obtained in part (*b*) of Problem 5, together with the special case of Duhamel's theorem described in Sec. 24, to derive the solution

$$u(x,t) = \frac{x}{2\sqrt{\pi k}} \int_0^t \frac{F(\tau)}{\sqrt{(t-\tau)^3}} \exp\left[-\frac{x^2}{4k(t-\tau)}\right] d\tau$$

of the temperature problem here.

7. (*a*) The face $x = 0$ of a semi-infinite solid $x \geq 0$ is insulated, and the initial temperature distribution is $f(x)$. Derive the temperature formula

$$u(x,t) = \frac{1}{\sqrt{\pi}} \int_{-x/(2\sqrt{kt})}^{\infty} f(x + 2\sigma\sqrt{kt})\, e^{-\sigma^2} d\sigma + \frac{1}{\sqrt{\pi}} \int_{x/(2\sqrt{kt})}^{\infty} f(-x + 2\sigma\sqrt{kt})\, e^{-\sigma^2} d\sigma.$$

(*b*) Show that if the function f in part (*a*) is defined by the equations

$$f(x) = \begin{cases} 1 & \text{when } 0 < x < c, \\ 0 & \text{when } \quad\;\; x > c, \end{cases}$$

then

$$u(x,t) = \frac{1}{2} \operatorname{erf}\left(\frac{c+x}{2\sqrt{kt}}\right) + \frac{1}{2} \operatorname{erf}\left(\frac{c-x}{2\sqrt{kt}}\right).$$

8. A semi-infinite string, with one end fixed at the origin, is stretched along the positive half of the x axis and released at rest from a position $y = f(x)$ $(x \geq 0)$. Derive the expression

$$y(x,t) = \frac{2}{\pi} \int_0^{\infty} \cos\alpha a t \sin\alpha x \int_0^{\infty} f(s) \sin\alpha s \, ds \, d\alpha$$

for the transverse displacements. Let $F(x)$ $(-\infty < x < \infty)$ denote the odd extension of $f(x)$, and show how this result reduces to the form

$$y(x,t) = \frac{1}{2}[F(x+at) + F(x-at)].$$

[Compare with solution (9), Sec. 39, of a string problem treated in that section.]

9. Find the function $u(x, y)$ that is bounded and harmonic in the horizontal semi-infinite strip $x > 0, 0 < y < 1$ and satisfies the boundary conditions

$$u_x(0,y) = 0, \qquad u_y(x,0) = 0, \qquad u(x,1) = e^{-x}.$$

Answer: $u(x,y) = \dfrac{2}{\pi} \displaystyle\int_0^{\infty} \frac{\cos\alpha x \cosh\alpha y}{(1+\alpha^2)\cosh\alpha} d\alpha.$

10. Find $u(x, y)$ when the boundary conditions in Problem 9 are replaced by the conditions

$$u_x(0,y) = 0, \qquad u_y(x,1) = -u(x,1), \qquad u(x,0) = f(x),$$

where

$$f(x) = \begin{cases} 1 & \text{when } 0 < x < 1, \\ 0 & \text{when } \quad x > 1. \end{cases}$$

Interpret this problem physically.

Answer: $u(x, y) = \dfrac{2}{\pi} \displaystyle\int_0^\infty \frac{\alpha \cosh \alpha(1-y) + \sinh \alpha(1-y)}{\alpha^2 \cosh \alpha + \alpha \sinh \alpha} \sin \alpha \cos \alpha x \, d\alpha.$

11. Find the bounded harmonic function $u(x, y)$ in the semi-infinite strip $0 < x < 1, y > 0$ that satisfies the conditions

$$u_y(x, 0) = 0, \qquad u(0, y) = 0, \qquad u_x(1, y) = f(y).$$

Answer: $u(x, y) = \dfrac{2}{\pi} \displaystyle\int_0^\infty \frac{\sinh \alpha x \cos \alpha y}{\alpha \cosh \alpha} \int_0^\infty f(s) \cos \alpha s \, ds \, d\alpha.$

12. Let a semi-infinite solid $x \geq 0$, which is initially at a uniform temperature, be cooled or heated by keeping its boundary at a uniform constant temperature (Sec. 50). Show that the times required for two interior points to reach the same temperature are proportional to the squares of the distances of those points from the boundary plane.

13. Solve the following boundary value problem for steady temperatures $u(x, y)$ in a thin plate in the shape of a semi-infinite strip when heat transfer to the surroundings at temperature zero takes place at the faces of the plate:

$$u_{xx}(x, y) + u_{yy}(x, y) - b\,u(x, y) = 0 \qquad (x > 0, 0 < y < 1),$$

$$u_x(0, y) = 0 \qquad (0 < y < 1),$$

$$u(x, 0) = 0, \qquad u(x, 1) = f(x) \qquad (x > 0),$$

where b is a positive constant and

$$f(x) = \begin{cases} 1 & \text{when } 0 < x < c, \\ 0 & \text{when } \quad x > c. \end{cases}$$

Answer: $u(x, y) = \dfrac{2}{\pi} \displaystyle\int_0^\infty \frac{\sin \alpha c \, \cos \alpha x \, \sinh\left(y\sqrt{\alpha^2 + b}\right)}{\alpha \sinh \sqrt{\alpha^2 + b}} \, d\alpha.$

14. Verify that for any constant C, the function

$$v(x, t) = Cxt^{-3/2} \exp\left(\frac{-x^2}{4kt}\right)$$

satisfies the heat equation $v_t = kv_{xx}$ when $x > 0$ and $t > 0$. Also, verify that for those values of x and t,

$$v(0+, t) = 0 \qquad \text{and} \qquad v(x, 0+) = 0.$$

Thus show that $v(x, t)$ can be added to the solution (9) found in Sec. 50 to form other solutions of the problem there if the temperature function is not required to be bounded. Note that v is unbounded as x and t tend to zero (this can be seen by letting x vanish while $t = x^2$).

15. Let $u = u(x, y, t)$ denote the bounded solution of the two-dimensional temperature problem indicated in Fig. 45, where

$$u_t = k(u_{xx} + u_{yy}) \qquad (x > 0, 0 < y < 1, t > 0),$$

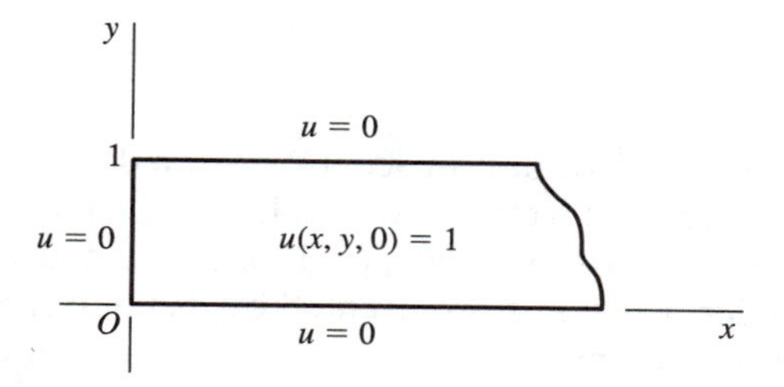

FIGURE 45

and let $v = v(x, t)$ and $w = w(y, t)$ denote the bounded solutions of the following one-dimensional temperature problems:

$$v_t = kv_{xx}, \qquad v(0, t) = 0, \qquad v(x, 0) = 1 \qquad (x > 0, t > 0),$$

$$w_t = kw_{yy}, \qquad w(0, t) = w(1, t) = 0, \qquad w(y, 0) = 1 \qquad (0 < y < 1, t > 0).$$

(*a*) With the aid of the result obtained in Problem 3, Sec. 42, show that $u = vw$.
(*b*) By referring to the solution (15), Sec. 50, of the temperature problem there and to the temperature function found in Example 1, Sec. 35, write explicit expressions for v and w. Then use the result in part (*a*) to show that

$$u(x, y, t) = \frac{4}{\pi}\operatorname{erf}\left(\frac{x}{2\sqrt{kt}}\right)\sum_{n=1}^{\infty}\frac{\sin(2n-1)\pi y}{2n-1}\exp[-(2n-1)^2\pi^2 kt].$$

16. Use the following method to show that

$$\int_0^\infty e^{-\sigma^2}\,d\sigma = \frac{\sqrt{\pi}}{2},$$

as stated in Sec. 50. Do this by denoting the integral by I, writing

$$I^2 = \int_0^\infty e^{-x^2}\,dx \int_0^\infty e^{-y^2}\,dy = \int_0^\infty\int_0^\infty e^{-(x^2+y^2)}\,dx\,dy,$$

and then evaluating this iterated integral by switching to polar coordinates.

17. Derive the integration formula (10), Sec. 50, by first writing

$$y(x) = \int_0^\infty e^{-\alpha^2 a}\cos\alpha x\,d\alpha \qquad (a > 0)$$

and differentiating the integral to find $y'(x)$. Then integrate the new integral by parts to show that $2ay'(x) = -xy(x)$, point out why

$$y(0) = \frac{1}{2}\sqrt{\frac{\pi}{a}}$$

(see Problem 16), and solve for $y(x)$. The desired result is the value of y when $x = b$.†

†Another derivation is indicated in the authors' book (2004, pp. 154–155), listed in the Bibliography.

51. TEMPERATURES IN AN UNLIMITED MEDIUM

For an application of the general Fourier integral formula in Sec. 44, we now derive expressions for the temperatures $u(x, t)$ in a medium that occupies all space, where the initial temperature distribution is $f(x)$. We assume that $f(x)$ is bounded and, moreover, that it satisfies conditions under which it is represented by its Fourier integral formula. The boundary value problem consists of a boundedness condition $|u(x, t)| < M$ and the conditions

$$u_t(x, t) = ku_{xx}(x, t) \qquad (-\infty < x < \infty, t > 0), \tag{1}$$

$$u(x, 0) = f(x) \qquad (-\infty < x < \infty). \tag{2}$$

Writing $u = X(x)T(t)$ and separating variables, we have the eigenvalue problem

$$X''(x) + \lambda X(x) = 0, \qquad |X(x)| < M_1 \qquad (-\infty < x < \infty),$$

whose eigenvalues are $\lambda = \alpha^2$ ($\alpha \geq 0$) and whose eigenfunctions are constant multiples of unity when $\alpha = 0$ and arbitrary linear combinations of $\cos \alpha x$ and $\sin \alpha x$ when $\alpha > 0$ (Problem 8, Sec. 48). The solutions of the differential equation

$$T'(t) + \lambda k T(t) = 0, \qquad |T(t)| < M_2 \qquad (t > 0)$$

that arise are constant multiples of

$$T(t) = e^{-\alpha^2 kt} \qquad (\alpha \geq 0).$$

Our generalized linear combination of the products $u = X(x)T(t)$ is then

$$u(x, t) = \int_0^\infty e^{-\alpha^2 kt} \left[A(\alpha) \cos \alpha x + B(\alpha) \sin \alpha x\right] d\alpha. \tag{3}$$

The coefficients $A(\alpha)$ and $B(\alpha)$ are to be determined so that the integral here represents $f(x)$ $(-\infty < x < \infty)$ when $t = 0$. According to equations (8) and (9) in Sec. 44 and our Fourier integral theorem (Sec. 47), the representation is valid if

$$A(\alpha) = \frac{1}{\pi} \int_{-\infty}^\infty f(x) \cos \alpha x \, dx, \qquad B(\alpha) = \frac{1}{\pi} \int_{-\infty}^\infty f(x) \sin \alpha x \, dx.$$

Thus

$$u(x, t) = \frac{1}{\pi} \int_0^\infty e^{-\alpha^2 kt} \int_{-\infty}^\infty f(s) \cos \alpha (s - x) \, ds \, d\alpha. \tag{4}$$

If we formally reverse the order of integration here, the integration formula (10) in Sec. 50 can be used to write equation (4) as

$$u(x, t) = \frac{1}{2\sqrt{\pi kt}} \int_{-\infty}^\infty f(s) \exp\left[-\frac{(s - x)^2}{4kt}\right] ds \qquad (t > 0). \tag{5}$$

An alternative form of this is

$$u(x, t) = \frac{1}{\sqrt{\pi}} \int_{-\infty}^\infty f(x + 2\sigma\sqrt{kt}\,) \, e^{-\sigma^2} \, d\sigma. \tag{6}$$

[Compare with expression (12), Sec. 50.]

Forms (5) and (6) can be verified by assuming only that f is piecewise continuous over some bounded interval $|x| < c$ and continuous and bounded over the rest of the x axis, or when $|x| \geq c$. If f is an odd function, $u(x, t)$ becomes the function found in Sec. 50 for positive values of x.

PROBLEMS

1. Let the initial temperature distribution $f(x)$ in the unlimited medium in Sec. 51 be defined by the equations

$$f(x) = \begin{cases} 0 & \text{when } x < 0, \\ 1 & \text{when } x > 0. \end{cases}$$

Show that

$$u(x, t) = \frac{1}{2} + \frac{1}{2} \operatorname{erf}\left(\frac{x}{2\sqrt{kt}}\right).$$

Verify this solution of the boundary value problem in Sec. 51 when f is this function.

2. Derive this solution of the wave equation $y_{tt} = a^2 y_{xx}$ $(-\infty < x < \infty, t > 0)$, which satisfies the conditions $y(x, 0) = f(x)$ and $y_t(x, 0) = 0$ when $-\infty < x < \infty$:

$$y(x, t) = \frac{1}{\pi} \int_0^\infty \cos \alpha a t \int_{-\infty}^\infty f(s) \cos \alpha (s - x)\, ds\, d\alpha.$$

Also, reduce the solution to the form obtained in the example in Sec. 27:

$$y(x, t) = \frac{1}{2} [f(x + at) + f(x - at)].$$

3. Find the bounded harmonic function $u(x, y)$ in the strip $-\infty < x < \infty, 0 < y < b$ such that $u(x, 0) = 0$ and $u(x, b) = f(x)$ $(-\infty < x < \infty)$, where f is bounded and represented by its Fourier integral.

Answer: $$u(x, y) = \frac{1}{\pi} \int_0^\infty \frac{\sinh \alpha y}{\sinh \alpha b} \int_{-\infty}^\infty f(s) \cos \alpha (s - x)\, ds\, d\alpha.$$

CHAPTER 7

ORTHONORMAL SETS

In this chapter, we provide a brief introduction to the theory of so-called orthonormal sets of functions. The chapter will not only clarify underlying concepts behind the several types of Fourier series that we have encountered but will also lay the foundation for finding other series representations that are needed in chapters to follow.

52. INNER PRODUCTS AND ORTHONORMAL SETS

Let f and g denote any two functions that are continuous on a closed bounded interval $a \leq x \leq b$. Dividing that interval into N closed subintervals of equal length $\Delta x = (b - a)/N$ and letting x_k denote any point in the kth subinterval, we recall from calculus that when N is large,

$$\int_a^b f(x)\, g(x)\, dx \doteq \sum_{k=1}^{N} f(x_k)\, g(x_k)\, \Delta x,$$

the symbol $\doteq$ here denoting approximate equality. That is,

$$\int_a^b f(x)\, g(x)\, dx \doteq \sum_{k=1}^{N} a_k\, b_k \tag{1}$$

where

$$a_k = f(x_k)\sqrt{\Delta x} \qquad \text{and} \qquad b_k = g(x_k)\sqrt{\Delta x}.$$

The left-hand side of expression (1) is, then, approximately equal to the inner product of two vectors in N-dimensional space when N is large. The approximation becomes exact in the limit as N tends to infinity.† This suggests defining an *inner product* of the functions f and g:

(2) $$(f, g) = \int_a^b f(x)g(x)\,dx.$$

The integral here is, of course, well defined when f and g are piecewise continuous on the *fundamental interval* $a < x < b$. Equation (2) can, therefore, be used to define an inner product of *any* two functions f and g in the function space $C_p(a, b)$ that was introduced in Sec. 1.

The function space $C_p(a, b)$, with inner product (2), is analogous to ordinary three-dimensional space. Indeed, the following counterparts of familiar properties of vectors in three-dimensional space hold for any functions f, g, and h in $C_p(a, b)$:

(3) $$(f, g) = (g, f),$$

(4) $$(f, g + h) = (f, g) + (f, h),$$

(5) $$(cf, g) = c(f, g),$$

where c is any constant, and

(6) $$(f, f) \geq 0.$$

The analogy is continued with the introduction of the *norm*

(7) $$\|f\| = (f, f)^{1/2}$$

of a function f in $C_p(a, b)$. It is evident from equation (2) that the norm of f can be written

(8) $$\|f\| = \left\{ \int_a^b [f(x)]^2\,dx \right\}^{1/2}.$$

The norm of the difference of two functions f and g,

(9) $$\|f - g\| = \left\{ \int_a^b [f(x) - g(x)]^2\,dx \right\}^{1/2},$$

is a measure of the area of the region between the graphs of $y = f(x)$ and $y = g(x)$ (Fig. 46). To be specific, the quotient $\|f - g\|^2/(b - a)$ is the mean, or average,

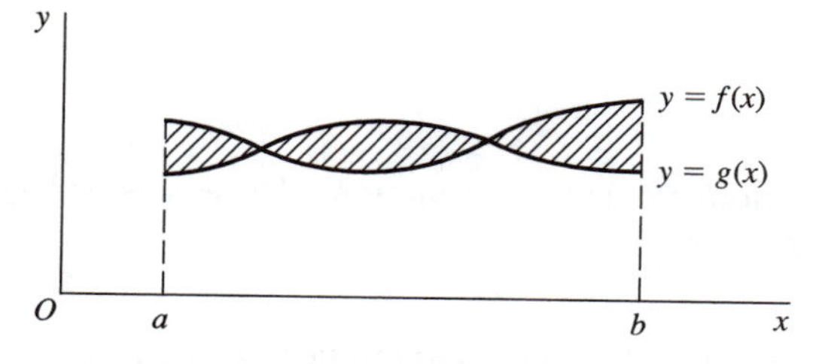

FIGURE 46

†See the book by Lanczos (1966, pp. 210ff), listed in the Bibliography, for an elaboration of this idea.

value of the squares of the vertical distances $|f(x) - g(x)|$ between points on those graphs over the interval $a < x < b$. The quantity $\|f - g\|^2$ is called the *mean square deviation* of one of the functions f and g from the other.

Two functions f and g in $C_p(a, b)$ are *orthogonal* when

$$(f, g) = 0,$$

or

$$\int_a^b f(x)g(x)\,dx = 0. \tag{10}$$

Also, if $\|f\| = 1$, the function f is said to be *normalized*. We have carried our analogy too far to preserve the original meaning of the geometric terminology. The orthogonality of two functions f and g signifies nothing about perpendicularity. It says only that the product fg assumes both positive and negative values on the fundamental interval in such a way that equation (10) holds.

A set of functions $\psi_n(x)$ $(n = 1, 2, \ldots)$ is orthogonal on an interval $a < x < b$ if $(\psi_m, \psi_n) = 0$ when $m \neq n$. Assuming that none of the functions ψ_n has zero norm (see Problem 4, Sec. 53), we define the related functions

$$\phi_n(x) = \frac{\psi_n(x)}{\|\psi_n\|} \qquad (n = 1, 2, \ldots). \tag{11}$$

Inasmuch as

$$(\phi_m, \phi_n) = \left(\frac{\psi_m}{\|\psi_m\|}, \frac{\psi_n}{\|\psi_n\|}\right) = \frac{(\psi_m, \psi_n)}{\|\psi_m\|\|\psi_n\|},$$

it follows that

$$(\phi_m, \phi_n) = \begin{cases} 0 & \text{when } m \neq n, \\ 1 & \text{when } m = n. \end{cases} \tag{12}$$

That is, the set $\{\phi_n(x)\}$ $(n = 1, 2, \ldots)$ is *orthonormal* in the sense that it is orthogonal and each $\phi_n(x)$ is normalized. The characterization (12) of an orthonormal set $\{\phi_n(x)\}$ is, of course, the same as

$$\int_a^b \phi_m(x)\,\phi_n(x)\,dx = \begin{cases} 0 & \text{when } m \neq n, \\ 1 & \text{when } m = n. \end{cases} \tag{13}$$

53. EXAMPLES

We present here three examples of orthonormal sets of functions. These sets will illustrate much of the theory in this chapter.

EXAMPLE 1. We recall from Sec. 4 that for positive integers m and n,

$$\int_0^\pi \sin mx \sin nx\,dx = \begin{cases} 0 & \text{when } m \neq n, \\ \pi/2 & \text{when } m = n. \end{cases} \tag{1}$$

Evidently, then, the set of sine functions

$$\psi_n(x) = \sin nx \qquad (n = 1, 2, \ldots)$$

is orthogonal on the interval of $0 < x < \pi$; and the norm $\|\psi_n\|$ of each of these functions is $\sqrt{\pi/2}$. Hence the corresponding orthonormal set $\{\phi_n(x)\}$ consists of the functions

$$\phi_n(x) = \sqrt{\frac{2}{\pi}} \sin nx \qquad (n = 1, 2, \ldots). \tag{2}$$

It is sometimes more convenient to index an infinite orthogonal or orthonormal set by starting with $n = 0$, rather than $n = 1$. This is the case in Examples 2 and 3.

EXAMPLE 2. The functions

$$\phi_0(x) = \frac{1}{\sqrt{\pi}}, \qquad \phi_n(x) = \sqrt{\frac{2}{\pi}} \cos nx \qquad (n = 1, 2, \ldots) \tag{3}$$

constitute a set $\{\phi_n(x)\}$ $(n = 0, 1, 2, \ldots)$ that is orthonormal on the same interval $0 < x < \pi$ as in Example 1.

To verify this, we start with the observations that

$$(\phi_0, \phi_n) = \frac{\sqrt{2}}{\pi} \int_0^\pi \cos nx\, dx = \frac{\sqrt{2}}{\pi} \left[\frac{\sin nx}{n} \right]_0^\pi = 0 \qquad (n = 1, 2, \ldots)$$

and

$$\|\phi_0\|^2 = (\phi_0, \phi_0) = \frac{1}{\pi} \int_0^\pi dx = 1.$$

Next, we let m and n denote positive integers and recall from Sec. 2 that

$$\int_0^\pi \cos mx \cos nx\, dx = \begin{cases} 0 & \text{when } m \neq n, \\ \pi/2 & \text{when } m = n. \end{cases} \tag{4}$$

This tells us that, for *distinct* positive integers m and n,

$$(\phi_m, \phi_n) = \frac{2}{\pi} \int_0^\pi \cos mx \cos nx\, dx = 0$$

and that

$$\|\phi_n\|^2 = (\phi_n, \phi_n) = \frac{2}{\pi} \int_0^\pi \cos nx \cos nx\, dx = 1 \qquad (n = 1, 2, \ldots).$$

The verification that the set (3) is orthonormal is now complete.

EXAMPLE 3. The set $\{\phi_n(x)\}$ $(n = 0, 1, 2, \ldots)$ consisting of the functions

$$\phi_0(x) = \frac{1}{\sqrt{2\pi}}, \qquad \phi_{2n-1}(x) = \frac{1}{\sqrt{\pi}} \cos nx, \qquad \phi_{2n}(x) = \frac{1}{\sqrt{\pi}} \sin nx \tag{5}$$

$$(n = 1, 2, \ldots)$$

is orthonormal on the interval $-\pi < x < \pi$.

Steps needed in the verification are as follows. Simple integration reveals that

$$(\phi_0, \phi_{2n-1}) = \frac{1}{\sqrt{2\pi}} \int_{-\pi}^{\pi} \cos nx \, dx = 0 \quad \text{and} \quad (\phi_0, \phi_{2n}) = \frac{1}{\sqrt{2\pi}} \int_{-\pi}^{\pi} \sin nx \, dx = 0$$

$$(n = 1, 2, \ldots)$$

and that

$$\|\phi_0\|^2 = \frac{1}{2\pi} \int_{-\pi}^{\pi} dx = 1.$$

The remaining steps depend on integration formulas (1) and (4), together with the observations that

$$\int_{-\pi}^{\pi} f(x)\, dx = 2 \int_0^{\pi} f(x)\, dx$$

when a given function f is even and

$$\int_{-\pi}^{\pi} f(x)\, dx = 0$$

when it is odd.

To be specific, for *distinct* positive integers m and n,

$$(\phi_{2m-1}, \phi_{2n-1}) = \frac{1}{\pi} \int_{-\pi}^{\pi} \cos mx \cos nx \, dx = \frac{2}{\pi} \int_0^{\pi} \cos mx \cos nx \, dx = 0$$

and

$$(\phi_{2m}, \phi_{2n}) = \frac{1}{\pi} \int_{-\pi}^{\pi} \sin mx \sin nx \, dx = \frac{2}{\pi} \int_0^{\pi} \sin mx \sin nx \, dx = 0.$$

Also, for *any* positive integers m and n,

$$(\phi_{2m-1}, \phi_{2n}) = \frac{1}{\pi} \int_{-\pi}^{\pi} \cos mx \sin nx \, dx = 0.$$

Finally, one needs to show that when $n = 1, 2, \ldots,$

$$\|\phi_{2n-1}\|^2 = \frac{1}{\pi} \int_{-\pi}^{\pi} \cos^2 nx \, dx = \frac{2}{\pi} \int_0^{\pi} \cos^2 nx \, dx = 1$$

and

$$\|\phi_{2n}\|^2 = \frac{1}{\pi} \int_{-\pi}^{\pi} \sin^2 nx \, dx = \frac{2}{\pi} \int_0^{\pi} \sin^2 nx \, dx = 1.$$

PROBLEMS

1. Show that the functions $\psi_1(x) = 1$ and $\psi_2(x) = x$ are orthogonal on the interval $-1 < x < 1$, and determine constants A and B such that the function

$$\psi_3(x) = 1 + Ax + Bx^2$$

is orthogonal to both ψ_1 and ψ_2 on that interval.

Answer: $A = 0, B = -3$.

2. Suppose that two continuous functions $f(x)$ and $\psi_1(x)$, with positive norms, are linearly independent on an interval $a \le x \le b$; that is, one is not a constant times the other. By determining the linear combination $f + A\psi_1$ of those functions that is orthogonal to ψ_1 on the fundamental interval $a < x < b$, obtain an orthogonal pair ψ_1, ψ_2 where

$$\psi_2(x) = f(x) - \frac{(f, \psi_1)}{\|\psi_1\|^2}\,\psi_1(x).$$

Interpret this expression geometrically when f, ψ_1, and ψ_2 represent vectors in three-dimensional space.

3. In Problem 2, suppose that the fundamental interval is $-\pi < x < \pi$ and that

$$f(x) = \cos nx + \sin nx \qquad \text{and} \qquad \psi_1(x) = \cos nx,$$

where n is a fixed positive integer. Show that the function $\psi_2(x)$ there turns out to be

$$\psi_2(x) = \sin nx.$$

Suggestion: One can avoid evaluating any integrals by using the fact that the set in Example 3, Sec. 53, is orthogonal on the interval $-\pi < x < \pi$.

4. Verify the following two statements, regarding functions f in the space $C_p(a, b)$.

(*a*) If $f(x) = 0$, except possibly at a finite number of points, in the interval $a < x < b$, then $\|f\| = 0$.

(*b*) Conversely, if $\|f\| = 0$, then $f(x) = 0$, except possibly at a finite number of points, in the interval $a < x < b$.

Suggestion: In part (*b*), use the fact that a definite integral of a nonnegative continuous function over a closed bounded interval has positive value if the function has a positive value somewhere in that interval.

5. Verify that for any two functions f and g in the space $C_p(a, b)$,

$$\frac{1}{2}\int_a^b\int_a^b [f(x)g(y) - g(x)f(y)]^2\,dx\,dy = \|f\|^2\|g\|^2 - (f, g)^2.$$

Thus establish the *Schwarz inequality*

$$|(f, g)| \le \|f\|\|g\|,$$

which is also valid when f and g denote vectors in three-dimensional space. In that case, it is known as Cauchy's inequality.

6. Let f and g denote any two functions in the space $C_p(a, b)$. Use the Schwarz inequality (Problem 5) to show that if either function has zero norm, then $(f, g) = 0$.

7. Prove that if f and g are functions in the space $C_p(a, b)$, then

$$\|f + g\| \le \|f\| + \|g\|.$$

If f and g denote, instead, vectors in three-dimensional space, this is the familiar triangle inequality, which states that the length of one side of a triangle is less than or equal to the sum of the lengths of the other two sides.

Suggestion: Start the proof by showing that

$$\|f + g\|^2 = \|f\|^2 + 2(f, g) + \|g\|^2,$$

and then use the Schwarz inequality (Problem 5).

54. GENERALIZED FOURIER SERIES

Let f be any given function in $C_p(a, b)$, the space of piecewise continuous functions defined on the interval $a < x < b$. When an orthonormal set of functions $\phi_n(x)$ $(n = 1, 2, \ldots)$ in $C_p(a, b)$ is specified, it *may* be possible to represent $f(x)$ by a linear combination of those functions, generalized to an infinite series that converges to $f(x)$ at all but possibly a finite number of points in the interval $a < x < b$:

$$f(x) = \sum_{n=1}^{\infty} c_n \phi_n(x) \qquad (a < x < b). \tag{1}$$

This is analogous to the expression for any vector in three-dimensional space in terms of three mutually orthogonal vectors of unit length, such as **i**, **j**, and **k**.

To discover an expression for the coefficients c_n in representation (1), if such a representation actually exists, we use the index of summation m, rather than n, to write

$$f(x) = \sum_{m=1}^{\infty} c_m \phi_m(x) \qquad (a < x < b). \tag{2}$$

We also assume that after each of the terms here is multiplied by a specific $\phi_n(x)$, the resulting series is integrable term by term over the interval $a < x < b$. This enables us to write

$$\int_a^b f(x)\, \phi_n(x)\, dx = \sum_{m=1}^{\infty} c_m \int_a^b \phi_m(x)\, \phi_n(x)\, dx,$$

or

$$(f, \phi_n) = \sum_{m=1}^{\infty} c_m\, (\phi_m, \phi_n). \tag{3}$$

But $(\phi_m, \phi_n) = 0$ for all values of m here except when $m = n$, in which case $(\phi_m, \phi_n) = \|\phi_n\|^2 = 1$. Hence equation (3) becomes $(f, \phi_n) = c_n$, and c_n is evidently the inner product of f and ϕ_n.

As indicated above, we cannot be certain that representation (1), with coefficients $c_n = (f, \phi_n)$, is actually valid for a specific f and a given orthonormal set $\{\phi_n\}$. Hence we write

$$f(x) \sim \sum_{n=1}^{\infty} c_n \phi_n(x) \qquad (a < x < b), \tag{4}$$

where the tilde symbol $\sim$ merely denotes correspondence when

$$c_n = (f, \phi_n) = \int_a^b f(x)\, \phi_n(x)\, dx \qquad (n = 1, 2, \ldots). \tag{5}$$

To strengthen the analogy with vectors, we recall that if a vector **A** in three-dimensional space is to be written in terms of the orthonormal set $\{\mathbf{i}, \mathbf{j}, \mathbf{k}\}$ as

$$\mathbf{A} = a_1\mathbf{i} + a_2\mathbf{j} + a_3\mathbf{k},$$

the components can be obtained by taking the inner product of $\mathbf{A}$ with each of the vectors of that set. That is, the inner product of $\mathbf{A}$ with $\mathbf{i}$ is a_1, etc.

The series in correspondence (4) is the *generalized Fourier series*, with respect to the orthonormal set $\{\phi_n\}$, for the function f on the interval $a < x < b$. The coefficients c_n are known as *Fourier constants*.

The generalized Fourier series that we shall encounter will always involve orthonormal sets and functions f in a space of the type $C_p(a, b)$, or subspaces of it; and we say that representation (1) is *valid* for functions f in a given space if equality holds everywhere except possibly at a finite number of points in the fundamental interval $a < x < b$. Representation (1) will not, however, always be valid even in very restricted function spaces.

We may anticipate this limitation in the following way. If just the two vectors $\mathbf{i}$ and $\mathbf{j}$ make up an orthonormal set in three-dimensional space, any vector $\mathbf{A}$ that is not parallel to the xy plane fails to have a representation of the form $\mathbf{A} = a_1\mathbf{i}+a_2\mathbf{j}$. In particular, the nonzero vector $\mathbf{A} = \mathbf{k}$ is orthogonal to both $\mathbf{i}$ and $\mathbf{j}$, in which case the components $a_1 = \mathbf{k}\cdot\mathbf{i}$ and $a_2 = \mathbf{k}\cdot\mathbf{j}$ would both be zero. Similarly, an orthonormal set $\{\phi_n(x)\}$ may not be large enough to write a generalized Fourier series. To be specific, if the function $f(x)$ in correspondence (4) is orthogonal to each function in the orthonormal set $\{\phi_n(x)\}$, we find that the Fourier constants $c_n = (f, \phi_n)$ are all zero. This means, of course, that the sum of the series is the zero function, whose norm is zero. Consequently, if f has a positive norm, the series is not a valid representation on the fundamental interval. [See Problem 4(*a*), Sec. 53.]

An orthonormal set is *closed* in $C_p(a, b)$, or a subspace of it, if there is no function in the space, with positive norm, that is orthogonal to each of the functions $\phi_n(x)$. Thus, according to the preceding paragraph, *if an orthonormal set* $\{\phi_n(x)\}$ *is not closed, then representation* (1) *cannot be valid for each function f in the space.*

EXAMPLE. Consider the orthonormal set (see Example 2, Sec. 53)

$$\phi_0(x) = \frac{1}{\sqrt{\pi}}, \quad \phi_n(x) = \sqrt{\frac{2}{\pi}}\cos nx \qquad (n = 1, 2, \ldots) \tag{6}$$

in the space $C_p'(0, \pi)$ of piecewise smooth functions (Sec. 9) on $0 < x < \pi$. If the function $\phi_0(x)$ is not included with the other functions, the resulting set is not closed because $\phi_0(x)$ is orthogonal to each of the functions in that smaller set.

In Example 2, Sec. 55, we shall see how the generalized Fourier series correspondence

$$f(x) \sim \sum_{n=0}^{\infty} c_n\,\phi_n(x) \qquad (0 < x < \pi) \tag{7}$$

for a function f in $C_p'(0, \pi)$ with respect to the orthonormal set (6) is, in fact, a Fourier cosine series correspondence

$$f(x) \sim \frac{a_0}{2} + \sum_{n=1}^{\infty} a_n \cos nx \qquad (0 < x < \pi), \tag{8}$$

where

$$a_0 = 2\frac{c_0}{\sqrt{\pi}}.$$

Since c_0 is the coefficient of $\phi_0(x)$ in correspondence (7) and since the set (6) is not closed without $\phi_0(x)$, it follows from the statement in italics preceding this example that correspondence (8) without a_0 does not provide a valid representation for all f in $C'_p(0, \pi)$. We do know (Sec. 13), however, that valid Fourier cosine series representations are obtained when a_0 is included.

Finally, we note that another way to write the statement in italics just before this example is that *if representation* (1), *with respect to a given orthonormal set* $\{\phi_n(x)\}$, *is valid for each function* f *in* $C_p(a, b)$, *or a subspace of it, then that set must be closed.*

55. EXAMPLES

This section is devoted to familiar examples of generalized Fourier series, namely Fourier sine and cosine series on the fundamental interval $0 < x < \pi$ and Fourier series involving both sines and cosines on $-\pi < x < \pi$. These special cases of series (4), Sec. 54, will be based on the examples in Sec. 53.

EXAMPLE 1. We saw in Example 1, Sec. 53, that the sine functions

$$\phi_n(x) = \sqrt{\frac{2}{\pi}} \sin nx \qquad (n = 1, 2, \ldots) \tag{1}$$

constitute an orthonormal set on the interval $0 < x < \pi$. The generalized Fourier series correspondence (4), Sec. 54, for a function $f(x)$ in $C_p(0, \pi)$ is then

$$f(x) \sim \sum_{n=1}^{\infty} c_n \sqrt{\frac{2}{\pi}} \sin nx \qquad (0 < x < \pi),$$

where

$$c_n = (f, \phi_n) = \sqrt{\frac{2}{\pi}} \int_0^{\pi} f(x) \sin nx \, dx \qquad (n = 1, 2, \ldots).$$

Upon writing

$$b_n = c_n \sqrt{\frac{2}{\pi}} \qquad (n = 1, 2, \ldots),$$

we have the Fourier sine series correspondence (Sec. 4)

$$f(x) \sim \sum_{n=1}^{\infty} b_n \sin nx \qquad (0 < x < \pi), \tag{2}$$

where

$$b_n = \frac{2}{\pi} \int_0^{\pi} f(x) \sin nx \, dx \qquad (n = 1, 2, \ldots). \tag{3}$$

EXAMPLE 2. Let f be any function in $C_p(0, \pi)$. We know from Example 2 in Sec. 53 that the set $\{\phi_n(x)\}$ $(n = 0, 1, 2, \ldots)$ consisting of the functions

$$\phi_0(x) = \frac{1}{\sqrt{\pi}}, \qquad \phi_n(x) = \sqrt{\frac{2}{\pi}} \cos nx \qquad (n = 1, 2, \ldots) \tag{4}$$

is orthonormal on the interval $0 < x < \pi$. The correspondence

$$f(x) \sim \sum_{n=0}^{\infty} c_n \, \phi_n(x) \qquad (0 < x < \pi),$$

which is correspondence (4), Sec. 54, with the summation starting from $n = 0$, becomes

$$f(x) \sim \frac{c_0}{\sqrt{\pi}} + \sum_{n=1}^{\infty} c_n \sqrt{\frac{2}{\pi}} \cos nx \qquad (0 < x < \pi).$$

Moreover,

$$c_0 = (f, \phi_0) = \frac{1}{\sqrt{\pi}} \int_0^{\pi} f(x)\, dx, \qquad c_n = (f, \phi_n) = \sqrt{\frac{2}{\pi}} \int_0^{\pi} f(x) \cos nx \, dx \qquad (n = 1, 2, \ldots).$$

By writing

$$a_0 = 2 \frac{c_0}{\sqrt{\pi}}, \qquad a_n = c_n \sqrt{\frac{2}{\pi}} \qquad (n = 1, 2, \ldots),$$

we thus arrive at the Fourier cosine series correspondence (Sec. 2)

$$f(x) \sim \frac{a_0}{2} + \sum_{n=1}^{\infty} a_n \cos nx \qquad (0 < x < \pi), \tag{5}$$

where

$$a_n = \frac{2}{\pi} \int_0^{\pi} f(x) \cos nx \, dx \qquad (n = 0, 1, 2, \ldots). \tag{6}$$

EXAMPLE 3. In Example 3, Sec. 53, we saw that the functions

$$\phi_0(x) = \frac{1}{\sqrt{2\pi}}, \qquad \phi_{2n-1}(x) = \frac{1}{\sqrt{\pi}} \cos nx, \qquad \phi_{2n}(x) = \frac{1}{\sqrt{\pi}} \sin nx \qquad (n = 1, 2, \ldots) \tag{7}$$

form an orthonormal set on the fundamental interval $-\pi < x < \pi$. The generalized Fourier series corresponding to a function $f(x)$ in $C_p(-\pi, \pi)$ is, therefore,

$$\sum_{n=0}^{\infty} c_n \, \phi_n(x) = c_0 \, \phi_0(x) + \sum_{n=1}^{\infty} [c_{2n-1} \, \phi_{2n-1}(x) + c_{2n} \, \phi_{2n}(x)].$$

That is,

$$f(x) \sim \frac{c_0}{\sqrt{2\pi}} + \sum_{n=1}^{\infty} \left(\frac{c_{2n-1}}{\sqrt{\pi}} \cos nx + \frac{c_{2n}}{\sqrt{\pi}} \sin nx \right) \qquad (-\pi < x < \pi) \tag{8}$$

where

$$c_0 = (f, \phi_0) = \frac{1}{\sqrt{2\pi}} \int_{-\pi}^{\pi} f(x)\, dx$$

and

$$c_{2n-1} = (f, \phi_{2n-1}) = \frac{1}{\sqrt{\pi}} \int_{-\pi}^{\pi} f(x) \cos nx\, dx \qquad (n = 1, 2, \ldots),$$

$$c_{2n} = (f, \phi_{2n}) = \frac{1}{\sqrt{\pi}} \int_{-\pi}^{\pi} f(x) \sin nx\, dx \qquad (n = 1, 2, \ldots).$$

So if we write

$$a_0 = 2\frac{c_0}{\sqrt{2\pi}}, \qquad a_n = \frac{c_{2n-1}}{\sqrt{\pi}}, \qquad b_n = \frac{c_{2n}}{\sqrt{\pi}} \qquad (n = 1, 2, \ldots),$$

correspondence (8) becomes (see Sec. 6)

$$f(x) \sim \frac{a_0}{2} + \sum_{n=1}^{\infty} (a_n \cos nx + b_n \sin nx) \qquad (-\pi < x < \pi), \tag{9}$$

where

$$a_n = \frac{1}{\pi} \int_{-\pi}^{\pi} f(x) \cos nx\, dx \qquad (n = 0, 1, 2, \ldots) \tag{10}$$

and

$$b_n = \frac{1}{\pi} \int_{-\pi}^{\pi} f(x) \sin nx\, dx \qquad (n = 1, 2, \ldots). \tag{11}$$

PROBLEMS

1. Let $\{\psi_n(x)\}$ $(n = 1, 2, \ldots)$ denote an orthogonal, but not necessarily orthonormal, set on a fundamental interval $a < x < b$. Show that the correspondence between a piecewise continuous function $f(x)$ and its generalized Fourier series with respect to the *orthonormal* set

$$\phi_n(x) = \frac{\psi_n(x)}{\|\psi_n\|} \qquad (n = 1, 2, \ldots)$$

can be written

$$f(x) \sim \sum_{n=1}^{\infty} \gamma_n \psi_n(x) \quad \text{where} \quad \gamma_n = \frac{(f, \psi_n)}{\|\psi_n\|^2}.$$

2. If we exclude the constant function $\phi_0(x)$ from the orthonormal set

$$\phi_0(x) = \frac{1}{\sqrt{\pi}}, \qquad \phi_n(x) = \sqrt{\frac{2}{\pi}} \cos nx \qquad (n = 1, 2, \ldots)$$

in Example 2, Sec. 53, we still have an orthonormal set. State why that smaller set is closed (Sec. 54) in the space of all functions f that are piecewise smooth on the interval $0 < x < \pi$ and satisfy the condition

$$\int_0^{\pi} f(x)\, dx = 0.$$

Suggestion: Refer to the statement in italics at the end of Sec. 54.

3. In the space of *continuous* functions on the interval $a \le x \le b$, prove that if two functions f and g have the same Fourier constants with respect to a *closed* (Sec. 54) orthonormal set $\{\phi_n(x)\}$, then f and g must be identical. Thus show that f is uniquely determined by its Fourier constants.

Suggestion: Show that the norm of the difference $f(x) - g(x)$ is zero. Then point out how it follows that $f(x) - g(x) \equiv 0$ (see also the suggestion with Problem 4, Sec. 53).

4. Let $\{\phi_n(x)\}$ be an orthonormal set in the space of *continuous* functions on the interval $a \le x \le b$, and suppose that the generalized Fourier series for a function $f(x)$ in that space converges *uniformly* (Sec. 16) to a sum $s(x)$ on that interval.

(*a*) Show that $s(x)$ and $f(x)$ have the same Fourier constants with respect to $\{\phi_n(x)\}$.
(*b*) Use results in part (*a*) and Problem 3 to show that if $\{\phi_n(x)\}$ is *closed* (Sec. 54), then $s(x) = f(x)$ on the interval $a \le x \le b$.

Suggestion: Recall from Sec. 16 that the sum of a uniformly convergent series of continuous functions is continuous and that such a series can be integrated term by term.

56. BEST APPROXIMATION IN THE MEAN

Since the material in this and the next two sections is not essential for subsequent chapters, the reader may at this time pass directly to Chap. 8 without loss of continuity.

Let f be a function in $C_p(a,b)$ and $\{\phi_n(x)\}$ $(n = 1, 2, \ldots)$ an orthonormal set in that space. We consider here the first N functions $\phi_1(x), \phi_2(x), \ldots, \phi_N(x)$ of the orthonormal set and let $\Phi_N(x)$ denote *any* linear combination of them:

$$\Phi_N(x) = \gamma_1\phi_1(x) + \gamma_2\phi_2(x) + \cdots + \gamma_N\phi_N(x). \tag{1}$$

The norm

$$\|f - \Phi_N\| = \left\{\int_a^b [f(x) - \Phi_N(x)]^2\,dx\right\}^{1/2}$$

is a measure of the deviation of the sum Φ_N from a given function f in $C_p(a,b)$ (see Sec. 52). Let us determine values of the constants γ_n $(n = 1, 2, \ldots, N)$ in expression (1) that make $\|f - \Phi_N\|$, or the quantity

$$E = \|f - \Phi_N\|^2 = \int_a^b [f(x) - \Phi_N(x)]^2\,dx, \tag{2}$$

as small as possible. The nonnegative number E represents the *mean square error* in the approximation by the function Φ_N to the function f; and we seek the *best approximation in the mean*.[†]

We start with the observation that

$$(f - \Phi_N)^2 = \left(f - \sum_{n=1}^N \gamma_n\phi_n\right)^2 = f^2 - 2f\sum_{n=1}^N \gamma_n\phi_n + \left(\sum_{n=1}^N \gamma_n\phi_n\right)^2.$$

[†]The approximation sought here is also called a *least squares approximation*.

But

$$\left(\sum_{n=1}^{N}\gamma_n\phi_n\right)^2 = \left(\sum_{m=1}^{N}\gamma_m\phi_m\right)\left(\sum_{n=1}^{N}\gamma_n\phi_n\right)$$

$$= \sum_{n=1}^{N}\left(\sum_{m=1}^{N}\gamma_m\phi_m\right)\gamma_n\phi_n$$

$$= \sum_{n=1}^{N}\left(\sum_{m=1}^{N}\gamma_m\gamma_n\phi_m\phi_n\right);$$

and this enables us to write

$$(f-\Phi_N)^2 = f^2 + \sum_{n=1}^{N}\left[\left(\sum_{m=1}^{N}\gamma_m\gamma_n\phi_m\phi_n\right) - 2\gamma_n f\phi_n\right].$$

Integrating each side here over the interval $a < x < b$ and then using the relations

$$\int_a^b \phi_m(x)\,\phi_n(x)\,dx = \begin{cases} 0 & \text{when } m \neq n, \\ 1 & \text{when } m = n \end{cases}$$

and

$$\int_a^b f(x)\,\phi_n(x)\,dx = c_n,$$

where c_n are Fourier constants (Sec. 54), we arrive at the following expression for the error E, defined above:

$$E = \|f\|^2 + \sum_{n=1}^{N}(\gamma_n^2 - 2\gamma_n c_n).$$

If we complete the squares in the terms being summed here and write

$$\gamma_n^2 - 2\gamma_n c_n = \left(\gamma_n^2 - 2c_n\gamma_n + c_n^2\right) - c_n^2 = (\gamma_n - c_n)^2 - c_n^2,$$

this expression for E takes the form

$$E = \|f\|^2 + \sum_{n=1}^{N}(\gamma_n - c_n)^2 - \sum_{n=1}^{N}c_n^2. \tag{3}$$

In view of the squares in the first summation appearing in equation (3), the smallest possible value of E is, then, obtained when $\gamma_n = c_n$ $(n = 1, 2, \ldots, N)$, that value being

$$E = \|f\|^2 - \sum_{n=1}^{N}c_n^2. \tag{4}$$

We state the result as a theorem.

Theorem. *Let c_n $(n = 1, 2, \ldots)$ be the Fourier constants for a function f in $C_p(a, b)$ with respect to an orthonormal set $\{\phi_n(x)\}$ $(n = 1, 2, \ldots)$ in that space. Then, of all possible linear combinations of the functions $\phi_1(x), \phi_2(x), \ldots, \phi_N(x)$, the combination*

$$c_1\phi_1(x) + c_2\phi_2(x) + \cdots + c_N\phi_N(x)$$

is the best approximation in the mean to $f(x)$ on the fundamental interval $a < x < b$. In that case, the mean square error E is given by equation (4).

This theorem is analogous to, and even suggested by, a corresponding result in three-dimensional space. Namely, suppose that we wish to approximate a vector $\mathbf{A} = a_1\mathbf{i} + a_2\mathbf{j} + a_3\mathbf{k}$ by a linear combination of just the two basis vectors $\mathbf{i}$ and $\mathbf{j}$. If we interpret $\mathbf{A}$ and any linear combination $\alpha_1\mathbf{i} + \alpha_2\mathbf{j}$ as radius vectors, it is geometrically evident that the shortest distance d between their tips occurs when $\alpha_1\mathbf{i} + \alpha_2\mathbf{j}$ is the vector projection of $\mathbf{A}$ onto the plane of $\mathbf{i}$ and $\mathbf{j}$. That projection is, of course, the vector $a_1\mathbf{i} + a_2\mathbf{j}$ (see Fig. 47), the components a_1 and a_2 being the inner products of $\mathbf{A}$ with $\mathbf{i}$ and $\mathbf{j}$, respectively.

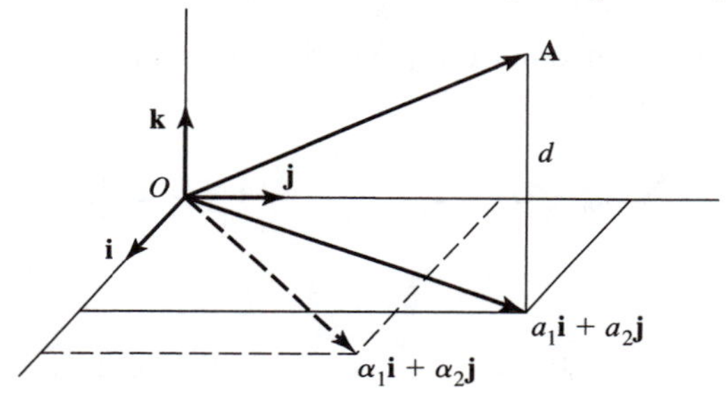

FIGURE 47

EXAMPLE. We recall from Example 3, Sec. 55, that when the orthonormal set of functions

$$\phi_0(x) = \frac{1}{\sqrt{2\pi}}, \qquad \phi_{2n-1}(x) = \frac{1}{\sqrt{\pi}}\cos nx, \qquad \phi_{2n}(x) = \frac{1}{\sqrt{\pi}}\sin nx$$

$$(n = 1, 2, \ldots)$$

in $C_p(-\pi, \pi)$ is used, the generalized Fourier series

$$(5) \quad \sum_{n=0}^{\infty} c_n\phi_n(x) = c_0\phi_0(x) + \sum_{n=1}^{\infty}[c_{2n-1}\phi_{2n-1}(x) + c_{2n}\phi_{2n}(x)] \qquad (-\pi < x < \pi)$$

corresponding to a function f in $C_p(-\pi, \pi)$ is the ordinary Fourier series

$$(6) \qquad \frac{a_0}{2} + \sum_{n=1}^{\infty}(a_n \cos nx + b_n \sin nx) \qquad (-\pi < x < \pi),$$

where

$$a_0 = 2\frac{c_0}{\sqrt{2\pi}}, \qquad a_n = \frac{c_{2n-1}}{\sqrt{\pi}}, \qquad b_n = \frac{c_{2n}}{\sqrt{\pi}} \qquad (n = 1, 2, \ldots).$$

The above theorem now tells us that of all possible linear combinations of the functions $\phi_n(x)$ $(n = 0, 1, 2, \ldots, 2N)$, the partial sum

$$\sum_{n=0}^{2N} c_n \phi_n(x) = c_0 \phi_0(x) + \sum_{n=1}^{N} [c_{2n-1} \phi_{2n-1}(x) + c_{2n} \phi_{2n}(x)]$$

of series (5) is the best approximation in the mean to f on the interval $-\pi < x < \pi$. That is, the partial sum

$$S_N(x) = \frac{a_0}{2} + \sum_{n=1}^{N} (a_n \cos nx + b_n \sin nx) \qquad (-\pi < x < \pi) \tag{7}$$

of series (6) is the best approximation of all linear combinations of the functions

$$1, \qquad \cos nx, \qquad \sin nx \qquad (n = 1, 2, \ldots, N).$$

57. BESSEL'S INEQUALITY AND PARSEVAL'S EQUATION

Because the mean square error E, defined by equation (2) in Sec. 56, is nonnegative, it follows from expression (4) there that

$$\|f\|^2 - \sum_{n=1}^{N} c_n^2 \geq 0 \qquad (N = 1, 2, \ldots),$$

or

$$\sum_{n=1}^{N} c_n^2 \leq \|f\|^2 \qquad (N = 1, 2, \ldots). \tag{1}$$

This is *Bessel's inequality* for the Fourier constants c_n. The following important theorem is an immediate consequence of it.

Theorem 1. *If c_n $(n = 1, 2, \ldots)$ are the Fourier constants for a function f in $C_p(a, b)$ with respect to an orthonormal set in that space, then*

$$\lim_{n \to \infty} c_n = 0. \tag{2}$$

Our proof of this theorem is similar to an argument in Sec. 10, based on a special case of inequality (1) (see Problem 3, Sec. 58), showing that the coefficients a_n in a Fourier cosine series on $0 < x < \pi$ tend to zero as n tends to infinity. We start here with the fact that since $\|f\|^2$ is independent of N, the sums of the squares c_n^2 on the left-hand side of inequality (1) form a sequence that is bounded and nondecreasing as N increases. Such a sequence must converge; and since it is the sequence of partial sums of the series whose terms are c_n^2 $(n = 1, 2, \ldots)$, the series must converge. Finally, because the nth term of a convergent series tends to zero as n tends to infinity, limit (2) is established.

We turn now to a modification of inequality (1) in which the inequality is actually an equality. A sequence of functions $s_N(x)$ $(N = 1, 2, \ldots)$ in $C_p(a, b)$

is said to *converge in the mean* to a function $f(x)$ in $C_p(a, b)$ if the mean square error (Sec. 56)

$$E = \|f - s_N\|^2 = \int_a^b [f(x) - s_N(x)]^2 \, dx \tag{3}$$

in the approximation by s_N to f tends to zero as N tends to infinity. That is, convergence in the mean occurs when

$$\lim_{N\to\infty} \|f - s_N\| = 0. \tag{4}$$

Sometimes condition (4) is also written

$$\underset{N\to\infty}{\text{l.i.m.}} \; s_N(x) = f(x), \tag{5}$$

where the abbreviation "l.i.m." stands for *limit in the mean*.

It should be emphasized that statement (5) is *not* the same as the statement

$$\lim_{N\to\infty} s_N(x) = f(x) \qquad (a < x < b), \tag{6}$$

even if a finite number of points in the interval are excepted.† In fact, neither of the statements (5) and (6) implies the other (see Problems 5 and 6, Sec. 58).

Suppose that the functions s_N are the partial sums of a generalized Fourier series (Sec. 54) corresponding to f on the fundamental interval $a < x < b$:

$$s_N(x) = \sum_{n=1}^{N} c_n \phi_n(x). \tag{7}$$

This is the linear combination $\Phi_N(x)$ in Sec. 56 when $\gamma_n = c_n$ there. If condition (4) is satisfied by each function f in our function space $C_p(a, b)$, or possibly a subspace containing the orthonormal set $\{\phi_n(x)\}$, we say that $\{\phi_n(x)\}$ is *complete* in that space or subspace.‡ Thus each function $f(x)$ can be approximated arbitrarily closely in the mean by some linear combination of the functions $\phi_n(x)$ of a complete set, namely the linear combination (7) when N is large enough.

According to equation (4), Sec. 56, the mean square error in the approximation by $s_N(x)$ to $f(x)$ is

$$\|f - s_N\|^2 = \|f\|^2 - \sum_{n=1}^{N} c_n^2. \tag{8}$$

Hence when $\{\phi_n(x)\}$ is complete, it is always true that

$$\sum_{n=1}^{\infty} c_n^2 = \|f\|^2. \tag{9}$$

---

†An example of a sequence of functions that converges in the mean to zero but *diverges at each point* of the interval is given in the book by Franklin (1964, p. 408), listed in the Bibliography.

‡In the mathematical literature, including some earlier editions of this text, the terms *complete* and *closed* are sometimes applied to sets that we have called *closed* (Sec. 54) and *complete*, respectively.

Equation (9) is known as *Parseval's equation*. It identifies the sum of the squares of the Fourier constants for f, with respect to the orthonormal set $\{\phi_n(x)\}$, as the square of the norm of f.

Conversely, if each function f in the space satisfies Parseval's equation, the set $\{\phi_n(x)\}$ is complete in the sense of mean convergence. This is because, in view of equation (8), the limit (4) is merely a restatement of equation (9). We now have a theorem that provides an alternative characterization of complete sets.

Theorem 2. *A necessary and sufficient condition for an orthonormal set $\{\phi_n(x)\}$ $(n = 1, 2, \ldots)$ to be complete is that for each function f in the space considered, Parseval's equation*

$$\sum_{n=1}^{\infty} c_n^2 = \| f \|^2, \tag{10}$$

where c_n are the Fourier constants $c_n = (f, \phi_n)$, be satisfied.

Each of the theorems in this section will be illustrated in Sec. 58, with applications to ordinary Fourier series.

58. APPLICATIONS TO FOURIER SERIES

In Example 3, Sec. 53, we saw that the functions

$$\phi_0(x) = \frac{1}{\sqrt{2\pi}}, \qquad \phi_{2n-1}(x) = \frac{1}{\sqrt{\pi}} \cos nx, \qquad \phi_{2n}(x) = \frac{1}{\sqrt{\pi}} \sin nx \qquad (n = 1, 2, \ldots) \tag{1}$$

form an orthonormal set on the fundamental interval $-\pi < x < \pi$. The Fourier constants c_n $(n = 0, 1, 2, \ldots)$ in the generalized Fourier series for a function f in $C_p(-\pi, \pi)$ with respect to this set were then used in Example 3, Sec. 55, to define the constants

$$a_0 = 2\frac{c_0}{\sqrt{2\pi}}, \qquad a_n = \frac{c_{2n-1}}{\sqrt{\pi}}, \qquad b_n = \frac{c_{2n}}{\sqrt{\pi}} \qquad (n = 1, 2, \ldots). \tag{2}$$

That gave rise to the Fourier series correspondence

$$f(x) \sim \frac{a_0}{2} + \sum_{n=1}^{\infty} (a_n \cos nx + b_n \sin nx) \qquad (-\pi < x < \pi), \tag{3}$$

where

$$a_n = \frac{1}{\pi} \int_{-\pi}^{\pi} f(x) \cos nx \, dx \qquad (n = 0, 1, 2, \ldots) \tag{4}$$

and

$$b_n = \frac{1}{\pi} \int_{-\pi}^{\pi} f(x) \sin nx \, dx \qquad (n = 1, 2, \ldots). \tag{5}$$

Bessel's inequality (1), Sec. 57, involving $2N+1$ terms in the sum there, is

$$c_0^2 + \sum_{n=1}^{N} (c_{2n-1}^2 + c_{2n}^2) \leq \| f \|^2 \qquad (N = 1, 2, \ldots);$$

and, in view of relations (2), this is the same as (compare with Problem 4, Sec. 11)

$$\frac{a_0^2}{2} + \sum_{n=1}^{N} (a_n^2 + b_n^2) \leq \frac{1}{\pi} \int_{-\pi}^{\pi} [f(x)]^2 \, dx \qquad (N = 1, 2, \ldots). \tag{6}$$

Theorem 1 in Sec. 57 tells us, moreover, that

$$\lim_{n\to\infty} a_n = 0 \quad \text{and} \quad \lim_{n\to\infty} b_n = 0. \tag{7}$$

Limits (7) were obtained directly in Sec. 10, where a_n and b_n were also the coefficients in the Fourier cosine and sine series for certain functions related to f. (See Problems 2 and 3, where the Bessel inequalities appearing in Sec. 10 are derived from orthonormal sets.)

We now prove a theorem that follows from Theorem 2 in Sec. 57 and states that the orthonormal set (1) is complete in the space consisting of functions satisfying the same conditions as in the lemma in Sec. 15, as well as in the theorems in Secs. 16 and 17.

Theorem. *The orthonormal set* (1) *is complete in the space in which each function f has these properties:*

(i) f *is continuous on the interval* $-\pi \leq x \leq \pi$;
(ii) $f(-\pi) = f(\pi)$;
(iii) its derivative f' *is piecewise continuous on the interval* $-\pi < x < \pi$.

Observe that, just as Bessel's inequality in Sec. 57 became inequality (6) when the set (1) was used, Parseval's equation (10) in Sec. 57 becomes

$$\frac{a_0^2}{2} + \sum_{n=1}^{\infty} (a_n^2 + b_n^2) + \frac{1}{\pi} \int_{-\pi}^{\pi} [f(x)]^2 \, dx. \tag{8}$$

Hence once we show that the coefficients (4) and (5) actually satisfy equation (8), the theorem here is proved.

The fact that equation (8) is satisfied is an easy consequence of the theorem in Sec. 16, which tells us that for the functions f in the space considered here the series in correspondence (3) converges uniformly to $f(x)$ on the interval $-\pi \leq x \leq \pi$:

$$f(x) = \frac{a_0}{2} + \sum_{n=1}^{\infty} (a_n \cos nx + b_n \sin nx) \qquad (-\pi \leq x \leq \pi). \tag{9}$$

Now a uniformly convergent series of continuous functions can be integrated term by term (Sec. 16). Hence we may multiply each term in equation (9) by $f(x)$ itself, thus leaving the series still uniformly convergent, and then integrate over the fundamental interval:

$$\int_{-\pi}^{\pi} [f(x)]^2\,dx$$
$$= \frac{a_0}{2}\int_{-\pi}^{\pi} f(x)\,dx + \sum_{n=1}^{\infty}\left[a_n\int_{-\pi}^{\pi} f(x)\cos nx\,dx + b_n\int_{-\pi}^{\pi} f(x)\sin nx\,dx\right].$$

In view of expressions (4) and (5), the integrals on the right here can be written in terms of a_n and b_n; and we find that

$$\int_{-\pi}^{\pi} [f(x)]^2\,dx = \pi\left[\frac{a_0^2}{2} + \sum_{n=1}^{\infty}(a_n^2 + b_n^2)\right].$$

Since this is the same as Parseval's equation (8), the proof is finished.

The theorem above is readily modified so as to apply to the orthonormal sets leading to Fourier cosine and sine series on the interval $0 < x < \pi$. More specifically, the set of normalized cosine functions in Example 2, Sec. 53, is complete in the space of continuous functions f, on the interval $0 \le x \le \pi$, whose derivatives f' are piecewise continuous. When the normalized sine functions in Example 1, Sec. 53, are used to obtain a sine series, the conditions $f(0) = f(\pi) = 0$ are also needed in order for the set to be complete.

The function space in the theorem is quite restricted. It can be shown that Parseval's equation (8) holds for any function f whose square is integrable over the interval $-\pi < x < \pi$.[†]

PROBLEMS

1. Use Theorem 2 in Sec. 57 to show that *an orthonormal set* $\{\phi_n(x)\}$ *is closed* (Sec. 54) *in a given function space if it is complete in that space.*
2. Apply Bessel's inequality (1), Sec. 57, to the orthonormal set (Example 1, Sec. 53)

$$\phi_n(x) = \sqrt{\frac{2}{\pi}}\sin nx \qquad (n = 1, 2, \ldots)$$

in $C_p(0, \pi)$ to show that

$$\sum_{n=1}^{N} b_n^2 \le \frac{2}{\pi}\int_0^{\pi} [f(x)]^2\,dx \qquad (N = 1, 2, \ldots)$$

when f is in $C_p(0, \pi)$ and b_n are the coefficients in its Fourier sine series.

3. Let a_n $(n = 0, 1, 2, \ldots)$ be the usual coefficients in the Fourier cosine series for a function f in $C_p(0, \pi)$. By referring to the orthonormal set (Example 2, Sec. 53)

$$\phi_0(x) = \frac{1}{\sqrt{\pi}}, \qquad \phi_n(x) = \sqrt{\frac{2}{\pi}}\cos nx \qquad (n = 1, 2, \ldots)$$

[†] See, for instance, the book by Tolstov (1976, pp. 54–57 and 117–120), which is listed in the Bibliography.

and using Bessel's inequality (1), Sec. 57, show that

$$\frac{a_0^2}{2} + \sum_{n=1}^{N} a_n^2 \le \frac{2}{\pi} \int_0^{\pi} [f(x)]^2 \, dx \qquad (N = 1, 2, \ldots).$$

4. (*a*) Use the same steps as in Example 3, Sec. 53, to verify that the set of functions

$$\phi_0(x) = \frac{1}{\sqrt{2c}}, \qquad \phi_{2n-1}(x) = \frac{1}{\sqrt{c}} \cos \frac{n\pi x}{c}, \qquad \phi_{2n}(x) = \frac{1}{\sqrt{c}} \sin \frac{n\pi x}{c}$$
$$(n = 1, 2, \ldots)$$

is orthonormal on the interval $-c < x < c$. (This set becomes the one in that example when $c = \pi$.)

(*b*) By proceeding as in Example 3, Sec. 55, show that the generalized Fourier series corresponding to a function $f(x)$ in $C_p(-c, c)$ with respect to the orthonormal set in part (*a*) can be written as an ordinary Fourier series on $-c < x < c$ (Sec. 14), with the usual coefficients a_n and b_n.

(*c*) Derive Bessel's inequality

$$\frac{a_0^2}{2} + \sum_{n=1}^{N} \left(a_n^2 + b_n^2\right) \le \frac{1}{c} \int_{-c}^{c} [f(x)]^2 \, dx \qquad (N = 1, 2, \ldots)$$

for the coefficients a_n and b_n in part (*b*) from the general form (1), Sec. 57, of that inequality for Fourier constants. [Compare with inequality (6), Sec. 58.]

Suggestion: In part (*a*), some integrals to be used can be evaluated by writing

$$x = \frac{\pi}{c} s$$

in integrals (1) and (4), Sec. 53.

5. Consider the sequence of functions $s_N(x)$ $(N = 1, 2, \ldots)$ defined on the interval $0 \le x \le 1$ by the equations

$$s_N(x) = \begin{cases} 0 & \text{when } 0 \le x \le \dfrac{1}{N}, \\ \sqrt{N} & \text{when } \dfrac{1}{N} < x < \dfrac{2}{N}, \\ 0 & \text{when } \dfrac{2}{N} \le x \le 1. \end{cases}$$

Show that this sequence converges pointwise to the function $f(x) = 0\,(0 \le x \le 1)$ but that it does *not* converge in the mean to f in the space $C_p(0, 1)$ or any subspace of $C_p(0, 1)$.

6. Let $s_N(x)$ $(N = 1, 2, \ldots)$ be a sequence of functions defined on the interval $0 \le x \le 1$ by the equations

$$s_N(x) = \begin{cases} 0 & \text{when } x = 1, \dfrac{1}{2}, \ldots, \dfrac{1}{N}, \\ 1 & \text{when } x \ne 1, \dfrac{1}{2}, \ldots, \dfrac{1}{N}. \end{cases}$$

Show that this sequence converges in the mean to the function $f(x) = 1$ in $C_p(0, 1)$ but that for each positive integer p,

$$\lim_{N\to\infty} s_N\left(\frac{1}{p}\right) = 0.$$

Suggestion: Observe that

$$s_N\left(\frac{1}{p}\right) = 0 \qquad \text{when} \qquad N \geq p.$$

CHAPTER 8

STURM-LIOUVILLE PROBLEMS AND APPLICATIONS

We turn now to a careful presentation of the basic theory of Sturm-Liouville problems and their solutions. Once that is done, we shall illustrate the Fourier method in solving physical problems involving eigenfunctions not encountered in earlier chapters.

59. REGULAR STURM-LIOUVILLE PROBLEMS

In Chap. 5, we found solutions of various boundary value problems by the Fourier method. Except in Sec. 43, the method always led to the need for a Fourier cosine or sine series representation of a given function. The cosine and sine functions in the series were the eigenfunctions of one of the following two Sturm-Liouville problems on an interval $0 \le x \le c$:

$$X''(x) + \lambda X(x) = 0, \qquad X'(0) = 0, \qquad X'(c) = 0, \tag{1}$$

$$X''(x) + \lambda X(x) = 0, \qquad X(0) = 0, \qquad X(c) = 0. \tag{2}$$

When applied to many other boundary value problems in partial differential equations, the Fourier method continues to involve a Sturm-Liouville problem consisting of a linear homogeneous ordinary differential equation of the type

$$[r(x)X'(x)]' + [q(x) + \lambda p(x)]X(x) = 0 \qquad (a < x < b), \tag{3}$$

together with a pair of *separated* boundary conditions

$$a_1 X(a) + a_2 X'(a) = 0, \qquad b_1 X(b) + b_2 X'(b) = 0. \tag{4}$$

The interval $a < x < b$ is understood to be bounded. We agree, moreover, that a_1 and a_2 are not both zero and that the same is true of the constants b_1 and b_2. Values of the parameter λ and corresponding nontrivial solutions $X(x)$ are to be determined.

The parameter λ appears in a Sturm-Liouville problem only as indicated above. That is, the real-valued functions p, q, and r in the differential equation (3) are independent of λ, and the real numbers a_1, a_2, b_1, b_2 in boundary conditions (4) are also independent of λ. The Sturm-Liouville problem is said to be *regular* when[†]

(*i*) p, q, r, and r' are continuous on the closed interval $a \le x \le b$;
(*ii*) $p(x) > 0$ and $r(x) > 0$ when $a \le x \le b$.

EXAMPLES. Problems (1) and (2) are regular Sturm-Liouville problems. Other examples, to be solved later in this chapter, are

$$X''(x) + \lambda X(x) = 0 \qquad (0 < x < c),$$

$$X'(0) = 0, \qquad hX(c) + X'(c) = 0,$$

where h denotes a positive constant, and

$$[x^2 X'(x)]' + \lambda X(x) = 0 \qquad (1 < x < b),$$

$$X(1) = 0, \qquad X(b) = 0.$$

As was the case with problems (1) and (2), a value of λ for which problem (3)–(4) has a nontrivial solution is called an *eigenvalue*; and the nontrivial solution is called an *eigenfunction*. Note that if $X(x)$ is an eigenfunction, then so is $CX(x)$, where C is any nonzero constant. It is understood that for $X(x)$ to be an eigenfunction, $X(x)$ *and* $X'(x)$ *must be continuous on the closed interval* $a \le x \le b$. Such continuity conditions are usually required of solutions of boundary value problems in ordinary differential equations.

The set of eigenvalues of problem (3)–(4) is called the *spectrum* of the problem. The spectrum of a regular Sturm-Liouville problem consists of an infinite number of eigenvalues $\lambda_1, \lambda_2, \ldots$. We state this fact without proof, which is quite involved.[‡] In special cases, the eigenvalues will be found; and so their existence will not be in doubt. When eigenvalues are sought, however, it is useful to know that they are all real and hence that there is no possibility of discovering others in the complex plane. The proof that the eigenvalues must be real is given in Sec. 61; and we agree that they are to be arranged in ascending order of magnitude, so that $\lambda_n < \lambda_{n+1}$ $(n = 1, 2, \ldots)$. It can be shown that $\lambda_n \to \infty$ as $n \to \infty$.

[†]Papers by J. C. F. Sturm and J. Liouville giving the first extensive development of the theory of this problem appeared in vols. 1–3 of the *Journal de mathématique* (1836–1838).

[‡]For verification of statements in this section that we do not prove, see the book by Churchill (1972, chap. 9), which contains proofs when $a_2 = b_2 = 0$ in conditions (4), and the one by Birkhoff and Rota (1989). Also, extensive treatments of Sturm-Liouville theory appear in the books by Ince (1956) and Titchmarsh (1962). These references are all listed in the Bibliography.

60. MODIFICATIONS

Although we are mainly concerned in this chapter with the theory and application of *regular* Sturm-Liouville problems, described in Sec. 59, certain important modifications are also of interest in practice. We mention them here since some of their theory is conveniently included in the discussion of regular Sturm-Liouville problems in Sec. 61.

A Sturm-Liouville problem

$$[r(x)X'(x)]' + [q(x) + \lambda p(x)]X(x) = 0 \qquad (a < x < b), \tag{1}$$

$$a_1X(a) + a_2X'(a) = 0, \qquad b_1X(b) + b_2X'(b) = 0 \tag{2}$$

is *singular* when the interval on which it is defined is unbounded or when at least one of the regularity conditions stated in Sec. 59 fails to be satisfied. Eigenvalue problems on unbounded intervals have already been encountered in Chap. 6. On a bounded interval $a \le x \le b$, the function q might, for instance, have an infinite discontinuity at an endpoint of the interval. The problem is also singular if $p(x)$ or $r(x)$ vanishes at an endpoint. When $r(x)$ does this, we drop the boundary condition at the endpoint in question. Note that the dropping of the boundary condition at $x = a$ is the same as letting both of the coefficients a_1 and a_2 in that condition be zero; a similar remark can be made when the condition at $x = b$ is to be dropped.

EXAMPLE 1. One singular Sturm-Liouville problem to be studied in Chap. 9 consists of the differential equation

$$[xX'(x)]' + \left(-\frac{n^2}{x} + \lambda x\right)X(x) = 0 \qquad (0 < x < c),$$

where $n = 0, 1, 2, \ldots$, and the single boundary condition $X(c) = 0$. Observe that the functions $p(x) = x$ and $r(x) = x$ both vanish at $x = 0$ and that the function $q(x) = -n^2/x$ has an infinite discontinuity there when n is positive.

EXAMPLE 2. The differential equation

$$[(1 - x^2)X'(x)]' + \lambda X(x) = 0 \qquad (-1 < x < 1),$$

with no boundary conditions, constitutes a singular Sturm-Liouville problem. Here the function $r(x) = 1 - x^2$ vanishes at both ends $x = \pm 1$ of the interval $-1 \le x \le 1$. This problem is the main one that is solved and used in Chap. 10.

The singular problems in Chap. 6 had *continuous* spectra, consisting of either all nonnegative or all positive values of λ, and it will turn out that the problems in Examples 1 and 2 just above have the *discrete* spectra of regular Sturm-Liouville problems, where the eigenvalues may be indexed with nonnegative or positive integers. As already indicated in Sec. 59, the nature of the spectrum of any particular problem will be determined by actually finding the eigenvalues.

Finally, in addition to singular problems, another modification of problem (1)–(2) occurs when $r(a) = r(b)$ and conditions (2) are replaced by the *periodic* boundary conditions

$$X(a) = X(b), \qquad X'(a) = X'(b). \tag{3}$$

EXAMPLE 3. The problem

$$X''(x) + \lambda X(x) = 0, \qquad X(-\pi) = X(\pi), \qquad X'(-\pi) = X'(\pi),$$

already solved in Sec. 43, has periodic boundary conditions.

61. ORTHOGONALITY OF EIGENFUNCTIONS

As pointed out in Sec. 59, a regular Sturm-Liouville problem always has an infinite number of eigenvalues $\lambda_1, \lambda_2, \ldots$. In this section, we shall establish the orthogonality of eigenfunctions corresponding to *distinct* eigenvalues. The concept of orthogonality to be used here is, however, a slight generalization of the one originally introduced in Sec. 52. To be specific, a set $\{\psi_n(x)\}$ $(n = 1, 2, \ldots)$ is orthogonal on an interval $a < x < b$ with respect to a *weight function* $p(x)$, which is piecewise continuous and positive on that interval, if

$$\int_a^b p(x)\psi_m(x)\psi_n(x)\,dx = 0 \qquad \text{when } m \neq n.$$

The integral here represents an inner product (ψ_m, ψ_n) with respect to the weight function. The set is normalized by dividing each $\psi_n(x)$ by $\|\psi_n\|$, where

$$\|\psi_n\|^2 = (\psi_n, \psi_n) = \int_a^b p(x)[\psi_n(x)]^2\,dx$$

and where it is assumed that $\|\psi_n\| \neq 0$. This type of orthogonality can, of course, be reduced to that in Sec. 52 by using the products $\sqrt{p(x)}\psi_n(x)$ as functions of the set. In Sec. 64, we shall illustrate how expansions of arbitrary functions in series of such normalized eigenfunctions follow from our earlier discussion of generalized Fourier series (Sec. 54).

The following theorem states that eigenfunctions associated with distinct eigenvalues of a regular Sturm-Liouville problem

$$[r(x)X'(x)]' + [q(x) + \lambda p(x)]X(x) = 0 \qquad (a < x < b), \tag{1}$$

$$a_1 X(a) + a_2 X'(a) = 0, \qquad b_1 X(b) + b_2 X'(b) = 0 \tag{2}$$

are orthogonal on the interval $a < x < b$ with respect to the weight function $p(x)$, where $p(x)$ is the same function as in equation (1). In presenting the theorem, we relax the conditions of regularity on the coefficients in the differential equation (1) so that the result can also be applied to eigenfunctions that are found for some of the modifications of regular Sturm-Liouville problems mentioned in Sec. 60. We retain all the conditions for a regular problem, stated in Sec. 59, except that now q may be discontinuous at an endpoint of the interval $a \le x \le b$ and $p(x)$ and $r(x)$ may vanish at an endpoint. That is,

(*i*) p, r, and r' are continuous on the closed interval $a \le x \le b$, and q is continuous on the open interval $a < x < b$;

(*ii*) $p(x) > 0$ and $r(x) > 0$ when $a < x < b$.

Theorem. *If λ_m and λ_n are distinct eigenvalues of the Sturm-Liouville problem (1)–(2), then corresponding eigenfunctions $X_m(x)$ and $X_n(x)$ are orthogonal with respect to the weight function $p(x)$ on the interval $a < x < b$. The orthogonality also holds in each of the following cases:*

(*a*) *when $r(a) = 0$ and the first of boundary conditions (2) is dropped from the problem;*

(*b*) *when $r(b) = 0$ and the second of conditions (2) is dropped;*

(*c*) *when $r(a) = r(b)$ and conditions (2) are replaced by the conditions*

$$X(a) = X(b), \qquad X'(a) = X'(b).$$

Note that *both* cases (*a*) and (*b*) here may apply to a single Sturm-Liouville problem (see Example 2, Sec. 60).

To prove the theorem, we first observe that

$$(rX'_m)' + qX_m = -\lambda_m pX_m, \qquad (rX'_n)' + qX_n = -\lambda_n pX_n$$

since each eigenfunction satisfies equation (1) when λ is the eigenvalue to which it corresponds. We then multiply each side of these two equations by X_n and X_m, respectively, and subtract:

$$(\lambda_m - \lambda_n)pX_mX_n = X_m(rX'_n)' - X_n(rX'_m)'.$$

Since the right-hand side of this last equation can be written

$$[X_m(rX'_n)' + X'_m(rX'_n)] - [X_n(rX'_m)' + X'_n(rX'_m)],$$

or

$$\frac{d}{dx}[X_m(rX'_n) - X_n(rX'_m)],$$

it follows that

$$(\lambda_m - \lambda_n)pX_mX_n = \frac{d}{dx}\,[r(X_mX'_n - X_nX'_m)]. \tag{3}$$

The function q has been eliminated, and the continuity conditions on the remaining functions allow us to write

$$(\lambda_m - \lambda_n)\int_a^b pX_mX_n\,dx = [r(x)\,\Delta(x)]_a^b, \tag{4}$$

where $\Delta(x)$ is the determinant

$$\Delta(x) = \begin{vmatrix} X_m(x) & X'_m(x) \\ X_n(x) & X'_n(x) \end{vmatrix} = X_m(x)X'_n(x) - X_n(x)X'_m(x). \tag{5}$$

That is,

$$(\lambda_m - \lambda_n)\int_a^b pX_mX_n\,dx = r(b)\Delta(b) - r(a)\Delta(a). \tag{6}$$

The first of boundary conditions (2) requires that

$$a_1X_m(a) + a_2X'_m(a) = 0,$$
$$a_1X_n(a) + a_2X'_n(a) = 0;$$

and for this pair of linear homogeneous equations in a_1 and a_2 to be satisfied by numbers a_1 and a_2, not both zero, it is necessary that the determinant $\Delta(a)$ be zero. Similarly, from the second boundary condition, where b_1 and b_2 are not both zero, we see that $\Delta(b) = 0$. Thus, according to equation (6),

$$(\lambda_m - \lambda_n) \int_a^b pX_m X_n \, dx = 0; \tag{7}$$

and since $\lambda_m \neq \lambda_n$, the desired orthogonality property follows:

$$\int_a^b p(x) X_m(x) X_n(x) \, dx = 0. \tag{8}$$

If $r(a) = 0$, property (8) follows from equation (6) even when $\Delta(a) \neq 0$, or when $a_1 = a_2 = 0$, in which case the first of boundary conditions (2) disappears. Similarly, if $r(b) = 0$, the second of those conditions is not used.

If $r(a) = r(b)$ and the periodic boundary conditions

$$X(a) = X(b), \qquad X'(a) = X'(b)$$

are used in place of conditions (2), then $\Delta(a) = \Delta(b)$. Hence

$$r(b)\Delta(b) = r(a)\Delta(a);$$

and, again, property (8) follows. This completes the proof of the theorem.

EXAMPLE 1. The eigenfunctions of the regular Sturm-Liouville problem

$$X''(x) + \lambda X(x) = 0 \qquad (0 < x < c),$$

$$X(0) = 0, \qquad X(c) = 0$$

are (Sec. 32)

$$X_n(x) = \sin \frac{n\pi x}{c} \qquad (n = 1, 2, \ldots).$$

The theorem tells us that any two distinct eigenfunctions $X_m(x)$ and $X_n(x)$ are orthogonal on the interval $0 < x < c$ with weight function $p(x) = 1$:

$$\int_0^c \sin \frac{m\pi x}{c} \sin \frac{n\pi x}{c} \, dx = 0 \qquad (m \neq n). \tag{9}$$

We recall that the value of this integral was established directly when $c = \pi$ in Problem 9, Sec. 5.

EXAMPLE 2. Eigenfunctions corresponding to distinct eigenvalues of the regular Sturm-Liouville problem

$$[xX'(x)]' + \frac{\lambda}{x} X(x) = 0 \qquad (1 < x < b),$$

$$X(1) = 0, \qquad X(b) = 0$$

are, according to the theorem, orthogonal on the interval $1 < x < b$ with weight function $p(x) = 1/x$. In Problem 1 the eigenfunctions are actually found, and the orthogonality is verified.

The following corollary is an immediate consequence of the theorem.

Corollary. *If* λ *is an eigenvalue of the Sturm-Liouville problem* (1)–(2), *then it must be a real number; and the same is true in cases* (*a*), (*b*), *and* (*c*), *treated in the theorem.*

We begin the proof by writing the eigenvalue as $\lambda = \alpha + i\beta$, where α and β are real numbers. If X denotes a corresponding eigenfunction, which is nontrivial and may be complex-valued, conditions (1) and (2) are satisfied. Now the complex conjugate of λ is the number $\overline{\lambda} = \alpha - i\beta$; and $\overline{X} = u - iv$ and $X' = u' + iv'$ if $X = u + iv$. Also, the conjugate of a sum or product of two complex numbers is the sum or product, respectively, of the conjugates of those numbers. Hence by taking the conjugates of both sides of the equations in conditions (1) and (2) and keeping in mind that the functions p, q, and r are real-valued and that the coefficients in conditions (2) are real numbers, we see that

$$(r\overline{X}')' + (q + \overline{\lambda}p)\overline{X} = 0,$$
$$a_1\overline{X}(a) + a_2\overline{X}'(a) = 0, \qquad b_1\overline{X}(b) + b_2\overline{X}'(b) = 0.$$

Thus the nontrivial function $\overline{X}$ is an eigenfunction corresponding to $\overline{\lambda}$.

If we assume that $\beta \neq 0$, then $\overline{\lambda} \neq \lambda$; and the theorem tells us that X and $\overline{X}$ are orthogonal on the interval $a < x < b$ with respect to the weight function $p(x)$, even in cases (a), (b), and (c):

$$\int_a^b p(x)X(x)\overline{X}(x)\,dx = 0. \tag{10}$$

But $p(x) > 0$ when $a < x < b$. Moreover,

$$X\overline{X} = u^2 + v^2 = |X|^2 \geq 0$$

when $a \leq x \leq b$; and $|X|^2$ is not identically equal to zero since X is an eigenfunction. So integral (10) has positive value, and our assumption that $\beta \neq 0$ has led us to a contradiction. Hence we must conclude that $\beta = 0$, or that λ is real.

PROBLEMS

1. (*a*) After writing the differential equation in the regular Sturm-Liouville problem

$$[xX'(x)]' + \frac{\lambda}{x}X(x) = 0 \qquad (1 < x < b),$$
$$X(1) = 0, \qquad X(b) = 0$$

in Cauchy-Euler form (see Problem 1, Sec. 38), use the substitution $x = \exp s$ to transform the problem into one consisting of the differential equation

$$\frac{d^2X}{ds^2} + \lambda X = 0 \qquad (0 < s < \ln b)$$

and the boundary conditions

$$X = 0 \quad \text{when} \quad s = 0 \qquad \text{and} \qquad X = 0 \quad \text{when} \quad s = \ln b.$$

Then, by simply referring to the solutions of the Sturm-Liouville problem in Sec. 32, show that the eigenvalues and eigenfunctions of the original problem here are

$$\lambda_n = \alpha_n^2, \qquad X_n(x) = \sin(\alpha_n \ln x) \qquad (n = 1, 2, \ldots),$$

where $\alpha_n = n\pi / \ln b$.

(*b*) By making the substitution

$$s = \pi \frac{\ln x}{\ln b}$$

in the integral involved and then referring to Problem 9, Sec. 5, give a direct verification that the eigenfunctions $X_n(x)$ obtained in part (*a*) are orthogonal on the interval $1 < x < b$ with weight function $p(x) = 1/x$, as ensured by the theorem in Sec. 61.

2. Note that the differential equation

$$(rX')' + (q + \lambda p)X = 0$$

in a Sturm-Liouville problem can be put in the form

$$\mathcal{L}[X] + \lambda pX = 0,$$

where $\mathcal{L}$ is the differential operator defined by the equation

$$\mathcal{L}[X] = (rX')' + qX.$$

Show that the identity

$$X(rY')' - Y(rX')' = \frac{d}{dx}[r(XY' - YX')],$$

which is obtained by the steps leading up to equation (3) in Sec. 61, can be written

$$X\mathcal{L}[Y] - Y\mathcal{L}[X] = \frac{d}{dx}[r(XY' - YX')].$$

This is called *Lagrange's identity* for the operator $\mathcal{L}$.

3. (*a*) Suppose that the operator $\mathcal{L}$ in Problem 2 is defined on a space of functions satisfying the conditions

$$a_1 X(a) + a_2 X'(a) = 0, \qquad b_1 X(b) + b_2 X'(b) = 0,$$

where a_1 and a_2 are not both zero and where the same is true of b_1 and b_2. Use Lagrange's identity, obtained in that problem, to show that

$$(X, \mathcal{L}[Y]) = (\mathcal{L}[X], Y),$$

where these inner products are on the interval $a < x < b$ with weight function unity.

(*b*) Let λ_m and λ_n denote distinct eigenvalues of a regular Sturm-Liouville problem, whose differential equation is (see Problem 2)

$$\mathcal{L}[X] + \lambda pX = 0.$$

Use the result in part (*a*) to prove that if X_m and X_n are eigenfunctions corresponding to λ_m and λ_n, then

$$(pX_m, X_n) = 0.$$

Thus show that X_m and X_n are orthogonal on the interval $a < x < b$ with weight function p, as already demonstrated in Sec. 61.

62. REAL-VALUED EIGENFUNCTIONS AND NONNEGATIVE EIGENVALUES

In the study of ordinary differential equations, problems in which all boundary data are given at one point are called *initial value problems*. We begin here by stating without proof a fundamental result from the theory of such problems.[†]

Lemma. *Let P and Q denote functions of x that are continuous on an interval $a \le x \le b$. If x_0 is a point in that interval and A and B are prescribed constants, then there is one and only one function y, which is continuous together with its derivative y' when $a \le x \le b$, that satisfies the differential equation*

$$y''(x) + P(x)y'(x) + Q(x)y(x) = 0 \qquad (a < x < b)$$

and the two initial conditions

$$y(x_0) = A, \qquad y'(x_0) = B.$$

Note that $y'' = -Py' - Qy$, and so y'' is continuous when $a \le x \le b$. Also, since any values can be assigned to the constants A and B, the general solution of the differential equation has two arbitrary constants.

Suppose now that X and Y are two eigenfunctions corresponding to the same eigenvalue λ of the regular Sturm-Liouville problem

$$(rX')' + (q + \lambda p)X = 0 \qquad (a < x < b), \tag{1}$$

$$a_1 X(a) + a_2 X'(a) = 0, \qquad b_1 X(b) + b_2 X'(b) = 0. \tag{2}$$

As stated in Sec. 59, *the functions p, q, r, and r' are continuous on the interval $a \le x \le b$; also, $p(x) > 0$ and $r(x) > 0$ when $a \le x \le b$.* The above lemma enables us to prove the following theorem, which shows that X and Y can differ by at most a constant factor and that there is always a *real-valued* eigenfunction associated with λ.

Theorem 1. *If X and Y are eigenfunctions corresponding to the same eigenvalue of a regular Sturm-Liouville problem, then*

$$Y(x) = CX(x) \qquad (a \le x \le b) \tag{3}$$

where C is a nonzero constant. Also, each eigenfunction can be made real-valued by multiplying it by an appropriate nonzero constant.

According to this theorem, a regular Sturm-Liouville problem cannot have two linearly independent eigenfunctions corresponding to the same eigenvalue. For certain modifications of regular Sturm-Liouville problems, however, it is possible to have an eigenvalue with linearly independent eigenfunctions (see Sec. 43).

[†] A proof can be found in, for instance, the book by Coddington (1989, chap. 6), which is listed in the Bibliography.

We let $X(x)$ and $Y(x)$ be as stated in the hypothesis of the theorem and start the proof by observing that in view of the principle of superposition of solutions for linear homogeneous ordinary differential equations, the linear combination

$$Z(x) = Y'(a)X(x) - X'(a)Y(x) \tag{4}$$

satisfies the differential equation

$$(rZ')' + (q + \lambda p)Z = 0 \qquad (a < x < b); \tag{5}$$

in addition, $Z'(a) = 0$. Since X and Y satisfy the conditions

$$a_1 X(a) + a_2 X'(a) = 0,$$
$$a_1 Y(a) + a_2 Y'(a) = 0,$$

where a_1 and a_2 are not both zero, and since $Z(a)$ is the determinant of this pair of linear homogeneous equations in a_1 and a_2, we also know that $Z(a) = 0$. According to the lemma at the beginning of this section, then, $Z(x) = 0$ when $a \le x \le b$. That is,

$$Y'(a)X(x) - X'(a)Y(x) = 0 \qquad (a \le x \le b). \tag{6}$$

Since eigenfunctions cannot be identically equal to zero, it is clear from relation (6) that if either of the values $X'(a)$ or $Y'(a)$ is zero, then so is the other.

Relation (3) now follows from equation (6), provided that $X'(a)$ and $Y'(a)$ are nonzero. Suppose, on the other hand, that $X'(a) = Y'(a) = 0$. Then $X(a)$ and $Y(a)$ are nonzero since, otherwise, X and Y would be identically equal to zero, according to the lemma; and zero is not an eigenfunction. The procedure that we have applied to $Z(x)$ may now be used to show that the linear combination

$$W(x) = Y(a)X(x) - X(a)Y(x) \tag{7}$$

is zero when $a \le x \le b$ and hence that relation (3) still holds.

It follows immediately from relation (3) that except possibly for a nonzero constant factor, any eigenfunction X of problem (1)–(2) is real-valued. To show this, we first recall from the corollary in Sec. 61 that the eigenvalue λ to which X corresponds must be real. So if we make the substitution $X = U + iV$, where U and V are real-valued functions, in problem (1)–(2) and separate real and imaginary parts, we find that U and V are themselves eigenfunctions corresponding to λ. Hence there is a nonzero constant β such that $V = \beta U$. Here β is real since U and V are real-valued, and we may conclude that

$$X = U + i\beta U = (1 + i\beta)U.$$

That is, X can be expressed as a nonzero constant times a real-valued function. Since

$$U = \left(\frac{1}{1 + i\beta}\right)X,$$

the final statement in Theorem 1 is now proved.

The next theorem, which uses the fact that there is always a real-valued eigenfunction corresponding to a given eigenvalue of problem (1)–(2), is an additional aid in determining eigenvalues since it often eliminates the possibility that there are negative ones. We already know from the corollary in Sec. 61 that each eigenvalue must be real.

Theorem 2. *Let* λ *be an eigenvalue of the regular Sturm-Liouville problem* (1)–(2). *If the conditions*

$$q(x) \leq 0 \; (a \leq x \leq b) \qquad \text{and} \qquad a_1 a_2 \leq 0, \; b_1 b_2 \geq 0$$

are satisfied, then $\lambda \geq 0$.

To prove this, we let X denote a real-valued eigenfunction corresponding to the eigenvalue λ. Equation (1) is thus satisfied, and we multiply each term of that equation by X and integrate each of the resulting terms from $x = a$ to $x = b$:

$$\int_a^b X(rX')' \, dx + \int_a^b qX^2 \, dx + \lambda \int_a^b pX^2 \, dx = 0. \tag{8}$$

After applying integration by parts to the first of these integrals, one can write equation (8) in the form

$$\lambda \int_a^b pX^2 \, dx = \int_a^b (-qX^2) \, dx + \int_a^b r(X')^2 \, dx + r(a)X(a)X'(a) - r(b)X(b)X'(b). \tag{9}$$

Let us now assume that the conditions stated in Theorem 2 are satisfied. Since $-q(x) \geq 0$ and $r(x) > 0$ when $a \leq x \leq b$, the values of the two integrals on the right in equation (9) are clearly nonnegative. As for the third term on the right, we note that if $a_1 = 0$ or $a_2 = 0$ in the first of conditions (2), then $X'(a) = 0$ or $X(a) = 0$, respectively. In either case, the third term is zero. If, on the other hand, neither a_1 nor a_2 is zero, then

$$r(a)X(a)X'(a) = \frac{r(a)[a_1 X(a)]^2}{-a_1 a_2} \geq 0.$$

Similarly, $-r(b)X(b)X'(b) \geq 0$; and it follows that all the terms on the right-hand side of equation (9) are nonnegative. Consequently,

$$\lambda \int_a^b p(x)[X(x)]^2 \, dx \geq 0.$$

But this integral has a positive value, and so $\lambda \geq 0$.

63. METHODS OF SOLUTION

We turn now to two examples that illustrate methods to be used in finding eigenvalues and eigenfunctions. The basic method has already been touched on in Secs. 31 and 43, where simpler Sturm-Liouville problems were solved.

EXAMPLE 1. Let us solve the regular Sturm-Liouville problem

$$X'' + \lambda X = 0 \qquad (0 < x < c), \tag{1}$$

$$X'(0) = 0, \qquad hX(c) + X'(c) = 0, \tag{2}$$

where h is a positive constant.

From Theorem 2 in Sec. 62, we know that there are no negative eigenvalues. If $\lambda = 0$, the general solution of equation (1) is $X(x) = Ax + B$ where A and B are constants; and it follows from boundary conditions (2) that $A = 0$ and $B = 0$. But eigenfunctions cannot be identically equal to zero. Consequently, the number $\lambda = 0$ is not an eigenvalue. This leaves only the possibility that $\lambda > 0$.

If $\lambda > 0$, we write $\lambda = \alpha^2$ $(\alpha > 0)$. The general solution of equation (1) this time is

$$X(x) = C_1 \cos \alpha x + C_2 \sin \alpha x.$$

It reduces to

$$X(x) = C_1 \cos \alpha x \tag{3}$$

when the first of boundary conditions (2) is applied. The second boundary condition then requires that

$$C_1(h \cos \alpha c - \alpha \sin \alpha c) = 0. \tag{4}$$

If the function (3) is to be nontrivial, the constant C_1 must be nonzero. Hence the factor in parentheses in equation (4) must be equal to zero. That is, if there is an eigenvalue $\lambda = \alpha^2$ $(\alpha > 0)$, the number α must be a positive root of the equation

$$\tan \alpha c = \frac{h}{\alpha}. \tag{5}$$

Figure 48, where the graphs of

$$y = \tan \alpha c \qquad \text{and} \qquad y = \frac{h}{\alpha}$$

are plotted, shows that equation (5) has an infinite number of positive roots

$$\alpha_1, \alpha_2, \alpha_3, \ldots,$$

where

$$\alpha_n < \alpha_{n+1} \qquad (n = 1, 2, \ldots);$$

they are the positive values of α for which those graphs intersect. The eigenvalues are, then, the numbers $\lambda_n = \alpha_n^2$ $(n = 1, 2, \ldots)$. We identify them by simply writing

$$\lambda_n = \alpha_n^2, \qquad \text{where} \qquad \tan \alpha_n c = \frac{h}{\alpha_n} \qquad (\alpha_n > 0). \tag{6}$$

Note that the dashed vertical lines in Fig. 48 are equally spaced π/c units apart. Also, as n tends to infinity, the numbers α_n tend to be the positive roots of the equation $\tan \alpha c = 0$, which are the same as the positive roots of $\sin \alpha c = 0$. In fact, the numbers $\alpha_n c$ are approximately $(n - 1)\pi$ when n is large. The first few positive roots

$$x_1, x_2, x_3, \ldots$$

of the equation

$$\tan x = \frac{a}{x} \qquad (a = hc)$$

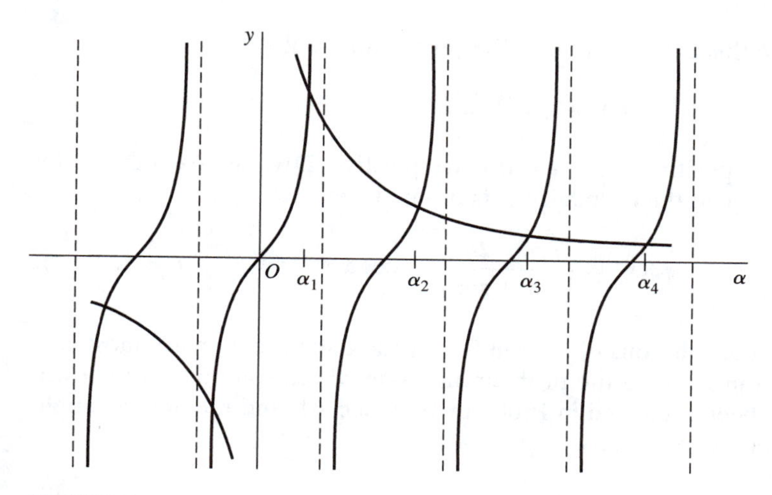

FIGURE 48

have been tabulated for various values of the constant a, and it follows from equation (5) that†

$$\alpha_1 = \frac{x_1}{c}, \qquad \alpha_2 = \frac{x_2}{c}, \qquad \alpha_3 = \frac{x_3}{c}, \qquad \ldots .$$

In view of the above remarks, the eigenvalues $\lambda_n = \alpha_n^2$ are approximately

$$\left[\frac{(n-1)\pi}{c}\right]^2$$

when n is large. This is in agreement with the statement, made earlier in Sec. 59, that if λ_n are the eigenvalues of a regular Sturm-Liouville problem, arranged in ascending order of magnitude, then it is always true that $\lambda_n \to \infty$ as $n \to \infty$.

Expression (3) now tells us that except for constant factors, the corresponding eigenfunctions are $X_n(x) = \cos \alpha_n x$ $(n = 1, 2, \ldots)$. Let us put these eigenfunctions in *normalized* form (Sec. 61), the form that we shall need in the applications. To accomplish this, we note that since the functions $X_n(x)$ are orthogonal on the interval $0 < x < c$ with weight function unity, according to the theorem in Sec. 61,

$$\|X_n\|^2 = \int_0^c \cos^2 \alpha_n x \, dx = \frac{1}{2}\int_0^c (1 + \cos 2\alpha_n x)\, dx = \frac{1}{2}\left(c + \frac{\sin 2\alpha_n c}{2\alpha_n}\right).$$

The relations

$$\sin 2\alpha_n c = 2 \sin \alpha_n c \cos \alpha_n c, \qquad \alpha_n = \frac{h}{\tan \alpha_n c}$$

†Roots of this and the related equation $\tan x = ax$, arising in some of the problems of this section, are tabulated in, for example, the handbook edited by Abramowitz and Stegun (1972, pp. 224–225), which is listed in the Bibliography.

enable us to write this expression for $\|X_n\|^2$ as the quotient

$$\|X_n\|^2 = \frac{hc + \sin^2 \alpha_n c}{2h}, \tag{7}$$

which is obviously positive since h and c are positive. Dividing each $X_n(x)$ by $\|X_n\|$, we then arrive at the normalized eigenfunctions

$$\phi_n(x) = \sqrt{\frac{2h}{hc + \sin^2 \alpha_n c}} \cos \alpha_n x \qquad (n = 1, 2, \ldots). \tag{8}$$

Sometimes the solutions of a given Sturm-Liouville problem are most easily obtained by transforming the problem into one whose solutions are known. This has already been indicated in Problem 1(a), Sec. 61, and the next example illustrates the method more fully.

EXAMPLE 2. We consider here the problem

$$(xX')' + \frac{\lambda}{x} X = 0 \qquad (1 < x < b), \tag{9}$$

$$X'(1) = 0, \qquad hX(b) + X'(b) = 0, \tag{10}$$

where h is a positive constant.

Since equation (9) can be put in the Cauchy-Euler form (see Problem 1, Sec. 38)

$$x^2X'' + xX' + \lambda X = 0,$$

the substitution $x = \exp s$ transforms it into the equation

$$\frac{d^2X}{ds^2} + \lambda X = 0 \qquad (0 < s < \ln b). \tag{11}$$

Also, since

$$\frac{dX}{dx} = \frac{dX}{ds} e^{-s},$$

the boundary conditions (10) become

$$\frac{dX}{ds} = 0 \text{ when } s = 0, \qquad (hb)X + \frac{dX}{ds} = 0 \text{ when } s = \ln b. \tag{12}$$

Hence, by referring to Example 1, we see immediately that the eigenvalues of problem (11)–(12), and therefore of problem (9)–(10), are the numbers

$$\lambda_n = \alpha_n^2, \qquad \text{where} \qquad \tan(\alpha_n \ln b) = \frac{hb}{\alpha_n} \qquad (\alpha_n > 0). \tag{13}$$

The corresponding eigenfunctions are evidently

$$X_n = \cos \alpha_n s = \cos(\alpha_n \ln x) \qquad (n = 1, 2, \ldots).$$

From equations (9) and (11), we know that the weight functions for the eigenfunctions $X_n = \cos(\alpha_n \ln x)$ and $X_n = \cos \alpha_n s$ are $1/x$ and 1, respectively. The value of the norm $\|X_n\|$ is, however, the same regardless of whether we think of X_n as a function of x or s. For the substitution $x = \exp s$ $(s = \ln x)$ shows that

$$\int_1^b \frac{1}{x}\cos^2(\alpha_n \ln x)\,dx = \int_0^{\ln b} \cos^2 \alpha_n s\,ds.$$

So, in view of expression (7), the normalized eigenfunctions of problem (9)–(10) are

$$\phi_n(x) = \sqrt{\frac{2hb}{hb\ln b + \sin^2(\alpha_n \ln b)}}\cos(\alpha_n \ln x) \qquad (n = 1, 2, \ldots). \tag{14}$$

PROBLEMS

In Problems 1 through 5, solve directly (without referring to any other problems) for the eigenvalues and normalized eigenfunctions.

1. $X'' + \lambda X = 0, \qquad X(0) = 0, \qquad X'(1) = 0.$

Answer: $\lambda_n = \alpha_n^2, \quad \phi_n(x) = \sqrt{2}\sin \alpha_n x \quad (n = 1, 2, \ldots); \quad \alpha_n = \dfrac{(2n-1)\pi}{2}.$

2. $X'' + \lambda X = 0, \qquad X(0) = 0, \qquad hX(1) + X'(1) = 0 \quad (h > 0).$

Answer: $\lambda_n = \alpha_n^2, \quad \phi_n(x) = \sqrt{\dfrac{2h}{h + \cos^2 \alpha_n}}\sin \alpha_n x \quad (n = 1, 2, \ldots);$
$\tan \alpha_n = -\dfrac{\alpha_n}{h} \quad (\alpha_n > 0).$

3. $X'' + \lambda X = 0, \qquad X'(0) = 0, \qquad X(c) = 0.$

Answer: $\lambda_n = \alpha_n^2, \quad \phi_n(x) = \sqrt{\dfrac{2}{c}}\cos \alpha_n x \quad (n = 1, 2, \ldots); \quad \alpha_n = \dfrac{(2n-1)\pi}{2c}.$

4. $X'' + \lambda X = 0, \qquad X(0) = 0, \qquad X(1) - X'(1) = 0.$

Suggestion: The trigonometric identity

$$\cos^2 A = \frac{1}{1 + \tan^2 A}$$

is useful in putting $\|X_n\|^2$ in a form that leads to the expression for $\phi_n(x)$ in the answer below.

Answer: $\lambda_0 = 0, \quad \lambda_n = \alpha_n^2, \quad \phi_0(x) = \sqrt{3}x, \quad \phi_n(x) = \dfrac{\sqrt{2(\alpha_n^2 + 1)}}{\alpha_n}\sin \alpha_n x$
$(n = 1, 2, \ldots)$; $\tan \alpha_n = \alpha_n$ $(\alpha_n > 0)$.

5. $X'' + \lambda X = 0, \qquad hX(0) - X'(0) = 0 \quad (h > 0), \qquad X(1) = 0.$

Answer: $\lambda_n = \alpha_n^2, \quad \phi_n(x) = \sqrt{\dfrac{2h}{h + \cos^2 \alpha_n}}\sin \alpha_n(1 - x) \quad (n = 1, 2, \ldots);$
$\tan \alpha_n = -\dfrac{\alpha_n}{h} \quad (\alpha_n > 0).$

6. In Problem 1(*a*), Sec. 61, the eigenvalues and eigenfunctions of the Sturm-Liouville problem

$$(xX')' + \frac{\lambda}{x}X = 0, \qquad X(1) = 0, \qquad X(b) = 0$$

were found to be

$$\lambda_n = \alpha_n^2, \qquad X_n(x) = \sin(\alpha_n \ln x) \qquad (n = 1, 2, \ldots),$$

where $\alpha_n = n\pi/\ln b$. Show that the *normalized* eigenfunctions are

$$\phi_n(x) = \sqrt{\frac{2}{\ln b}}\sin(\alpha_n \ln x) \qquad (n = 1, 2, \ldots).$$

Suggestion: The integral that arises can be evaluated by making the substitution

$$s = \pi\frac{\ln x}{\ln b}$$

and then referring to the integration formula established in Problem 9, Sec. 5.

7. Find the eigenvalues and normalized eigenfunctions of the Sturm-Liouville problem

$$X'' + \lambda X = 0, \qquad X(0) = 0, \qquad X'(c) = 0$$

by making the substitution $s = x/c$ and referring to the solutions of Problem 1.

Answer: $\lambda_n = \alpha_n^2, \quad \phi_n(x) = \sqrt{\frac{2}{c}}\sin\alpha_n x \quad (n = 1, 2, \ldots); \quad \alpha_n = \frac{(2n-1)\pi}{2c}.$

8. (*a*) Show that the solutions obtained in Problem 2 can be written

$$\lambda_n = \alpha_n^2, \qquad \phi_n(x) = \sqrt{\frac{2(\alpha_n^2 + h^2)}{\alpha_n^2 + h^2 + h}}\sin\alpha_n x \qquad (n = 1, 2, \ldots),$$

where $\alpha_n \cos\alpha_n = -h\sin\alpha_n$ $(\alpha_n > 0)$.

(*b*) By referring to the solutions of Problem 1, point out why the solutions in part (*a*) here are actually valid solutions of Problem 2 when $h \geq 0$, not just when $h > 0$.

Suggestion: The trigonometric identity in the suggestion with Problem 4 is useful in part (*a*) here as well.

9. Use the solutions obtained in Problem 3 to find the eigenvalues and normalized eigenfunctions of the Sturm-Liouville problem

$$(xX')' + \frac{\lambda}{x}X = 0, \qquad X'(1) = 0, \qquad X(b) = 0.$$

Answer:

$$\lambda_n = \alpha_n^2, \qquad \phi_n(x) = \sqrt{\frac{2}{\ln b}}\cos(\alpha_n \ln x) \quad (n = 1, 2, \ldots); \qquad \alpha_n = \frac{(2n-1)\pi}{2\ln b}.$$

10. By making an appropriate substitution and referring to the known solutions of the same problem on a different interval in the section indicated, find the eigenfunctions of the Sturm-Liouville problem

(*a*) $X'' + \lambda X = 0, \qquad X'(-\pi) = 0, \qquad X'(\pi) = 0$ (Sec. 31);

(*b*) $X'' + \lambda X = 0, \qquad X(-c) = X(c), \qquad X'(-c) = X'(c)$ (Sec. 43).

Answers: (*a*) $1, \cos\frac{n(x+\pi)}{2} \quad (n = 1, 2, \ldots);$

(*b*) $1, \cos\frac{n\pi x}{c}, \sin\frac{n\pi x}{c} \quad (n = 1, 2, \ldots).$

11. (*a*) By making the substitutions

$$X = \frac{Y}{\sqrt{x}} \qquad \text{and} \qquad \lambda = \frac{1}{4} + \mu,$$

transform the regular Sturm-Liouville problem

$$(x^2X')' + \lambda X = 0, \qquad X(1) = 0, \qquad X(b) = 0,$$

where $b > 1$, into the problem

$$(xY')' + \frac{\mu}{x}Y = 0, \qquad Y(1) = 0, \qquad Y(b) = 0.$$

(*b*) Obtain the eigenvalues and normalized eigenfunctions of the new problem in part (*a*) by referring to Problem 6. Then substitute back to show that for the original problem in part (*a*), the eigenvalues and normalized eigenfunctions are

$$\lambda_n = \frac{1}{4} + \alpha_n^2, \qquad \phi_n(x) = \sqrt{\frac{2}{x \ln b}} \sin(\alpha_n \ln x) \qquad (n = 1, 2, \ldots),$$

where $\alpha_n = n\pi / \ln b$.

12. Find the eigenfunctions of each of these Sturm-Liouville problems:

(*a*) $X'' + \lambda X = 0, \quad X(0) = 0, \quad hX(1) + X'(1) = 0 \quad (h < -1);$

(*b*) $(x^3 X')' + \lambda x X = 0, \quad X(1) = 0, \quad X(e) = 0.$

Answers: (*a*) $X_0(x) = \sinh \alpha_0 x$, where $\tanh \alpha_0 = -\dfrac{\alpha_0}{h} \quad (\alpha_0 > 0)$,

$X_n(x) = \sin \alpha_n x \quad (n = 1, 2, \ldots)$, where $\tan \alpha_n = -\dfrac{\alpha_n}{h} \quad (\alpha_n > 0)$;

(*b*) $X_n(x) = \dfrac{1}{x} \sin(n\pi \ln x) \qquad (n = 1, 2, \ldots).$

13. Give details showing that the function $W(x)$ defined by equation (7), Sec. 62, is identically equal to zero on the interval $a \le x \le b$.

64. EXAMPLES OF EIGENFUNCTION EXPANSIONS

We now illustrate how generalized Fourier series representations (Sec. 54)

$$f(x) = \sum_{n=1}^{\infty} c_n \phi_n(x) \qquad (a < x < b) \tag{1}$$

are obtained when the functions $\phi_n(x)$ $(n = 1, 2, \ldots)$ are the normalized eigenfunctions of specific Sturm-Liouville problems. We have, of course, already illustrated the method when the eigenfunctions are the ones leading to Fourier cosine and sine series on the interval $0 < x < \pi$, as well as Fourier series on $-\pi < x < \pi$ (see Sec. 55). For other sets of normalized eigenfunctions $\phi_n(x)$ $(n = 1, 2, \ldots)$, we shall use the expression

$$c_n = (f, \phi_n) = \int_a^b p(x) f(x) \phi_n(x)\, dx \qquad (n = 1, 2, \ldots) \tag{2}$$

for the coefficients c_n in expansion (1) that was found in Sec. 54 when the weight function $p(x)$ was unity.

Except for a few cases in which it is easy to establish the validity of an expansion by transforming it into a known Fourier series representation, in this book we do not treat the convergence of series (1). We merely accept the fact that results analogous to the Fourier theorem and its corollary in Sec. 12 exist when specific eigenfunctions are used. Such results are often obtained with the aid of the theory of functions of a complex variable.† Proofs are complicated by the fact that explicit solutions of the Sturm-Liouville differential equation with arbitrary coefficients cannot be written.

†The theory of eigenfunction expansions is extensively developed in the volumes by Titchmarsh (1962, 1958), listed in the Bibliography.

EXAMPLE 1. According to Problem 6, Sec. 63, the Sturm-Liouville problem

$$(xX')' + \frac{\lambda}{x}X = 0, \qquad X(1) = 0, \qquad X(b) = 0$$

has eigenvalues and normalized eigenfunctions

$$\lambda_n = \alpha_n^2, \qquad \phi_n(x) = \sqrt{\frac{2}{\ln b}}\sin(\alpha_n \ln x) \qquad (n = 1, 2, \ldots),$$

where $\alpha_n = n\pi/\ln b$. Since the orthogonality of the set $\{\phi_n(x)\}$ $(n = 1, 2, \ldots)$ is with respect to the weight function $p(x) = 1/x$, the coefficients in the expansion

$$1 = \sum_{n=1}^{\infty} c_n \phi_n(x) \qquad (1 < x < b) \tag{3}$$

are

$$c_n = (f, \phi_n) = \sqrt{\frac{2}{\ln b}}\int_1^b \frac{1}{x}\sin(\alpha_n \ln x)\,dx.$$

Making the substitution $s = \ln x$ here and noting that

$$\cos(\alpha_n \ln b) = \cos n\pi = (-1)^n,$$

we readily see that

$$\int_1^b \frac{1}{x}\sin(\alpha_n \ln x)\,dx = \int_0^{\ln b} \sin \alpha_n s\,ds = \frac{1-(-1)^n}{\alpha_n}.$$

Thus

$$c_n = \sqrt{\frac{2}{\ln b}} \cdot \frac{1-(-1)^n}{\alpha_n} \qquad (n = 1, 2, \ldots),$$

and expansion (3) becomes

$$1 = \frac{4}{\ln b}\sum_{n=1}^{\infty}\frac{\sin(\alpha_{2n-1}\ln x)}{\alpha_{2n-1}} \qquad (1 < x < b). \tag{4}$$

The validity of this representation is evident if we make the substitution

$$(2n-1)s = \alpha_{2n-1}\ln x$$

and note that $\alpha_{2n-1} = (2n-1)\pi/\ln b$. For the result,

$$1 = \frac{4}{\pi}\sum_{n=1}^{\infty}\frac{\sin(2n-1)s}{2n-1} \qquad (0 < s < \pi),$$

is a known [Problem 1(*b*), Sec. 5] Fourier sine series representation on the indicated interval.

EXAMPLE 2. The eigenvalues and normalized eigenfunctions of the Sturm-Liouville problem

$$X'' + \lambda X = 0, \qquad X(0) = 0, \qquad X'(c) = 0$$

are (Problem 7, Sec. 63)

$$\lambda_n = \alpha_n^2, \qquad \phi_n(x) = \sqrt{\frac{2}{c}} \sin \alpha_n x \qquad (n = 1, 2, \ldots),$$

where

$$\alpha_n = \frac{(2n-1)\pi}{2c}.$$

The weight function is $p(x) = 1$, and we may find the coefficients in the expansion

$$x = \sum_{n=1}^{\infty} c_n \phi_n(x) \qquad (0 < x < c)$$

by writing

$$c_n = (f, \phi_n) = \sqrt{\frac{2}{c}} \int_0^c x \sin \alpha_n x \, dx = \sqrt{\frac{2}{c}} \left[-\frac{x \cos \alpha_n x}{\alpha_n} + \frac{\sin \alpha_n x}{\alpha_n^2} \right]_0^c .$$

Since $\cos \alpha_n c = 0$ and $\sin \alpha_n c = (-1)^{n+1}$, this expression for c_n reduces to

$$c_n = \sqrt{\frac{2}{c}} \cdot \frac{(-1)^{n+1}}{\alpha_n^2} \qquad (n = 1, 2, \ldots).$$

Hence

$$x = \frac{2}{c} \sum_{n=1}^{\infty} \frac{(-1)^{n+1}}{\alpha_n^2} \sin \alpha_n x \qquad (0 < x < c). \tag{5}$$

After putting expansion (5) in the form

$$x = \frac{8c}{\pi^2} \sum_{n=1}^{\infty} \frac{(-1)^{n+1}}{(2n-1)^2} \sin \frac{(2n-1)\pi x}{2c} \qquad (0 < x < c),$$

we see from Problem 7, Sec. 14, that it is actually valid on the closed interval $-c \le x \le c$. Furthermore, since $\sin \alpha_n(x + 2c) = -\sin \alpha_n x$ $(n = 1, 2, \ldots)$, series (5) converges for all x; and if $H(x)$ denotes the sum of that series for each value of x, it is clear that $H(x)$ represents the triangular wave function defined by the equations (see Fig. 49)

$$\begin{aligned} H(x) &= x && (-c \le x \le c), \\ H(x + 2c) &= -H(x) && (-\infty < x < \infty). \end{aligned} \tag{6}$$

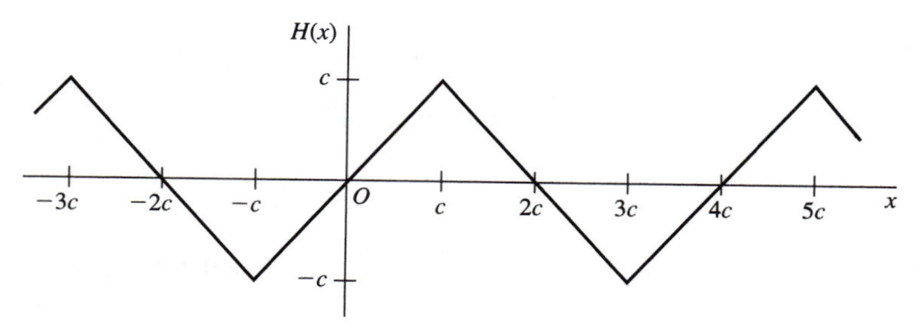

FIGURE 49

Thus $H(x)$ is an *antiperiodic* function. It is also periodic, with period $4c$, as is seen by writing

$$H(x+4c) = H(x+2c+2c) = -H(x+2c) = H(x).$$

Note, too, that

$$H(2c-x) = -H(x-2c) = -H(x+2c) = H(x).$$

We conclude with an example in which the series obtained is a sine series that cannot be transformed into an ordinary Fourier sine series. We must, therefore, accept the representation without verification.

EXAMPLE 3. We consider here the eigenvalues and normalized eigenfunctions of the Sturm-Liouville problem

$$X'' + \lambda X = 0, \qquad X(0) = 0, \qquad X(1) - X'(1) = 0.$$

According to Problem 4, Sec. 63, they are

$$\lambda_0 = 0, \qquad \lambda_n = \alpha_n^2 \qquad (n = 1, 2, \ldots)$$

and

$$\phi_0(x) = \sqrt{3}\,x, \qquad \phi_n(x) = \frac{\sqrt{2(\alpha_n^2+1)}}{\alpha_n} \sin \alpha_n x \qquad (n = 1, 2, \ldots),$$

where $\tan \alpha_n = \alpha_n$ ($\alpha_n > 0$), the weight function being unity. The coefficients in the representation

$$f(x) = c_0\,\phi_0(x) + \sum_{n=1}^{\infty} c_n\,\phi_n(x) \qquad (0 < x < 1)$$

of a piecewise smooth function $f(x)$ are

$$c_0 = (f, \phi_0) = \sqrt{3} \int_0^1 x f(x)\,dx$$

and

$$c_n = (f, \phi_n) = \frac{\sqrt{2(\alpha_n^2+1)}}{\alpha_n} \int_0^1 f(x) \sin \alpha_n x\,dx \qquad (n = 1, 2, \ldots).$$

Consequently,

$$f(x) = B_0\,x + \sum_{n=1}^{\infty} B_n \sin \alpha_n x, \tag{7}$$

where

$$B_0 = 3 \int_0^1 x f(x)\,dx \qquad \text{and} \qquad B_n = \frac{2(\alpha_n^2+1)}{\alpha_n^2} \int_0^1 f(x) \sin \alpha_n x\,dx \tag{8}$$

$$(n = 1, 2, \ldots).$$

PROBLEMS

1. Use the normalized eigenfunctions in Problem 3, Sec. 63, to derive the representation

$$1 = \frac{2}{c}\sum_{n=1}^{\infty}\frac{(-1)^{n+1}}{\alpha_n}\cos\alpha_n x \qquad (0 < x < c),$$

where

$$\alpha_n = \frac{(2n-1)\pi}{2c}.$$

2. Derive the expansion

$$1 = \frac{2}{c}\sum_{n=1}^{\infty}\frac{\sin\alpha_n x}{\alpha_n} \qquad (0 < x < c),$$

where

$$\alpha_n = \frac{(2n-1)\pi}{2c},$$

using the normalized eigenfunctions in Problem 7, Sec. 63.

3. Use the normalized eigenfunctions in Problem 2, Sec. 63, to derive the expansion

$$1 = 2h\sum_{n=1}^{\infty}\frac{1-\cos\alpha_n}{\alpha_n(h+\cos^2\alpha_n)}\sin\alpha_n x \qquad (0 < x < 1),$$

where $\tan\alpha_n = -\alpha_n/h$ $(\alpha_n > 0)$.

4. Using the normalized eigenfunctions in Problem 3, Sec. 63, when $c = \pi$, show that

$$\pi^2 - x^2 = \frac{4}{\pi}\sum_{n=1}^{\infty}\frac{(-1)^{n+1}}{\alpha_n^3}\cos\alpha_n x \qquad (0 < x < \pi),$$

where $\alpha_n = (2n-1)/2$.

5. (*a*) Use the normalized eigenfunctions in Problem 7, Sec. 63, to obtain the expansion

$$x(2c - x) = \frac{4}{c}\sum_{n=1}^{\infty}\frac{\sin\alpha_n x}{\alpha_n^3} \qquad (0 < x < c),$$

where

$$\alpha_n = \frac{(2n-1)\pi}{2c}.$$

(*b*) Show how it follows from the result in Problem 5, Sec. 8, that the series found in part (*a*) converges for all x and that its sum is the antiperiodic function (see Example 2, Sec. 64) $Q(x)$ that can be described by means of the equations

$$Q(x) = x(2c - x) \quad (0 \le x \le 2c), \qquad Q(x + 2c) = -Q(x) \quad (-\infty < x < \infty).$$

6. Using the normalized eigenfunctions in Problem 2, Sec. 63, derive the representation

$$x\left(\frac{2+h}{1+h} - x\right) = 4h\sum_{n=1}^{\infty}\frac{1-\cos\alpha_n}{\alpha_n^3(h+\cos^2\alpha_n)}\sin\alpha_n x \qquad (0 < x < 1),$$

where $\tan\alpha_n = -\alpha_n/h$ $(\alpha_n > 0)$.

Suggestion: In the simplifications, it is useful to note that

$$-h \sin \alpha_n = \alpha_n \cos \alpha_n.$$

7. Use the normalized eigenfunctions in Problem 1, Sec. 63, to show that

$$\sin \omega x = 2\omega \cos \omega \sum_{n=1}^{\infty} \frac{(-1)^n}{\omega^2 - \omega_n^2} \sin \omega_n x \qquad (0 < x < 1),$$

where

$$\omega_n = \frac{(2n-1)\pi}{2} \qquad \text{and} \qquad \omega \neq \omega_n \text{ for any value of } n.$$

Suggestion: The trigonometric identity

$$2 \sin A \sin B = \cos(A - B) - \cos(A + B)$$

is useful in evaluating the integrals that arise.

8. Find the Fourier constants c_n for the function $f(x) = x$ $(1 < x < b)$ with respect to the normalized eigenfunctions in Problem 6, Sec. 63, and reduce them to the form

$$c_n = \sqrt{2 \ln b}\, \frac{n\pi[1 + (-1)^{n+1} b]}{(\ln b)^2 + (n\pi)^2} \qquad (n = 1, 2, \ldots).$$

Suggestion: The integration formula

$$\int e^x \sin ax\, dx = \frac{e^x(\sin ax - a \cos ax)}{1 + a^2},$$

derived in calculus, is useful here.

9. Let f be a piecewise smooth function defined on the interval $1 < x < b$.

(*a*) Use the normalized eigenfunctions in Problem 6, Sec. 63, to show formally that if $\alpha_n = n\pi / \ln b$, then

$$f(x) = \sum_{n=1}^{\infty} B_n \sin(\alpha_n \ln x) \qquad (1 < x < b),$$

where

$$B_n = \frac{2}{\ln b} \int_1^b \frac{1}{x} f(x) \sin(\alpha_n \ln x)\, dx \qquad (n = 1, 2, \ldots).$$

(*b*) By making the substitution $x = \exp s$ in the series and integral in part (*a*) and then referring to Theorem 2 in Sec. 14, verify that the series representation in part (*a*) is valid for all points in the interval $1 < x < b$ at which f is continuous. (Compare with Example 1, Sec. 64.)

10. Suppose that a function f, defined on the interval $0 < x < c$, is piecewise smooth there.

(*a*) Use the normalized eigenfunctions (Problem 7, Sec. 63)

$$\phi_n(x) = \sqrt{\frac{2}{c}} \sin \alpha_n x \qquad (n = 1, 2, \ldots),$$

where

$$\alpha_n = \frac{(2n-1)\pi}{2c},$$

to show formally that

$$f(x) = \sum_{n=1}^{\infty} B_n \sin \alpha_n x \qquad (0 < x < c),$$

where

$$B_n = \frac{2}{c}\int_0^c f(x)\sin\alpha_n x\,dx \qquad (n = 1, 2, \ldots).$$

(*b*) Note that according to Problem 6, Sec. 14, the series in part (*a*) is actually a Fourier sine series for an extension of f on the interval $0 < x < 2c$. Then, with the aid of Theorem 2 in Sec. 14, state why the representation in part (*a*) is valid for each point x $(0 < x < c)$ at which f is continuous.

11. (*a*) Use the normalized eigenfunctions

$$\phi_n(x) = \sqrt{2}\sin\alpha_n x \qquad (n = 1, 2, \ldots),$$

where

$$\alpha_n = \frac{(2n-1)\pi}{2},$$

in Problem 1, Sec. 63, to show formally that

$$x\left(1 - \frac{1}{3}x^2\right) = 4\sum_{n=1}^{\infty}\frac{(-1)^{n+1}}{\alpha_n^4}\sin\alpha_n x \qquad (0 < x < 1).$$

(*b*) According to Problem 10, the series in part (*a*) here is a Fourier sine series on the interval $0 < x < 2$. With the aid of Theorem 2 in Sec. 14, show that the series just obtained in part (*a*) converges for all x and that its sum is the antiperiodic function (see Example 2, Sec. 64) $Q(x)$ that is described by the equations

$$Q(x) = x\left(1 - \frac{1}{3}x^2\right) \quad (-1 \le x \le 1), \qquad Q(x+2) = -Q(x) \quad (-\infty < x < \infty).$$

65. A TEMPERATURE PROBLEM IN RECTANGULAR COORDINATES

In this and the following section, we shall apply the Fourier method to temperature problems in rectangular coordinates where regular Sturm-Liouville problems other than those used in Chap. 5 arise. In these and the remaining sections of the chapter, we seek only *formal* solutions of our boundary value problems.

Let $u(x, t)$ denote temperatures in a slab $0 \le x \le 1$, initially at temperatures $f(x)$, when the face $x = 0$ is insulated and surface heat transfer takes place at the face $x = 1$ into a medium at temperature zero (Fig. 50). The condition on u at

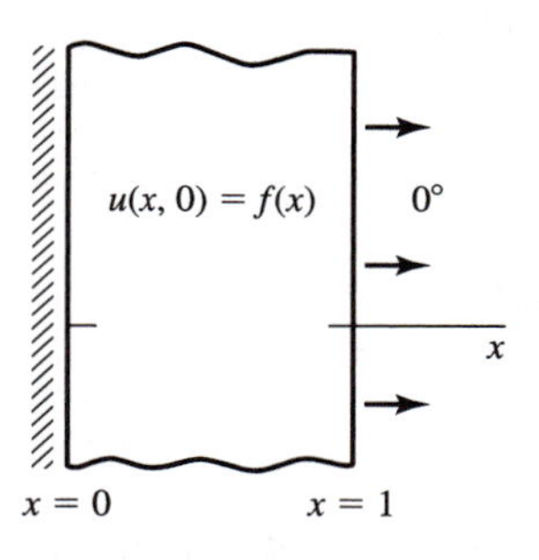

FIGURE 50

the face $x = 1$ is stated immediately below, in accordance with Newton's law of cooling (Sec. 24). The boundary value problem to be solved is evidently

$$u_t(x,t) = ku_{xx}(x,t) \qquad (0 < x < 1, t > 0), \tag{1}$$

$$u_x(0,t) = 0, \qquad u_x(1,t) = -hu(1,t), \qquad u(x,0) = f(x), \tag{2}$$

where h is a positive constant.

Writing $u = X(x)T(t)$ and separating variables, we arrive at the Sturm-Liouville problem

$$X''(x) + \lambda X(x) = 0, \qquad X'(0) = 0, \qquad hX(1) + X'(1) = 0, \tag{3}$$

along with the condition $T'(t) + \lambda kT(t) = 0$. The eigenvalues and normalized eigen functions of problem (3) are, according to Example 1, Sec. 63,

$$\lambda_n = \alpha_n^2, \qquad X_n = \phi_n(x) = \sqrt{\frac{2h}{h + \sin^2 \alpha_n}} \cos \alpha_n x \qquad (n = 1, 2, \ldots),$$

where $\tan \alpha_n = h/\alpha_n$ $(\alpha_n > 0)$. The corresponding functions of t are, moreover, constant multiples of

$$T_n(t) = \exp(-\alpha_n^2 kt) \qquad (n = 1, 2, \ldots).$$

Hence the formal solution of our temperature problem is

$$u(x,t) = \sum_{n=1}^{\infty} c_n \exp(-\alpha_n^2 kt)\, \phi_n(x), \tag{4}$$

where, in order that $u(x, 0) = f(x)$ $(0 < x < 1)$,

$$c_n = (f, \phi_n) = \int_0^1 f(x)\, \phi_n(x)\, dx = \sqrt{\frac{2h}{h + \sin^2 \alpha_n}} \int_0^1 f(x) \cos \alpha_n x\, dx \qquad (n = 1, 2, \ldots). \tag{5}$$

Observe that series (4), when the expression for $\phi_n(x)$ is substituted, is

$$u(x,t) = \sum_{n=1}^{\infty} \left(\sqrt{\frac{2h}{h + \sin^2 \alpha_n}}\, c_n \right) \exp(-\alpha_n^2 kt) \cos \alpha_n x.$$

Hence the solution just obtained can be written

$$u(x,\, t) = \sum_{n=1}^{\infty} A_n \exp(-\alpha_n^2 kt) \cos \alpha_n x, \tag{6}$$

where

$$A_n = \frac{2h}{h + \sin^2 \alpha_n} \int_0^1 f(x) \cos \alpha_n x\, dx \qquad (n = 1, 2, \ldots). \tag{7}$$

It is easy to show that solution (6), with coefficients (7), also satisfies the boundary value problem

$$u_t(x,t) = ku_{xx}(x,t) \qquad (-1 < x < 1, t > 0), \tag{8}$$

$$u_x(-1,t) = hu(-1,t), \qquad u_x(1,t) = -hu(1,t) \qquad (t > 0), \tag{9}$$

$$u(x,0) = f(x) \qquad (-1 < x < 1) \tag{10}$$

when f is an *even* function, or when $f(-x) = f(x)$ $(-1 < x < 1)$. For we already know that u satisfies the heat equation and the second of boundary conditions (9). Since the cosine function is even, it is clear from expression (6) that u is even in x; and its partial derivative u_x is odd in x. Hence the first of boundary conditions (9) is also satisfied:

$$u_x(-1,t) = -u_x(1,t) = hu(1,t) = hu(-1,t).$$

Finally, we already know that $u(x,0) = f(x)$ when $0 < x < 1$; furthermore, when $-1 < x < 0$, the fact that u and f are even in x enables us to write

$$u(x,0) = u(-x,0) = f(-x) = f(x).$$

The boundary value problem (8)–(10) is, of course, a temperature problem for a slab $-1 \leq x \leq 1$ with initial temperatures (10) and with surface heat transfer at both faces into a medium at temperature zero (Fig. 51). The solution of problem (8)–(10) when f is not necessarily even is obtained in the problems.

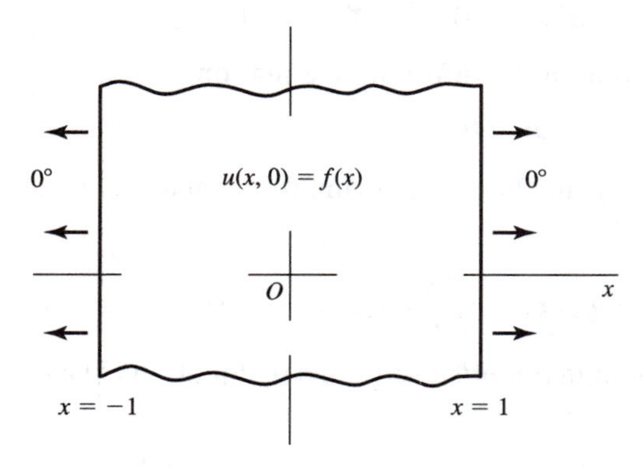

FIGURE 51

66. ANOTHER PROBLEM

Let $u(x,y)$ denote the bounded steady temperatures in a semi-infinite slab bounded by the planes $x=0$, $x=\pi$, and $y=0$ (Fig. 52), whose faces are subject to the following conditions. The face in the plane $x=0$ is insulated, the face in the plane $x=\pi$ is kept at temperature zero, and the flux inward through the face in the plane $y=0$ (see Sec. 24) is a prescribed function $f(x)$. The boundary value problem for this slab is

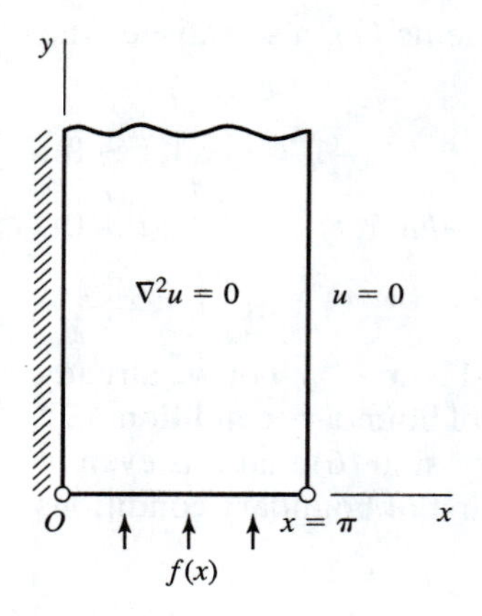

FIGURE 52

$$u_{xx}(x, y) + u_{yy}(x, y) = 0 \qquad (0 < x < \pi, y > 0), \tag{1}$$

$$u_x(0, y) = 0, \qquad u(\pi, y) = 0 \qquad (y > 0), \tag{2}$$

$$-Ku_y(x, 0) = f(x) \qquad (0 < x < \pi), \tag{3}$$

where K is a positive constant.

By assuming a product solution $u = X(x)Y(y)$ of conditions (1) and (2) and separating variables, we find that

$$X''(x) + \lambda X(x) = 0, \qquad X'(0) = 0, \qquad X(\pi) = 0 \tag{4}$$

and that $Y(y)$ is to be a bounded solution of the differential equation

$$Y''(y) - \lambda Y(y) = 0. \tag{5}$$

According to Problem 3, Sec. 63, the eigenvalues and normalized eigenfunctions of the Sturm-Liouville problem (4) are

$$\lambda_n = \alpha_n^2, \qquad X_n = \phi_n(x) = \sqrt{\frac{2}{\pi}} \cos \alpha_n x \qquad (n = 1, 2, \ldots),$$

where $\alpha_n = (2n - 1)/2$. The corresponding *bounded* solutions of equation (5) are constant multiples of the functions

$$Y_n(y) = \exp(-\alpha_n y) \qquad (n = 1, 2, \ldots).$$

Consequently,

$$u(x, y) = \sum_{n=1}^{\infty} c_n \exp(-\alpha_n y)\, \phi_n(x). \tag{6}$$

Applying the nonhomogeneous condition (3) to this expression, we see that the constants c_n must be such that

$$f(x) = \sum_{n=1}^{\infty} (Kc_n\alpha_n)\, \phi_n(x) \qquad (0 < x < \pi).$$

That is,

$$(7)\qquad Kc_n\alpha_n = (f, \phi_n) = \sqrt{\frac{2}{\pi}}\int_0^\pi f(x)\cos\alpha_n x\,dx \qquad (n = 1, 2, \ldots).$$

Finally, it follows from expressions (6) and (7) that

$$(8)\qquad u(x, y) = \frac{1}{K}\sum_{n=1}^{\infty} A_n \frac{\exp(-\alpha_n y)}{\alpha_n}\cos\alpha_n x,$$

where

$$(9)\qquad A_n = \frac{2}{\pi}\int_0^\pi f(x)\cos\alpha_n x\,dx \qquad (n = 1, 2, \ldots).$$

Since $\alpha_n = (2n-1)/2$, equations (8) and (9) can, of course, be written in the form

$$(10)\qquad u(x, y) = \frac{2}{K}\sum_{n=1}^{\infty}\frac{A_n}{2n-1}\exp\left[-\frac{(2n-1)y}{2}\right]\cos\frac{(2n-1)x}{2},$$

where

$$(11)\qquad A_n = \frac{2}{\pi}\int_0^\pi f(x)\cos\frac{(2n-1)x}{2}\,dx \qquad (n = 1, 2, \ldots).$$

PROBLEMS†

1. Show that when $f(x) = 1$ $(0 < x < 1)$ in the boundary value problem in Sec. 65, the solution (6)–(7) there reduces to

$$u(x, t) = 2h\sum_{n=1}^{\infty}\frac{\sin\alpha_n}{\alpha_n(h + \sin^2\alpha_n)}\exp\left(-\alpha_n^2 kt\right)\cos\alpha_n x,$$

where $\tan\alpha_n = h/\alpha_n$ $(\alpha_n > 0)$.

2. Use the normalized eigenfunctions of the Sturm-Liouville problem

$$X'' + \lambda X = 0, \qquad X(0) = 0, \qquad X'(\pi) = 0$$

to solve the boundary value problem

$$u_t(x, t) = ku_{xx}(x, t) \qquad (0 < x < \pi, t > 0),$$

$$u(0, t) = 0, \qquad u_x(\pi, t) = 0, \qquad u(x, 0) = f(x).$$

Show that the solution can be written

$$u(x, t) = \sum_{n=1}^{\infty} B_{2n-1}\exp\left[-\frac{(2n-1)^2k}{4}t\right]\sin\frac{(2n-1)x}{2},$$

---

†The eigenvalues and (normalized) eigenfunctions of any Sturm-Liouville problem that arises have already been found in Sec. 63 or in the problem set with that section.

where

$$B_{2n-1} = \frac{2}{\pi}\int_0^{\pi} f(x)\sin\frac{(2n-1)x}{2}\,dx \qquad (n = 1, 2, \ldots).$$

(The solution in this form was obtained in another way in Example 2, Sec. 35.)

3. Solve the boundary value problem

$$u_{xx}(x, y) + u_{yy}(x, y) = 0 \qquad (0 < x < a, 0 < y < b),$$

$$u_x(0, y) = 0, \qquad u_x(a, y) = -hu(a, y) \qquad (0 < y < b),$$

$$u(x, 0) = 0, \qquad u(x, b) = f(x) \qquad (0 < x < a),$$

where h is a positive constant, and interpret $u(x, y)$ physically.

Answer: $$u(x, y) = 2h\sum_{n=1}^{\infty}\frac{\cos\alpha_n x}{ha + \sin^2\alpha_n a}\cdot\frac{\sinh\alpha_n y}{\sinh\alpha_n b}\int_0^a f(s)\cos\alpha_n s\,ds,$$

where $\tan\alpha_n a = h/\alpha_n$ $(\alpha_n > 0)$.

4. A bounded harmonic function $u(x, y)$ in the semi-infinite strip $x > 0, 0 < y < 1$ is to satisfy the boundary conditions

$$u(x, 0) = 0, \qquad u_y(x, 1) = -hu(x, 1), \qquad u(0, y) = u_0,$$

where h $(h > 0)$ and u_0 are constants. Derive the expression

$$u(x, y) = 2hu_0\sum_{n=1}^{\infty}\frac{1 - \cos\alpha_n}{\alpha_n(h + \cos^2\alpha_n)}\exp(-\alpha_n x)\sin\alpha_n y,$$

where $\tan\alpha_n = -\alpha_n/h$ $(\alpha_n > 0)$. Interpret $u(x, y)$ physically.

5. Find the bounded harmonic function $u(x, y)$ in the semi-infinite strip $0 < x < 1, y > 0$ that satisfies the boundary conditions

$$u_x(0, y) = 0, \qquad u_x(1, y) = -hu(1, y), \qquad u(x, 0) = f(x),$$

where h is a positive constant, and interpret $u(x, y)$ physically.

Answer: $$u(x, y) = \sum_{n=1}^{\infty} A_n\exp(-\alpha_n y)\cos\alpha_n x,$$

where $\tan\alpha_n = h/\alpha_n$ $(\alpha_n > 0)$ and

$$A_n = \frac{2h}{h + \sin^2\alpha_n}\int_0^1 f(x)\cos\alpha_n x\,dx \qquad (n = 1, 2, \ldots).$$

6. Find the bounded solution of this boundary value problem, where b and h are positive constants:

$$u_{xx}(x, y) + u_{yy}(x, y) - bu(x, y) = 0 \qquad (0 < x < 1, y > 0),$$

$$u(0, y) = 0, \qquad u_x(1, y) = -hu(1, y), \qquad u(x, 0) = f(x).$$

Answer: $$u(x, y) = \sum_{n=1}^{\infty} B_n\frac{\sin\alpha_n x}{\exp\left(y\sqrt{b + \alpha_n^2}\right)},$$

where $\tan\alpha_n = -\alpha_n/h$ $(\alpha_n > 0)$ and

$$B_n = \frac{2h}{h + \cos^2\alpha_n}\int_0^1 f(x)\sin\alpha_n x\,dx \qquad (n = 1, 2, \ldots).$$

7. Give a full physical interpretation of the following temperature problem, involving a time-dependent diffusivity, and derive its solution:

$$(1+t)u_t(x,t) = u_{xx}(x,t) \qquad (0 < x < 1, t > 0),$$
$$u(0,t) = 0, \qquad u_x(1,t) = 0, \qquad u(x,0) = 1.$$

Answer: $u(x,t) = 2\sum_{n=1}^{\infty} \frac{\sin \alpha_n x}{\alpha_n}(1+t)^{-\alpha_n^2}$, where $\alpha_n = \frac{(2n-1)\pi}{2}$.

8. (*a*) Give a physical interpretation of the boundary value problem

$$u_t(x,t) = ku_{xx}(x,t) \qquad (0 < x < 1, t > 0),$$
$$u(0,t) = 0, \qquad u_x(1,t) = -hu(1,t), \qquad u(x,0) = f(x),$$

where h is a positive constant. Then derive the solution

$$u(x,t) = \sum_{n=1}^{\infty} B_n \exp\left(-\alpha_n^2 kt\right) \sin \alpha_n x,$$

where $\tan \alpha_n = -\alpha_n/h$ ($\alpha_n > 0$) and

$$B_n = \frac{2h}{h+\cos^2\alpha_n}\int_0^1 f(x)\sin \alpha_n x\, dx \qquad (n = 1, 2, \ldots).$$

(*b*) Use an argument similar to the one at the end of Sec. 65 to show that the solution found in part (*a*) formally satisfies the boundary value problem (8)–(10) in that section when the function f there is *odd*, or when

$$f(-x) = -f(x) \qquad (-1 < x < 1).$$

9. Use the following method to solve the temperature problem (see Fig. 51 in Sec. 65)

$$u_t(x,t) = ku_{xx}(x,t) \qquad (-1 < x < 1, t > 0),$$
$$u_x(-1,t) = hu(-1,t), \qquad u_x(1,t) = -hu(1,t) \qquad (t > 0),$$
$$u(x,0) = f(x) \qquad (-1 < x < 1)$$

when the function f is not necessarily even or odd, as it was in Sec. 65 and Problem 8(*b*), respectively.

(*a*) Show that if $v(x,t)$ is the solution of the problem when $f(x)$ is replaced by the function

$$G(x) = \frac{f(x)+f(-x)}{2}$$

and if $w(x,t)$ is the solution when $f(x)$ is replaced by

$$H(x) = \frac{f(x)-f(-x)}{2},$$

then the sum $u(x,t) = v(x,t) + w(x,t)$ satisfies the above boundary value problem.

(*b*) After noting that the functions G and H in part (*a*) are even and odd, respectively, apply the result there, together with results in Sec. 65 and Problem 8, to show that

$$u(x,t) = \sum_{n=1}^{\infty} A_n \exp\left(-\alpha_n^2 kt\right)\cos \alpha_n x + \sum_{n=1}^{\infty} B_n \exp\left(-\beta_n^2 kt\right)\sin \beta_n x,$$

where

$$\tan \alpha_n = h/\alpha_n \quad (\alpha_n > 0), \qquad \tan \beta_n = -\beta_n/h \quad (\beta_n > 0)$$

and

$$A_n = \frac{h}{h + \sin^2 \alpha_n} \int_{-1}^{1} f(x) \cos \alpha_n x \, dx, \qquad B_n = \frac{h}{h + \cos^2 \beta_n} \int_{-1}^{1} f(x) \sin \beta_n x \, dx.$$

Suggestion: In part (*b*), write

$$\int_0^1 G(x) \cos \alpha_n x \, dx = \frac{1}{2} \left[\int_0^1 f(x) \cos \alpha_n x \, dx + \int_0^1 f(-s) \cos \alpha_n s \, ds \right]$$

and

$$\int_0^1 H(x) \sin \beta_n x \, dx = \frac{1}{2} \left[\int_0^1 f(x) \sin \beta_n x \, dx - \int_0^1 f(-s) \sin \beta_n s \, ds \right],$$

where $G(x)$ and $H(x)$ are the functions in part (*a*). Then make the substitution $x = -s$ in the second integrals on the right-hand sides of these equations.

67. OTHER COORDINATES

To illustrate the methods of this chapter in other than rectangular coordinates, we next consider a problem involving polar coordinates and, in the problems immediately following this section, one that involves spherical coordinates.

We seek here a function $u(\rho, \phi)$ that satisfies a Dirichlet problem (Sec. 28) consisting of Laplace's equation

$$(1) \qquad \rho^2 u_{\rho\rho}(\rho, \phi) + \rho u_\rho(\rho, \phi) + u_{\phi\phi}(\rho, \phi) = 0 \qquad (1 < \rho < b, 0 < \phi < \pi)$$

and the boundary conditions (Fig. 53)

$$(2) \qquad u(\rho, 0) = 0, \qquad u(\rho, \pi) = u_0 \qquad (1 < \rho < b),$$

$$(3) \qquad u(1, \phi) = 0, \qquad u(b, \phi) = 0 \qquad (0 < \phi < \pi),$$

where u_0 is a constant.

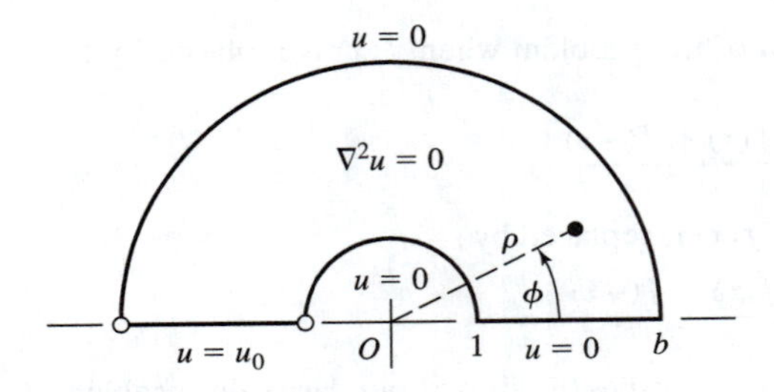

FIGURE 53

Substitution of the product $u = R(\rho)\Phi(\phi)$ into the homogeneous conditions here yields these conditions on R and Φ:

$$(4) \qquad [\rho R'(\rho)]' + \frac{\lambda}{\rho} R(\rho) = 0, \qquad R(1) = 0, \qquad R(b) = 0,$$

$$(5) \qquad \Phi''(\phi) - \lambda \Phi(\phi) = 0, \qquad \Phi(0) = 0.$$

Conditions (4) on R make up a Sturm-Liouville problem whose eigenvalues are $\lambda_n = \alpha_n^2$ $(n = 1, 2, \ldots)$, where $\alpha_n = n\pi/\ln b$, and whose normalized eigenfunctions are

$$R_n = \phi_n(\rho) = \sqrt{\frac{2}{\ln b}}\sin(\alpha_n \ln \rho) \qquad (n = 1, 2, \ldots).$$

(See Problem 6, Sec. 63.) Note that the weight function for these eigenfunctions is $1/\rho$. Except for constant factors, the corresponding functions of ϕ, arising from conditions (5), are

$$\Phi_n(\phi) = \sinh \alpha_n \phi \qquad (n = 1, 2, \ldots).$$

Hence

$$u(\rho, \phi) = \sum_{n=1}^{\infty}(c_n \sinh \alpha_n \phi)\, \phi_n(\rho). \tag{6}$$

Turning to the nonhomogeneous condition $u(\rho, \pi) = u_0$, we set $\phi = \pi$ in expression (6) and write

$$u_0 = \sum_{n=1}^{\infty}(c_n \sinh \alpha_n \pi)\, \phi_n(\rho) \qquad (1 < \rho < b).$$

Evidently, then,

$$c_n \sinh \alpha_n \pi = (u_0, \phi_n) = u_0 \sqrt{\frac{2}{\ln b}} \int_1^b \frac{1}{\rho} \sin(\alpha_n \ln \rho)\, d\rho.$$

This integral is readily evaluated by making the substitution $\rho = \exp s$; and by recalling that $\alpha_n = n\pi/\ln b$, one can simplify the result to show that

$$c_n \sinh \alpha_n \pi = \frac{u_0 \sqrt{2 \ln b}}{\pi} \cdot \frac{1 - (-1)^n}{n}. \tag{7}$$

So, in view of expressions (6) and (7),

$$u(\rho, \phi) = \frac{2u_0}{\pi} \sum_{n=1}^{\infty} \frac{1 - (-1)^n}{n} \cdot \frac{\sinh \alpha_n \phi}{\sinh \alpha_n \pi} \sin(\alpha_n \ln \rho).$$

That is,

$$u(\rho, \phi) = \frac{4u_0}{\pi} \sum_{n=1}^{\infty} \frac{\sinh \alpha_{2n-1} \phi}{\sinh \alpha_{2n-1} \pi} \cdot \frac{\sin(\alpha_{2n-1} \ln \rho)}{2n - 1}. \tag{8}$$

It is interesting to contrast this solution with the one obtained in Example 1, Sec. 38, for a Dirichlet problem involving the same region but with the nonhomogeneous condition $u = u_0$ occurring when $\rho = b$ instead of when $\phi = \pi$.

PROBLEMS†

1. Show that if the condition $u(\rho, \pi) = u_0$ $(1 < \rho < b)$ in Sec. 67 is replaced by the condition $u(\rho, \pi) = \rho$ $(1 < \rho < b)$, then

$$u(\rho, \phi) = 2\pi \sum_{n=1}^{\infty} \frac{n[1 + (-1)^{n+1} b]}{(\ln b)^2 + (n\pi)^2} \cdot \frac{\sinh \alpha_n \phi}{\sinh \alpha_n \pi} \sin(\alpha_n \ln \rho),$$

where $\alpha_n = n\pi / \ln b$.

Suggestion: The Fourier constants found in Problem 8, Sec. 64, can be used here.

2. Let ρ, ϕ, z denote cylindrical coordinates, and solve the following boundary value problem in the region $1 \le \rho \le b, 0 \le \phi \le \pi$ of the plane $z = 0$:

$$\rho^2 u_{\rho\rho}(\rho, \phi) + \rho u_\rho(\rho, \phi) + u_{\phi\phi}(\rho, \phi) = 0 \qquad (1 < \rho < b, 0 < \phi < \pi),$$

$$u_\rho(1, \phi) = 0, \qquad u_\rho(b, \phi) = -hu(b, \phi) \qquad (0 < \phi < \pi),$$

$$u_\phi(\rho, 0) = 0, \qquad u(\rho, \pi) = u_0 \qquad (1 < \rho < b),$$

where h $(h > 0)$ and u_0 are constants. Interpret the function $u(\rho, \phi)$ physically.

Answer: $$u(\rho, \phi) = 2hbu_0 \sum_{n=1}^{\infty} \frac{\cosh \alpha_n \phi}{\cosh \alpha_n \pi} \cdot \frac{\sin(\alpha_n \ln b) \cos(\alpha_n \ln \rho)}{\alpha_n [hb \ln b + \sin^2(\alpha_n \ln b)]},$$

where $\tan(\alpha_n \ln b) = hb/\alpha_n$ $(\alpha_n > 0)$.

3. The boundary $r = 1$ of a *solid sphere* is kept insulated and that solid is initially at temperatures $f(r)$, where r is the spherical coordinate described in Sec. 22. If $u(r, t)$ denotes subsequent temperatures, then

$$\frac{\partial u}{\partial t} = \frac{k}{r} \frac{\partial^2}{\partial r^2}(ru), \qquad u_r(1, t) = 0, \qquad u(r, 0) = f(r).$$

By writing $v(r, t) = r u(r, t)$ and requiring that u be continuous when $r = 0$ (compare with Example 3, Sec. 35), set up a boundary value problem in v, involving the boundary conditions

$$v(0, t) = 0, \qquad v(1, t) = v_r(1, t), \qquad v(r, 0) = rf(r).$$

Then derive the temperature formula

$$u(r, t) = B_0 + \sum_{n=1}^{\infty} B_n \frac{\sin \alpha_n r}{\alpha_n r} \exp(-\alpha_n^2 kt),$$

where $\tan \alpha_n = \alpha_n$ $(\alpha_n > 0)$ and

$$B_0 = 3 \int_0^1 r^2 f(r)\, dr, \qquad B_n = 2\left(\alpha_n + \frac{1}{\alpha_n}\right) \int_0^1 rf(r) \sin \alpha_n r\, dr \qquad (n = 1, 2, \ldots).$$

(An eigenfunction expansion similar to the one required here was found in Example 3, Sec. 64.)

---

†The footnote with the problem set for Sec. 66 applies here too.

68. A MODIFICATION OF THE METHOD

We now illustrate a certain modification of the Fourier method that can be used when generalized Fourier series arise. Another such modification is illustrated in Sec. 69. Both types of modifications were used in Chap. 5 when ordinary Fourier cosine and sine series were involved.

Assume that heat is introduced through the face $x = 1$ of a slab $0 \le x \le 1$ at a uniform rate $A\,(A > 0)$ per unit area (Sec. 24), while the face $x = 0$ is kept at the initial temperature zero of the slab. The temperature function $u(x, t)$ must satisfy the conditions

$$(1)\qquad u_t(x,t) = ku_{xx}(x,t) \qquad (0 < x < 1, t > 0),$$

$$(2)\qquad u(0,t) = 0, \qquad Ku_x(1,t) = A \qquad (t > 0),$$

$$(3)\qquad u(x,0) = 0 \qquad (0 < x < 1).$$

Because the second of conditions (2) is nonhomogeneous, we do not have two-point boundary conditions leading to a Sturm-Liouville problem. But, by writing

$$(4)\qquad u(x,t) = U(x,t) + \Phi(x)$$

(compare with Example 2, Sec. 34), we find that conditions (1)–(3) become

$$U_t(x,t) = k[U_{xx}(x,t) + \Phi''(x)],$$
$$U(0,t) + \Phi(0) = 0, \qquad K[U_x(1,t) + \Phi'(1)] = A,$$

and

$$U(x,0) + \Phi(x) = 0.$$

Hence, if we require that

$$(5)\qquad \Phi''(x) = 0 \quad \text{and} \quad \Phi(0) = 0, \qquad K\Phi'(1) = A,$$

we have the boundary value problem

$$(6)\qquad U_t(x,t) = kU_{xx}(x,t), \quad U(0,t) = 0, \quad U_x(1,t) = 0, \quad U(x,0) = -\Phi(x)$$

for $U(x, t)$ that does have two-point boundary conditions leading to a Sturm-Liouville problem.

It follows readily from conditions (5) that

$$(7)\qquad \Phi(x) = \frac{A}{K}x.$$

Also, by assuming a product solution $U = X(x)T(t)$ of the homogeneous conditions in problem (6), we see that

$$(8)\qquad X''(x) + \lambda X(x) = 0, \qquad X(0) = 0, \qquad X'(1) = 0$$

and $T'(t) + \lambda kT(t) = 0$. According to Problem 1, Sec. 63, the Sturm-Liouville problem (8) has the eigenvalues and normalized eigenfunctions

$$\lambda_n = \alpha_n^2, \qquad X_n = \phi_n(x) = \sqrt{2}\sin\alpha_n x \qquad (n = 1, 2, \ldots),$$

where $\alpha_n = (2n-1)\pi/2$; and the corresponding functions of t are

$$T_n(t) = \exp(-\alpha_n^2 kt) \qquad (n = 1, 2, \ldots).$$

Hence

$$U(x,t) = \sum_{n=1}^{\infty} c_n \exp(-\alpha_n^2 kt)\, \phi_n(x) \tag{9}$$

where, in view of the last of conditions (6),

$$-\frac{A}{K}x = \sum_{n=1}^{\infty} c_n \phi_n(x) \qquad (0 < x < 1).$$

Now the Fourier constants for x $(0 < x < 1)$ with respect to the normalized eigenfunctions here are already known to us (see Example 2, Sec. 64, when $c = 1$), and that earlier result tells us that

$$c_n = \sqrt{2}\frac{A}{K} \cdot \frac{(-1)^n}{\alpha_n^2}.$$

After substituting these values of c_n into expression (9) and then simplifying and combining the result with expression (7), as indicated in equation (4), we arrive at the desired temperature function:

$$u(x,t) = \frac{A}{K}\left[x + 2\sum_{n=1}^{\infty} \frac{(-1)^n}{\alpha_n^2} \exp(-\alpha_n^2 kt) \sin \alpha_n x\right], \tag{10}$$

where $\alpha_n = (2n-1)\pi/2$.

PROBLEMS†

1. With the aid of representation (5) in Example 2, Sec. 64, show that the temperature function (10) in Sec. 68 can be written in the form

$$u(x,t) = \frac{2A}{K} \sum_{n=1}^{\infty} \frac{(-1)^{n+1}}{\alpha_n^2} [1 - \exp(-\alpha_n^2 kt)] \sin \alpha_n x,$$

where $\alpha_n = (2n-1)\pi/2$.

2. Heat transfer takes place at the surface $x = 0$ of a slab $0 \le x \le 1$ into a medium at temperature zero, according to the linear law of surface heat transfer, so that (Sec. 24)

$$u_x(0,t) = hu(0,t) \qquad (h > 0).$$

The other boundary conditions are as indicated in Fig. 54, and the unit of time is chosen so that $k = 1$ in the heat equation. By proceeding as in Sec. 68, derive the temperature formula

$$u(x,t) = \frac{hx+1}{h+1} - 2h \sum_{n=1}^{\infty} \frac{\sin[\alpha_n(1-x)]}{\alpha_n(h + \cos^2 \alpha_n)} \exp(-\alpha_n^2 t),$$

where $\tan \alpha_n = -\alpha_n / h$ $(\alpha_n > 0)$.

† The footnote with the problem set for Sec. 66 also applies here.

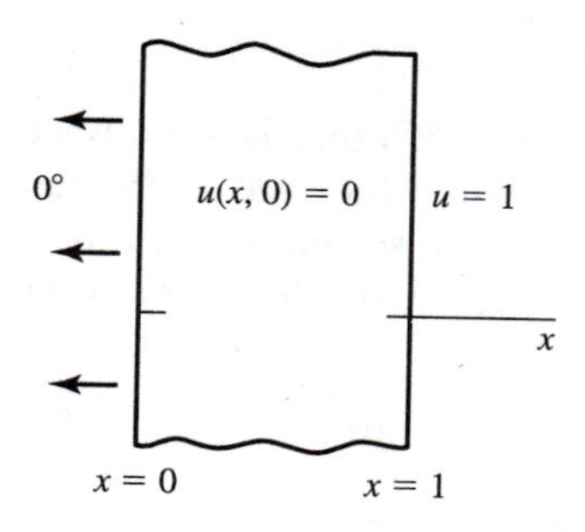

FIGURE 54

Suggestion: In simplifying the expression for the Fourier constants that arise, it is useful to note that

$$-\frac{h\sin\alpha_n}{\alpha_n^2} = \frac{\cos\alpha_n}{\alpha_n}.$$

3. Give a physical interpretation of the boundary value problem

$$u_t(x,t) = ku_{xx}(x,t) \qquad (0 < x < 1, t > 0),$$
$$u_x(0,t) = 0, \qquad u_x(1,t) = h[T - u(1,t)] \qquad (h > 0),$$
$$u(x,0) = 0,$$

where T is a constant (see Sec. 24). By making the substitution

$$u(x,t) = U(x,t) + \Phi(x)$$

and referring to the solution (6)–(7) in Sec. 65 that was found for another boundary value problem, derive the solution

$$u(x,t) = T\left[1 - 2h\sum_{n=1}^{\infty}\frac{\sin\alpha_n\cos\alpha_n x}{\alpha_n(h+\sin^2\alpha_n)}\exp\left(-\alpha_n^2 kt\right)\right],$$

where $\tan\alpha_n = h/\alpha_n$ $(\alpha_n > 0)$, of the boundary value problem here.

4. Use the same substitution as in Problem 3 and the same solution of an earlier boundary value problem to solve the temperature problem

$$u_t(x,t) = ku_{xx}(x,t) + q_0 \qquad (0 < x < 1, t > 0),$$
$$u_x(0,t) = 0, \qquad u_x(1,t) = -hu(1,t) \qquad (h > 0),$$
$$u(x,0) = 0,$$

where q_0 is a constant.

Answer: $u(x,t) = \dfrac{q_0}{2k}\left[\dfrac{2}{h} + 1 - x^2 - 4h\displaystyle\sum_{n=1}^{\infty}\frac{\sin\alpha_n\cos\alpha_n x}{\alpha_n^3(h+\sin^2\alpha_n)}\exp\left(-\alpha_n^2 kt\right)\right],$

where $\tan\alpha_n = h/\alpha_n$ $(\alpha_n > 0)$.

5. Use the method in Sec. 68 to solve the boundary value problem

$$(1+t)\,u_t(x,t) = u_{xx}(x,t) \qquad (0 < x < 1, t > 0),$$
$$u_x(0,t) = -1, \qquad u(1,t) = 0, \qquad u(x,0) = 0.$$

Interpret this problem physically (compare with Problem 7, Sec. 66).

Answer: $u(x,t) = 1 - x - 2\displaystyle\sum_{n=1}^{\infty}\frac{\cos\alpha_n x}{\alpha_n^2}(1+t)^{-\alpha_n^2}$, where $\alpha_n = \dfrac{(2n-1)\pi}{2}$.

69. ANOTHER MODIFICATION

Let $u(x, t)$ denote temperatures in a slab $0 \le x \le \pi$ (Fig. 55) that is initially at temperature zero and whose face $x = 0$ is insulated, while the face $x = \pi$ has temperatures $u(\pi, t) = t$ $(t \ge 0)$. If the unit of time is chosen so that the thermal diffusivity k in the heat equation is unity, the boundary value problem for $u(x, t)$ is

$$u_t(x, t) = u_{xx}(x, t) \qquad (0 < x < \pi, t > 0), \tag{1}$$

$$u_x(0, t) = 0, \qquad u(\pi, t) = t, \qquad u(x, 0) = 0. \tag{2}$$

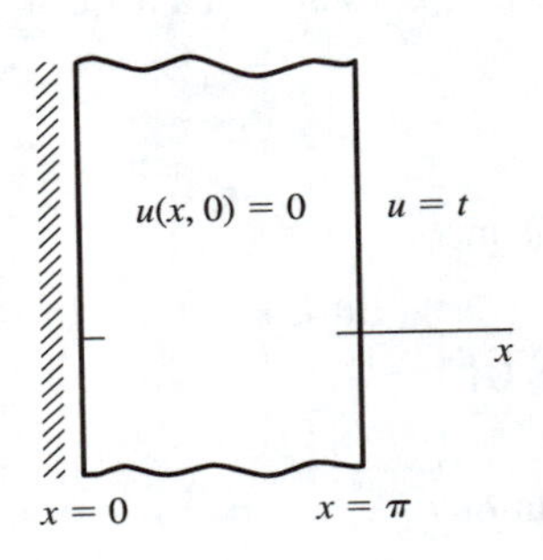

FIGURE 55

Observe that if $u(x, t)$ satisfies the first two of conditions (2), then the related function $U(x, t) = u(x, t) - t$ satisfies the conditions

$$U_x(0, t) = 0 \quad \text{and} \quad U(\pi, t) = 0, \tag{3}$$

both of which are homogeneous. In fact, by writing

$$u(x, t) = U(x, t) + t,$$

we have the related boundary value problem consisting of the differential equation

$$U_t(x, t) = U_{xx}(x, t) - 1 \tag{4}$$

and conditions (3), along with the condition

$$U(x, 0) = 0. \tag{5}$$

The nonhomogeneity in the second of conditions (2) is now transferred to the differential equation in the new boundary value problem; and this suggests applying the method of variation of parameters, first used in Sec. 36.

We begin by noting that when the method of separation of variables is applied to the homogeneous differential equation $U_t(x, t) = U_{xx}(x, t)$, which is equation (4) with the term -1 deleted, and to conditions (3), the Sturm-Liouville problem

$$X''(x) + \lambda X(x) = 0, \qquad X'(0) = 0, \qquad X(\pi) = 0$$

arises. Furthermore, from Problem 3, Sec. 63, we know that the eigenfunctions of this problem are the cosine functions

$$X_n(x) = \cos \alpha_n x \qquad \text{where} \qquad \alpha_n = \frac{2n-1}{2} \qquad (n = 1, 2, \ldots).$$

We thus seek a solution of the boundary value problem (3)–(5) having the form

$$U(x,t) = \sum_{n=1}^{\infty} A_n(t) \cos \alpha_n x. \tag{6}$$

By substituting series (6) into equation (4) and referring to Problem 1, Sec. 64, for the expansion

$$1 = \frac{2}{\pi} \sum_{n=1}^{\infty} \frac{(-1)^{n+1}}{\alpha_n} \cos \alpha_n x \qquad (0 < x < \pi),$$

we find that

$$\sum_{n=1}^{\infty} \left[A_n'(t) + \alpha_n^2 A_n(t)\right] \cos \alpha_n x = \sum_{n=1}^{\infty} \frac{2(-1)^n}{\pi \alpha_n} \cos \alpha_n x.$$

Then, by identifying the coefficients in the series on each side here, we have the differential equation

$$A_n'(t) + \alpha_n^2 A_n(t) = \frac{2(-1)^n}{\pi \alpha_n} \qquad (n = 1, 2, \ldots). \tag{7}$$

Also, condition (5) tells us that

$$\sum_{n=1}^{\infty} A_n(0) \cos \alpha_n x = 0,$$

or $A_n(0) = 0$ $(n = 1, 2, \ldots)$.

Now an integrating factor for the linear first-order differential equation (7) is

$$\exp \int \alpha_n^2 \, dt = \exp \alpha_n^2 t.$$

Hence if we multiply through the differential equation by this integrating factor, we have

$$\frac{d}{dt}\left[(\exp \alpha_n^2 t) A_n(t)\right] = \frac{2(-1)^n}{\pi \alpha_n} \exp \alpha_n^2 t.$$

By replacing the variable t here by τ, integrating the result from $\tau = 0$ to $\tau = t$, and keeping in mind the requirement that $A_n(0) = 0$, we see that

$$(\exp \alpha_n^2 t) A_n(t) = \frac{2(-1)^n}{\pi \alpha_n^3} (\exp \alpha_n^2 t - 1),$$

or

$$A_n(t) = \frac{2(-1)^n}{\pi\alpha_n^3}[1 - \exp(-\alpha_n^2 t)] \qquad (n = 1, 2, \ldots).$$

Finally, by substituting this expression for $A_n(t)$ into equation (6) and then recalling that $u(x, t) = U(x, t) + t$, we obtain the solution of the original boundary value problem:

$$(8) \qquad u(x, t) = t + \frac{2}{\pi}\sum_{n=1}^{\infty}\frac{(-1)^n}{\alpha_n^3}[1 - \exp(-\alpha_n^2 t)]\cos\alpha_n x,$$

where

$$(9) \qquad \alpha_n = \frac{2n-1}{2} \qquad (n = 1, 2, \ldots).$$

Note that in view of the representation

$$\pi^2 - x^2 = \frac{4}{\pi}\sum_{n=1}^{\infty}\frac{(-1)^{n+1}}{\alpha_n^3}\cos\alpha_n x \qquad (0 < x < \pi),$$

found in Problem 4, Sec. 64, this solution can be written

$$(10) \qquad u(x, t) = t - \frac{\pi^2 - x^2}{2} - \frac{2}{\pi}\sum_{n=1}^{\infty}\frac{(-1)^n}{\alpha_n^3}\exp(-\alpha_n^2 t)\cos\alpha_n x.$$

PROBLEMS†

1. Use the method of variation of parameters to solve the boundary value problem

$$u_t(x, t) = k u_{xx}(x, t) + q(t) \qquad (0 < x < 1, t > 0),$$

$$u_x(0, t) = 0, \qquad u(1, t) = 0, \qquad u(x, 0) = 0$$

for temperatures in an internally heated slab.

Suggestion: The representation, with $c = 1$, that was found in Problem 1, Sec. 64, is needed here.

Answer: $u(x, t) = 2\sum_{n=1}^{\infty}\frac{(-1)^{n+1}}{\alpha_n}\cos\alpha_n x\int_0^t \exp[-\alpha_n^2 k(t-\tau)]\,q(\tau)\,d\tau,$

where $\alpha_n = (2n-1)\pi/2$.

2. Solve the temperature problem

$$u_t(x, t) = u_{xx}(x, t) \qquad (0 < x < 1, t > 0),$$

$$u_x(0, t) = 0, \qquad u(1, t) = F(t), \qquad u(x, 0) = 0,$$

where F is continuous and $F(0) = 0$. [Compare with the boundary value problem (1)–(2) in Sec. 69.] Express the answer in the form

$$u(x, t) = F(t) + 2\sum_{n=1}^{\infty}\frac{(-1)^n}{\alpha_n}\cos\alpha_n x\int_0^t \exp[-\alpha_n^2(t-\tau)]\,F'(\tau)\,d\tau,$$

where $\alpha_n = (2n-1)\pi/2$.

---

† See the footnote with the problem set for Sec. 66.

3. By using the method of variation of parameters, derive the bounded solution of this problem:

$$u_{xx}(x, y) + u_{yy}(x, y) + q_0 = 0 \qquad (0 < x < 1, y > 0),$$

$$u(0, y) = 0, \qquad u_x(1, y) = -hu(1, y) \quad (h > 0), \qquad u(x, 0) = 0,$$

where q_0 and h are constants. Interpret the problem physically.

Suggestion: The representation found in Problem 3, Sec. 64, is needed here. Also, for general comments on solving the nonhomogeneous linear second-order differential equation that arises, see the suggestion with Problem 5, Sec. 37.

Answer: $u(x, y) = 2q_0 h \sum_{n=1}^{\infty} \frac{1 - \cos\alpha_n}{\alpha_n^3(h + \cos^2\alpha_n)} [1 - \exp(-\alpha_n y)] \sin\alpha_n x,$

where $\tan\alpha_n = -\alpha_n/h$ $(\alpha_n > 0)$.

4. With the aid of the representation found in Problem 6, Sec. 64, write the solution in Problem 3 above as

$$u(x, t) = \frac{q_0}{2}\left[x\left(\frac{2+h}{1+h} - x\right) - 4h \sum_{n=1}^{\infty} \frac{1 - \cos\alpha_n}{\alpha_n^3(h + \cos^2\alpha_n)} \exp(-\alpha_n y) \sin\alpha_n x\right],$$

where $\tan\alpha_n = -\alpha_n/h$ $(\alpha_n > 0)$. Then observe how it follows that

$$\lim_{y\to\infty} u(x, y) = \frac{q_0}{2} x\left(\frac{2+h}{1+h} - x\right).$$

70. A VERTICALLY HUNG ELASTIC BAR

An unstrained elastic bar, or heavy coiled spring, is clamped along its length c so as to prevent longitudinal displacements and then hung from its end $x=0$ (Fig. 56). At the instant $t=0$, the clamp is released and the bar vibrates longitudinally because of its own weight. If $y(x, t)$ denotes longitudinal displacements in the bar once it is released, then $y(x, t)$ satisfies the modified form

$$y_{tt}(x, t) = a^2 y_{xx}(x, t) + g \qquad (0 < x < c, t > 0) \tag{1}$$

of the wave equation, where g is acceleration due to gravity. The stated conditions at the ends of the bar tell us that (see Sec. 26)

$$y(0, t) = 0, \qquad y_x(c, t) = 0, \tag{2}$$

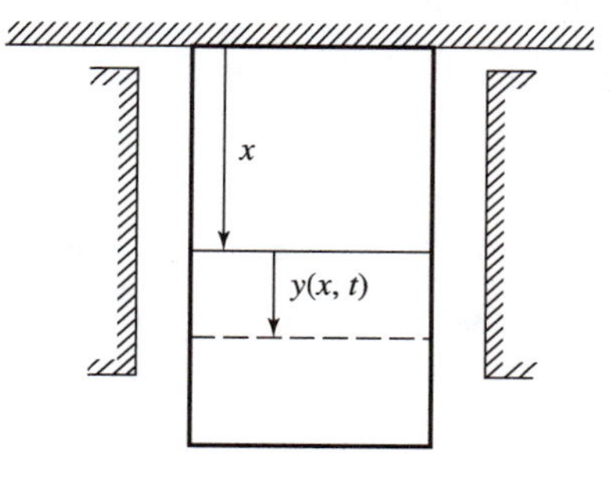

FIGURE 56

the initial conditions being

$$y(x,0) = 0, \qquad y_t(x,0) = 0. \tag{3}$$

The fact that equation (1) is nonhomogeneous suggests that we use the method of variation of parameters. More precisely, we seek a solution of our boundary value problem having the form

$$y(x,t) = \sum_{n=1}^{\infty} B_n(t) \sin \alpha_n x, \tag{4}$$

where

$$\alpha_n = \frac{(2n-1)\pi}{2c}.$$

We have chosen the sine functions $\sin \alpha_n x$ $(n = 1, 2, \ldots)$ here since they are the eigenfunctions (Problem 7, Sec. 63) of the Sturm-Liouville problem

$$X''(x) + \lambda X(x) = 0, \qquad X(0) = 0, \qquad X'(c) = 0,$$

which arises when the method of separation of variables is applied to the homogeneous wave equation $y_{xx}(x,t) = a^2 y_{tt}(x,t)$ and conditions (2). Substituting series (4) into equation (1) and recalling the representation (Problem 2, Sec. 64)

$$1 = \frac{2}{c} \sum_{n=1}^{\infty} \frac{\sin \alpha_n x}{\alpha_n} \qquad (0 < x < c),$$

we find that

$$\sum_{n=1}^{\infty} [B_n''(t) + (\alpha_n a)^2 B_n(t)] \sin \alpha_n x = \sum_{n=1}^{\infty} \frac{2g}{c\alpha_n} \sin \alpha_n x.$$

That is,

$$B_n''(t) + (\alpha_n a)^2 B_n(t) = \frac{2g}{c\alpha_n} \qquad (n = 1, 2, \ldots). \tag{5}$$

It follows, moreover, from conditions (3) that

$$B_n(0) = 0, \qquad B_n'(0) = 0 \qquad (n = 1, 2, \ldots). \tag{6}$$

Now the general solution of the complementary equation

$$B_n''(t) + (\alpha_n a)^2 B_n(t) = 0$$

is

$$B_n(t) = C_1 \cos \alpha_n a t + C_2 \sin \alpha_n a t,$$

where C_1 and C_2 are arbitrary constants, and it is easy to see that a particular solution of equation (5) is

$$B_n(t) = \frac{2g}{a^2 c \alpha_n^3}.$$

Hence the general solution of equation (5) is

$$B_n(t) = C_1 \cos \alpha_n at + C_2 \sin \alpha_n at + \frac{2g}{a^2 c \alpha_n^3}. \tag{7}$$

The constants C_1 and C_2 are readily determined by imposing conditions (6) on expression (7). The result is

$$B_n(t) = \frac{2g}{a^2 c \alpha_n^3}(1 - \cos \alpha_n at);$$

and in view of equation (4), it follows that

$$y(x, t) = \frac{2g}{a^2 c} \sum_{n=1}^{\infty} \frac{\sin \alpha_n x}{\alpha_n^3}(1 - \cos \alpha_n at). \tag{8}$$

This solution can actually be written in closed form in the following way. We first recall from Problem 5, Sec. 64, that

$$\frac{4}{c} \sum_{n=1}^{\infty} \frac{\sin \alpha_n x}{\alpha_n^3} = Q(x) \qquad (-\infty < x < \infty), \tag{9}$$

where $Q(x)$ is the antiperiodic function described by the equations

$$\begin{aligned} Q(x) &= x(2c - x) && (0 \le x \le 2c), \\ Q(x + 2c) &= -Q(x) && (-\infty < x < \infty). \end{aligned} \tag{10}$$

Thus we can put expression (8) in the form

$$y(x, t) = \frac{g}{2a^2}\left[x(2c - x) - \frac{4}{c} \sum_{n=1}^{\infty} \frac{\sin \alpha_n x \cos \alpha_n at}{\alpha_n^3}\right]. \tag{11}$$

As for the remaining series here, the trigonometric identity

$$2 \sin A \cos B = \sin(A + B) + \sin(A - B)$$

enables us to write

$$2 \sum_{n=1}^{\infty} \frac{\sin \alpha_n x \cos \alpha_n at}{\alpha_n^3} = \sum_{n=1}^{\infty} \frac{\sin \alpha_n (x + at)}{\alpha_n^3} + \sum_{n=1}^{\infty} \frac{\sin \alpha_n (x - at)}{\alpha_n^3},$$

or

$$\frac{4}{c} \sum_{n=1}^{\infty} \frac{\sin \alpha_n x \cos \alpha_n at}{\alpha_n^3} = \frac{Q(x + at) + Q(x - at)}{2}.$$

Finally, then,

$$y(x, t) = \frac{g}{2a^2}\left[x(2c - x) - \frac{Q(x + at) + Q(x - at)}{2}\right]. \tag{12}$$

PROBLEMS†

1. A horizontal elastic bar, with its end $x = 0$ kept fixed, is initially stretched so that its longitudinal displacements are $y(x, 0) = bx$ $(0 \le x \le c)$. It is released from rest in that position at the instant $t = 0$; and its end $x = c$ is kept free, so that $y_x(c, t) = 0$. Derive this expression for the displacements:
$$y(x,t) = b\left[\frac{H(x+at) + H(x-at)}{2}\right],$$
where $H(x)$ is the triangular wave function (6) in Example 2, Sec. 64. (Except for the condition at $x = 0$, the boundary value problem here is the same as the one solved in Sec. 41.)

2. Suppose that the end $x = 0$ of a horizontal elastic bar of length c is kept fixed and that a constant force F_0 per unit area acts parallel to the bar at the end $x = c$. Let all parts of the bar be initially unstrained and at rest. The displacements $y(x, t)$ then satisfy the boundary value problem
$$y_{tt}(x,t) = a^2 y_{xx}(x,t) \qquad (0 < x < c, t > 0),$$
$$y(0,t) = 0, \qquad Ey_x(c,t) = F_0,$$
$$y(x,0) = 0, \qquad y_t(x,0) = 0,$$
where $a^2 = E/\delta$, the constant E is Young's modulus of elasticity, and δ is the mass per unit volume of the material (see Sec. 26).
   (*a*) Write $y(x, t) = Y(x, t) + \Phi(x)$ (compare with Sec. 68) and determine $\Phi(x)$ such that $Y(x, t)$ satisfies a boundary value problem whose solution is obtained by simply referring to the solution in Problem 1. Thus show that
$$y(x,t) = \frac{F_0}{E}\left[x - \frac{H(x+at) + H(x-at)}{2}\right],$$
   where $H(x)$ is the same triangular wave function as in Problem 1.
   (*b*) Use the expression for $y(x, t)$ in part (*a*) to show that those displacements are periodic in t, with period
$$T_0 = \frac{4c}{a} = 4c\sqrt{\frac{\delta}{E}}.$$
   That is, show that $y(x, t + T_0) = y(x, t)$.

3. Show that the displacements at the end $x = c$ of the bar in Problem 2 are
$$y(c,t) = \frac{F_0}{E}[c + H(at - c)]$$
and that the graph of this function is as shown in Fig. 57.

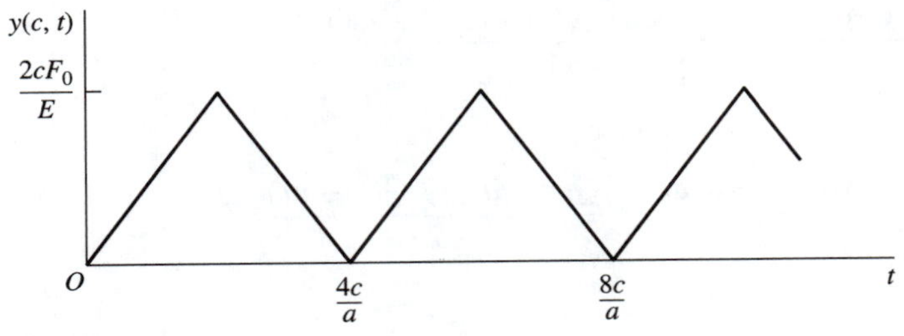

FIGURE 57

†The footnote with the problem set for Sec. 66 applies here as well.

4. Show that the force per unit area exerted by the bar in Problem 2 on the support at the end $x = 0$ is the function (see Sec. 26)

$$Ey_x(0, t) = F_0\,[1 - H'(at)]$$

and that the graph of this function is as shown in Fig. 58. (Note that this force becomes twice the applied force during regularly spaced intervals of time.)

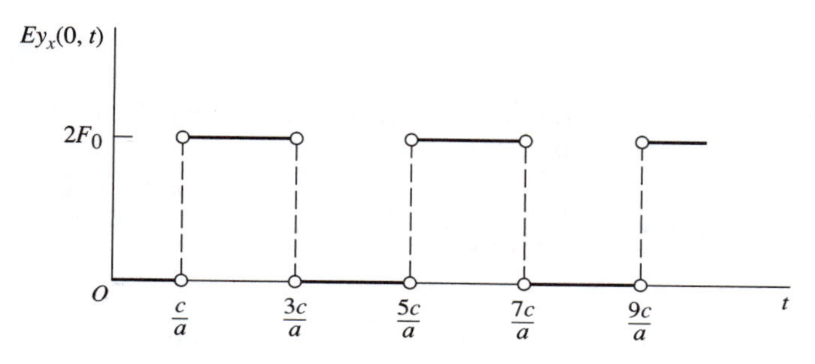

FIGURE 58

5. Let the constant F_0 in Problem 2 be replaced by a finite impulse of duration $4c/a$:

$$F(t) = \begin{cases} F_0 & \text{when } 0 < t < \dfrac{4c}{a}, \\ 0 & \text{when } \quad t > \dfrac{4c}{a}. \end{cases}$$

(*a*) State why the displacements $y(x, t)$ between the times $t = 0$ and $t = 4c/a$ are the same as in Problem 2. Then, after showing that

$$y\left(x, \frac{4c}{a}\right) = 0 \qquad \text{and} \qquad y_t\left(x, \frac{4c}{a}\right) = 0$$

when $y(x, t)$ is the solution in Problem 2, state why there is no motion in the bar here after time $t = 4c/a$.

(*b*) Use results in part (*a*) and Problem 3 to show that if

$$t_0 = \frac{2c}{a} \qquad \text{and} \qquad v_0 = \frac{aF_0}{E},$$

the end $x = c$ of the bar moves with velocity

$$y_t(c, t) = \begin{cases} v_0 & \text{when } 0 < t < t_0, \\ -v_0 & \text{when } t_0 < t < 2t_0, \end{cases}$$

and that it remains stationary after time $t = 2t_0$.

6. The end $x = 1$ of a stretched string is elastically supported (Fig. 59), so that the transverse displacements $y(x, t)$ satisfy the condition $y_x(1, t) = -hy(1, t)$ where h is a positive constant. Also,

$$y(0, t) = 0, \qquad y(x, 0) = bx, \qquad y_t(x, 0) = 0,$$

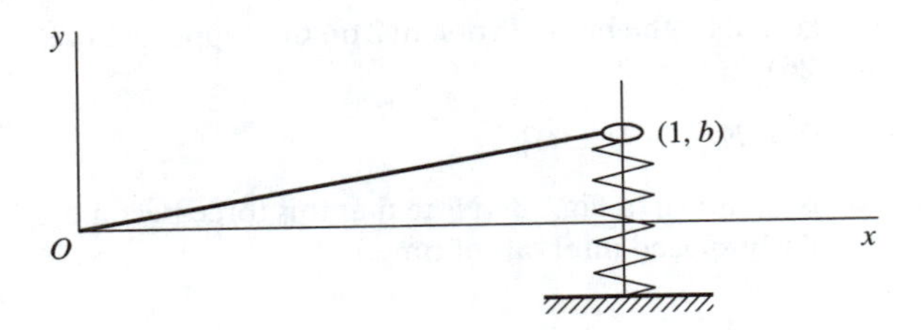

FIGURE 59

where b is a positive constant; and the wave equation $y_{tt} = y_{xx}$ is satisfied. Derive the following expression for the displacements:

$$y(x,t) = 2bh(h+1)\sum_{n=1}^{\infty} \frac{\sin\alpha_n \sin\alpha_n x}{\alpha_n^2(h+\cos^2\alpha_n)} \cos\alpha_n t,$$

where $\tan\alpha_n = -\alpha_n/h$ ($\alpha_n > 0$).

Suggestion: In simplifying the solution to the form given here, note that

$$-\frac{\cos\alpha_n}{\alpha_n} = \frac{h\sin\alpha_n}{\alpha_n^2}.$$

7. An unstrained elastic bar of length c, whose cross sections have area A and whose modulus of elasticity (Sec. 26) is E, is moving lengthwise with velocity v_0 when at the instant $t = 0$ its right-hand end $x = c$ meets and adheres to a rigid support (Fig. 60). The displacements $y(x,t)$ thus satisfy the wave equation $y_{tt} = a^2 y_{xx}$ and the end conditions $y_x(0,t) = y(c,t) = 0$, as well as the initial conditions

$$y(x,0) = 0, \qquad y_t(x,0) = v_0.$$

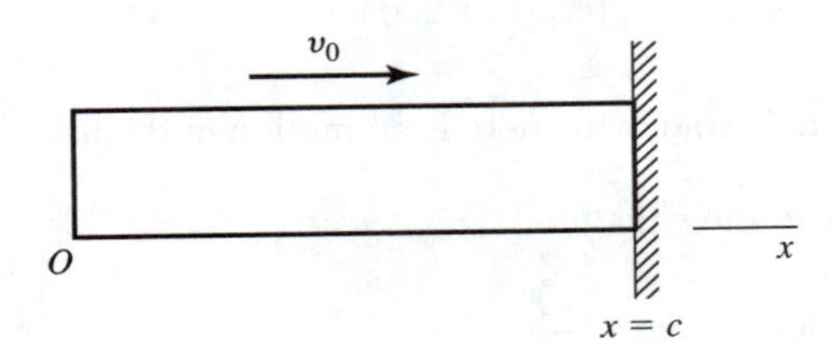

FIGURE 60

(*a*) Derive this expression for the displacements:

$$y(x,t) = \frac{2v_0}{ac}\sum_{n=1}^{\infty} \frac{(-1)^{n+1}}{\alpha_n^2} \cos\alpha_n x \sin\alpha_n at,$$

where

$$\alpha_n = \frac{(2n-1)\pi}{2c} \qquad (n = 1, 2, \ldots).$$

(*b*) Use the expression for $y(x, t)$ in part (*a*) to show that

$$y\left(x, \frac{2c}{a}\right) = 0 \qquad \text{and} \qquad y_t\left(x, \frac{2c}{a}\right) = -v_0 \qquad (0 < x < c).$$

According to these two equations, if the end $x = c$ of the bar is suddenly freed from the support at time $t = 2c/a$, the bar will move after that time as a rigid unstrained body with velocity $-v_0$.

(*c*) Show how it follows from the expression in part (*a*) that as long as the end of the bar continues to adhere to the support, the force on the support can be written

$$-AEy_x(c, t) = \frac{AEv_0}{a} M\left(\frac{2c}{a}, t\right),$$

where $M(c, t)$ $(t > 0)$ is the square wave represented by the series (see Problem 3, Sec. 14)

$$M(c, t) = \frac{4}{\pi} \sum_{n=1}^{\infty} \frac{1}{2n-1} \sin \frac{(2n-1)\pi t}{c} \qquad (t \neq c, 2c, 3c, \ldots).$$

8. Let $y(x, t)$ denote longitudinal displacements in an elastic bar of length unity whose end $x = 0$ is fixed and at whose end $x = 1$ a force proportional to t^2 acts longitudinally (Fig. 61), so that

$$y(0, t) = 0 \qquad \text{and} \qquad y_x(1, t) = At^2 \qquad (A \neq 0).$$

The bar is initially unstrained and at rest, and the unit of time is such that $a = 1$ in the wave equation.

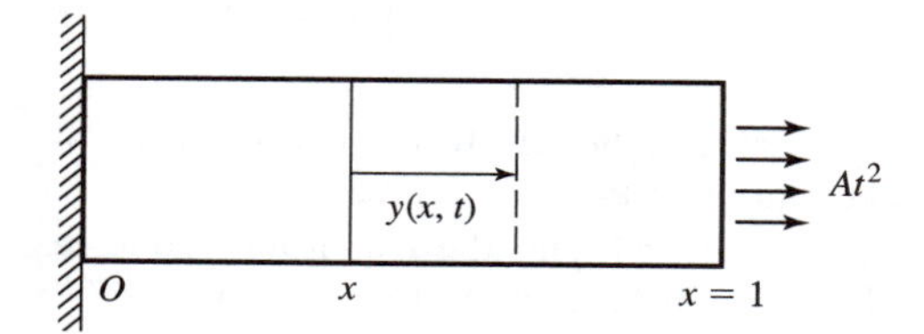

FIGURE 61

(*a*) Write out the complete boundary value problem for $y(x, t)$ and observe that if $Y(x, t) = y(x, t) - At^2x$, then

$$Y(0, t) = 0 \qquad \text{and} \qquad Y_x(1, t) = 0.$$

Set up the complete boundary value problem for $Y(x, t)$, the differential equation being

$$Y_{tt}(x, t) = Y_{xx}(x, t) - 2Ax \qquad (0 < x < 1, t > 0).$$

Then, with the aid of representation (5) in Example 2, Sec. 64, apply the method of variation of parameters to solve the boundary value problem for $Y(x, t)$ and thus derive this solution of the original problem:

$$y(x, t) = A\left[xt^2 - 4\sum_{n=1}^{\infty} \frac{(-1)^{n+1}}{\alpha_n^4}(1 - \cos \alpha_n t) \sin \alpha_n x\right],$$

where

$$\alpha_n = \frac{(2n-1)\pi}{2} \qquad (n = 1, 2, \ldots).$$

(*b*) Use the result in Problem 11(*b*), Sec. 64, to write the solution in part (*a*) here in the form

$$y(x,t) = A\left[x(t^2-1) + \frac{1}{3}x^3 + \frac{Q(x+t)+Q(x-t)}{2}\right],$$

where $Q(x)$ is the antiperiodic function described by the equations

$$Q(x) = x\left(1 - \frac{1}{3}x^2\right) \qquad (-1 \le x \le 1),$$

$$Q(x+2) = -Q(x) \qquad (-\infty < x < \infty).$$

9. Consider the same boundary value problem as in Problem 8 except that the condition at the end $x = 1$ of the bar is now replaced by the condition

$$y_x(1,t) = \sin \omega t.$$

(*a*) By proceeding in the same manner as in Problem 8, show that if

$$\omega_n = \frac{(2n-1)\pi}{2} \qquad (n = 1, 2, \ldots)$$

and $\omega \neq \omega_n$ for any value of n, then

$$y(x,t) = x \sin \omega t + 2\omega \sum_{n=1}^{\infty} \frac{(-1)^n}{\omega_n\left(\omega^2 - \omega_n^2\right)} \left(\frac{\omega}{\omega_n} \sin \omega t - \sin \omega_n t\right) \sin \omega_n x.$$

(*b*) Modify part (*a*) to show that resonance (Sec. 40) occurs when $\omega = \omega_N$ for any value of N.

Suggestion: In each part of this problem, it is helpful to refer to the general solution of a certain ordinary differential equation in Problem 2, Sec. 40.

10. By referring to expansion (5) in Example 2, Sec. 64, and the expansion found in Problem 7, Sec. 64, write the solution in Problem 9(*a*) just above in the form

$$y(x,t) = \frac{\sin \omega x \sin \omega t}{\omega \cos \omega} + 2\omega \sum_{n=1}^{\infty} \frac{(-1)^{n+1}}{\omega_n(\omega^2 - \omega_n^2)} \sin \omega_n x \sin \omega_n t,$$

where

$$\omega_n = \frac{(2n-1)\pi}{2} \qquad (n = 1, 2, \ldots).$$

CHAPTER 9

BESSEL FUNCTIONS AND APPLICATIONS

In boundary value problems that involve the laplacian $\nabla^2 u$ expressed in cylindrical or polar coordinates, the process of separating variables often produces a differential equation of the form

$$\rho^2 \frac{d^2 y}{d\rho^2} + \rho \frac{dy}{d\rho} + (\lambda\rho^2 - \nu^2)y = 0, \tag{1}$$

where y is a function of the coordinate ρ. In such a problem, $-\lambda$ is a separation constant and the values of λ are the eigenvalues of a singular Sturm-Liouville problem involving equation (1). The parameter ν is a nonnegative number determined by other aspects of the boundary value problem. Usually, ν is either zero or a positive integer.

In our applications, it turns out that $\lambda \geq 0$; and when $\lambda > 0$, the substitution $x = \sqrt{\lambda}\,\rho$ can be used to transform equation (1) into a form that is free of λ:

$$x^2 y''(x) + xy'(x) + (x^2 - \nu^2)y(x) = 0. \tag{2}$$

This differential equation is known as *Bessel's equation*. Its solutions are called *Bessel functions,* or sometimes cylindrical functions.

Equation (2) is an ordinary differential equation of the second order that is linear and homogeneous; and, upon comparing it with the standard form

$$y''(x) + P(x)y'(x) + Q(x)y(x) = 0 \tag{3}$$

of such equations, we see that

$$P(x) = \frac{1}{x} \qquad \text{and} \qquad Q(x) = \frac{x^2 - \nu^2}{x^2}.$$

Since these quotients do not have Maclaurin series representations with positive radii of convergence but the products

$$xP(x) \qquad \text{and} \qquad x^2Q(x)$$

do, the origin $x = 0$ is a *regular singular point* of Bessel's equation (2). From the theory of ordinary differential equations, it is known that when $x = 0$ is such a point, equation (3) always has a solution of the form

$$y = x^c \sum_{j=0}^{\infty} a_j x^j = \sum_{j=0}^{\infty} a_j x^{c+j} \qquad (a_0 \neq 0). \tag{4}$$

The determination of the constant c and the coefficients a_j in the case of Bessel's equation (2) is the subject of Sec. 71, where we limit our attention to the cases in which $\nu = n = 0, 1, 2, \ldots$.[†] Solutions when the nonnegative parameter ν has other values will be touched on later, but they will not arise in most of our applications.

71. BESSEL FUNCTIONS $J_n(x)$

We let n denote any fixed nonnegative integer and seek a solution of Bessel's equation

$$x^2y''(x) + xy'(x) + (x^2 - n^2)y(x) = 0 \qquad (n = 0, 1, 2, \ldots) \tag{1}$$

in the form

$$y = \sum_{j=0}^{\infty} a_j x^{c+j} \qquad (\alpha_0 \neq 0). \tag{2}$$

Assume for the moment that series (2) is differentiable and note that

$$y' = \sum_{j=0}^{\infty} (c+j)a_j x^{c+j-1}$$

and

$$y'' = \sum_{j=0}^{\infty} (c+j)(c+j-1)a_j x^{c+j-2}.$$

Substituting the function (2) and these derivatives into equation (1), we have

$$\sum_{j=0}^{\infty} (c+j)(c+j-1)a_j x^{c+j} + \sum_{j=0}^{\infty} (c+j)a_j x^{c+j}$$
$$+ \sum_{j=0}^{\infty} a_j x^{c+j+2} - \sum_{j=0}^{\infty} n^2 a_j x^{c+j} = 0,$$

[†]The series method used to solve equation (2) is often referred to as the *method of Frobenius* and is treated in introductory texts on ordinary differential equations, such as the one by Boyce and DiPrima (2005) or the one by Rainville, Bedient, and Bedient (1997). Both are listed in the Bibliography.

or

$$\sum_{j=0}^{\infty}[(c+j)(c+j-1)+(c+j)-n^2]a_j x^{c+j}+\sum_{j=0}^{\infty}a_j x^{c+j+2}=0.$$

But

$$(c+j)(c+j-1)+(c+j)=(c+j)[(c+j-1)+1]=(c+j)^2$$

in the first series here, and the second series can be written

$$\sum_{j=2}^{\infty}a_{j-2}x^{c+j}.$$

Hence

$$\sum_{j=0}^{\infty}[(c+j)^2-n^2]a_j x^{c+j}+\sum_{j=2}^{\infty}a_{j-2}x^{c+j}=0.$$

Multiplying through this equation by x^{-c} and then writing out the $j=0$ and $j=1$ terms of the first series separately, we have

$$(c^2-n^2)a_0+[(c+1)^2-n^2]a_1 x+\sum_{j=2}^{\infty}\{[(c+j)^2-n^2]a_j+a_{j-2}\}x^j=0. \tag{3}$$

Equation (3) is an identity in x if the coefficient of each power of x vanishes. In particular, the condition $a_0 \neq 0$ with series (2) tells us that $c=n$ or $c=-n$ if the constant term is to vanish, and we make the choice $c=n$. Then $a_1=0$, since

$$(n+1)^2-n^2=2n+1\neq 0.$$

Furthermore,

$$[(n+j)^2-n^2]a_j+a_{j-2}=0 \qquad (j=2,3,\ldots);$$

and since $(n+j)^2-n^2=j(2n+j)$, the *recurrence relation*

$$a_j=\frac{-1}{j(2n+j)}a_{j-2} \qquad (j=2,3,\ldots) \tag{4}$$

is obtained, giving each coefficient a_j $(j=2,3,\ldots)$ in terms of the second coefficient preceding it in the series. Note that when the nonnegative integer n is actually positive, the choice $c=-n$ does not lead to a well-defined relation of the type (4) in which the denominator on the right is never zero.

Since $a_1=0$, relation (4) requires that $a_3=0$; then $a_5=0$, etc. That is,

$$a_{2k+1}=0 \qquad (k=0,1,2,\ldots). \tag{5}$$

To obtain the remaining coefficients, we let k denote any positive integer and use relation (4) to write the following k equations:

$$a_2 = \frac{-1}{1(n+1)2^2}\, a_0,$$

$$a_4 = \frac{-1}{2(n+2)2^2}\, a_2,$$

$$\vdots$$

$$a_{2k} = \frac{-1}{k(n+k)2^2}\, a_{2k-2}.$$

Upon equating the product of the left-hand sides of these equations to the product of their right-hand sides, and then canceling the common factors $a_2, a_4, \ldots, a_{2k-2}$ on each side of the resulting equation, we arrive at the expression

$$(6) \qquad a_{2k} = \frac{(-1)^k}{k!(n+1)(n+2)\cdots(n+k)2^{2k}}\, a_0 \qquad (k = 1, 2, \ldots).$$

In view of identity (5) and since $c = n$, series (2) now takes the form

$$(7) \qquad y = a_0 x^n + \sum_{k=1}^{\infty} a_{2k} x^{n+2k},$$

where the coefficients a_{2k} $(k = 1, 2, \ldots)$ are those in expression (6). This series is absolutely convergent for all x, according to the ratio test:

$$\lim_{k\to\infty} \left| \frac{a_{2(k+1)} x^{n+2(k+1)}}{a_{2k} x^{n+2k}} \right| = \lim_{k\to\infty} \frac{1}{(k+1)(n+k+1)} \left(\frac{|x|}{2} \right)^2 = 0.$$

Hence it represents a continuous function and is differentiable with respect to x any number of times. Since it is differentiable and its coefficients satisfy the recurrence relation needed to make its sum satisfy Bessel's equation (1), series (7) is, indeed, a solution of that equation.

The coefficient a_0 in series (7) may have any nonzero value. If we substitute expression (6) into that series and write

$$y = a_0 x^n \left[1 + \sum_{k=1}^{\infty} \frac{(-1)^k}{k!(n+1)(n+2)\cdots(n+k)} \left(\frac{x}{2} \right)^{2k} \right],$$

we see that the choice

$$a_0 = \frac{1}{n!\,2^n}$$

simplifies our solution of Bessel's equation (1) to $y = J_n(x)$ where

$$(8) \qquad J_n(x) = \frac{1}{n!} \left(\frac{x}{2} \right)^n + \sum_{k=1}^{\infty} \frac{(-1)^k}{k!(n+k)!} \left(\frac{x}{2} \right)^{n+2k}$$

and where the convention $0! = 1$ is needed when $n = 0$. The function $J_n(x)$ is known as the *Bessel function of the first kind of order* n and can be written more compactly as

$$(9) \qquad J_n(x) = \sum_{k=0}^{\infty} \frac{(-1)^k}{k!(n+k)!} \left(\frac{x}{2} \right)^{n+2k} \qquad (n = 0, 1, 2, \ldots).$$

From expression (9), we note that

$$J_n(-x) = (-1)^n J_n(x) \qquad (n = 0, 1, 2, \ldots); \tag{10}$$

that is, J_n is an even function if $n = 0, 2, 4, \ldots$ but odd if $n = 1, 3, 5, \ldots$. Also, it is clear from expression (8) that $J_n(0) = 0$ when $n = 1, 2, \ldots$ but that $J_0(0) = 1$.

The case in which $n = 0$ will be of special interest to us in the applications. Bessel's equation (1) then becomes

$$xy''(x) + y'(x) + xy(x) = 0, \tag{11}$$

and expression (9) reduces to

$$J_0(x) = \sum_{k=0}^{\infty} \frac{(-1)^k}{(k!)^2} \left(\frac{x}{2}\right)^{2k}. \tag{12}$$

Since

$$(k!)^2 2^{2k} = [(1)(2)(3)\cdots(k)2^k]^2 = [(2)(4)(6)\cdots(2k)]^2 = 2^2 4^2 6^2 \cdots (2k)^2$$

when $k \geq 1$, another form is

$$J_0(x) = 1 + \sum_{k=1}^{\infty} (-1)^k \frac{x^{2k}}{2^2 4^2 6^2 \cdots (2k)^2} = 1 - \frac{x^2}{2^2} + \frac{x^4}{2^2 4^2} - \frac{x^6}{2^2 4^2 6^2} + \cdots. \tag{13}$$

Expressions (12) and (13) bear some resemblance to the Maclaurin series for the even function $\cos x$. There is also a similarity between our power series for $J_n(x)$ when $n=1$ and the Maclaurin series for the odd function $\sin x$. Similarities between the properties of those functions include, as we shall see, the differentiation formula $J_0'(x) = -J_1(x)$, corresponding to the formula for the derivative of $\cos x$. Graphs of $y = J_0(x)$ and $y = J_1(x)$ are shown in Fig. 62. More details regarding these graphs, especially the nature of the zeros of $J_0(x)$ and $J_1(x)$, will be developed later on in the chapter.

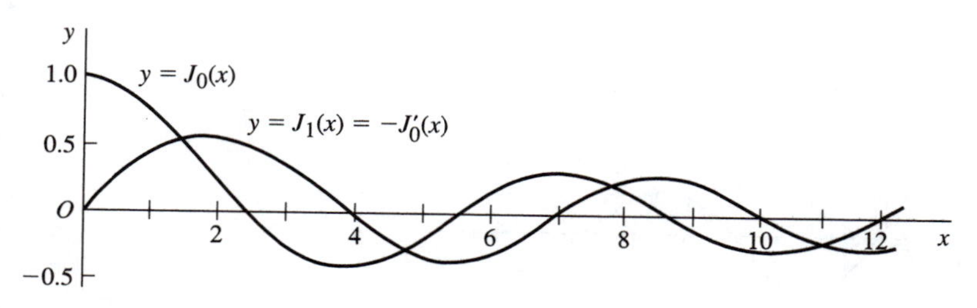

FIGURE 62

72. GENERAL SOLUTIONS OF BESSEL'S EQUATION

A function linearly independent of J_n that satisfies Bessel's equation

$$x^2 y''(x) + xy'(x) + (x^2 - n^2)y(x) = 0 \qquad (n = 0, 1, 2, \ldots) \tag{1}$$

can be obtained by various methods of a fairly elementary nature. In particular, the series procedure used in Sec. 71 can be extended so as to apply to equation (1). We do not give further details here but only state the results.

When $n = 0$, the general solution is found to be

$$y = AJ_0(x) + B\left[J_0(x)\ln x + \frac{x^2}{2^2} - \frac{x^4}{2^2 4^2}\left(1 + \frac{1}{2}\right) + \frac{x^6}{2^2 4^2 6^2}\left(1 + \frac{1}{2} + \frac{1}{3}\right) - \cdots\right], \tag{2}$$

where A and B are arbitrary constants and $x > 0$. Observe that as long as $B \neq 0$, any choice of A and B yields a solution which is unbounded as x tends to zero through positive values. Such a solution cannot, therefore, be expressed as a constant times $J_0(x)$, which tends to unity as x tends to zero. So $J_0(x)$ and the solution (2) are linearly independent when $B \neq 0$. It is most common to use *Euler's constant* $\gamma = 0.5772\cdots$, which is defined as the limit of the sequence

$$s_n = 1 + \frac{1}{2} + \frac{1}{3} + \cdots + \frac{1}{n} - \ln n \qquad (n = 1, 2, \ldots), \tag{3}$$

and to write

$$A = \frac{2}{\pi}(\gamma - \ln 2) \qquad \text{and} \qquad B = \frac{2}{\pi}.$$

When A and B are assigned those values, the second solution that arises is *Weber's Bessel function of the second kind of order zero.*[†]

$$Y_0(x) = \frac{2}{\pi}\left[\left(\ln\frac{x}{2} + \gamma\right)J_0(x) + \frac{x^2}{2^2} - \frac{x^4}{2^2 4^2}\left(1 + \frac{1}{2}\right) + \frac{x^6}{2^2 4^2 6^2}\left(1 + \frac{1}{2} + \frac{1}{3}\right) - \cdots\right]. \tag{4}$$

More generally, when n has any one of the values $n = 0, 1, 2, \ldots$, equation (1) has a solution $Y_n(x)$ that is valid when $x > 0$ and is unbounded as x tends to zero. (See Fig. 63, where $n = 0$.) Since $J_n(x)$ is continuous at $x = 0$, then, $J_n(x)$ and

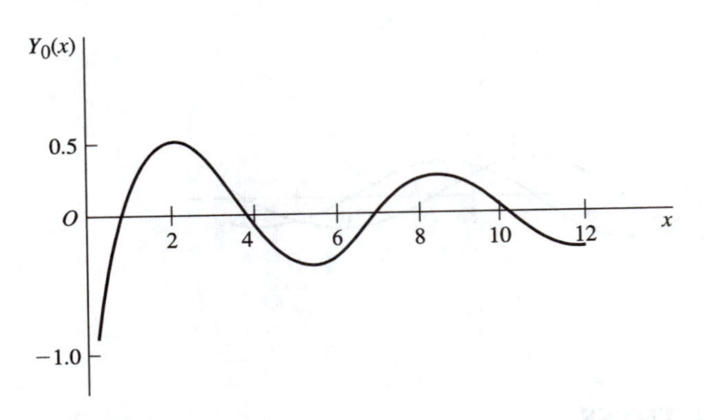

FIGURE 63

[†]There are other Bessel functions, and the notation varies widely throughout the literature. The treatise by Watson (1995) that is listed in the Bibliography is, however, usually regarded as the standard reference.

$Y_n(x)$ are linearly independent; and when $x > 0$, the general solution of equation (1) can be written

$$y = C_1 J_n(x) + C_2 Y_n(x) \qquad (n = 0, 1, 2, \ldots), \tag{5}$$

where C_1 and C_2 are arbitrary constants. The theory of the second solution $Y_n(x)$ is considerably more involved than that of $J_n(x)$, and we shall limit our applications to problems in which it is only necessary to know that $Y_n(x)$ is discontinuous at $x = 0$.

To write the general solution of Bessel's equation

$$x^2 y''(x) + xy'(x) + (x^2 - \nu^2)y(x) = 0 \qquad (\nu > 0;\ \nu \neq 1, 2, \ldots), \tag{6}$$

where ν is any positive number other than $1, 2, \ldots$, we need some elementary properties of the *gamma function*, defined when $\nu > 0$ by means of the equation[†]

$$\Gamma(\nu) = \int_0^\infty e^{-t} t^{\nu-1}\, dt \qquad (\nu > 0). \tag{7}$$

An integration by parts shows that

$$\Gamma(\nu + 1) = \int_0^\infty t^\nu e^{-t} dt = \nu \int_0^\infty e^{-t} t^{\nu-1} dt.$$

That is,

$$\Gamma(\nu + 1) = \nu\Gamma(\nu) \tag{8}$$

when $\nu > 0$. Property (8), in the form

$$\Gamma(\nu) = \frac{\Gamma(\nu + 1)}{\nu}, \tag{9}$$

is often used to define $\Gamma(\nu)$ when ν is negative but not an integer. To be specific, if $-1 < \nu < 0$, the inequalities $0 < \nu + 1 < 1$ enable us to use equation (9) to define $\Gamma(\nu)$ when $-1 < \nu < 0$. If $-2 < \nu < -1$, so that $-1 < \nu + 1 < 0$, the fact that $\Gamma(\nu)$ is now defined for $-1 < \nu < 0$ allows us to use equation (9) once again, this time to define $\Gamma(\nu)$ when $-2 < \nu < -1$. Continuing in this way, we now have $\Gamma(\nu)$ defined for every negative value of ν that is not an integer (Fig. 64).

We find from equation (7) that $\Gamma(1) = 1$. Also, it can be shown that $\Gamma(\nu)$ is continuous and positive when $\nu > 0$. So it follows from relation (9) that $\Gamma(0+) = \infty$ and, furthermore, that $|\Gamma(\nu)|$ becomes infinite as $\nu \to -n$ $(n = 0, 1, 2, \ldots)$. This means that $1/\Gamma(\nu)$ tends to zero as ν tends to $-n$ $(n = 0, 1, 2, \ldots)$; and, for brevity, we write

$$\frac{1}{\Gamma(-n)} = 0 \qquad \text{when } n = 0, 1, 2, \ldots.$$

Note that the reciprocal $1/\Gamma(\nu)$ is then continuous for all ν.

When its argument is a positive integer, the gamma function becomes a factorial:

$$\Gamma(n + 1) = n! \qquad (n = 0, 1, 2, \ldots). \tag{10}$$

[†]Thorough developments of the gamma function appear in the books by Lebedev (1972, chap. 1) and Rainville (1972, chap. 2), which are listed in the Bibliography.

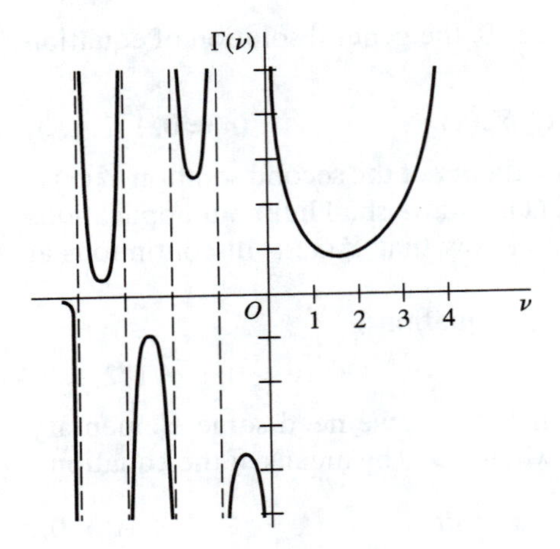

FIGURE 64

The verification of property (10) and the derivation of the property

$$\Gamma\left(\frac{1}{2}\right) = \sqrt{\pi} \tag{11}$$

are left to the problems.

The *Bessel function of the first kind of order* ν ($\nu > 0$) is defined as

$$J_\nu(x) = \sum_{k=0}^{\infty} \frac{(-1)^k}{k!\Gamma(\nu + k + 1)} \left(\frac{x}{2}\right)^{\nu+2k}. \tag{12}$$

When $\nu = n = 1, 2, \ldots$, this reduces to the expression

$$J_n(x) = \sum_{k=0}^{\infty} \frac{(-1)^k}{k!(n + k)!} \left(\frac{x}{2}\right)^{n+2k} \qquad (n = 1, 2, \ldots)$$

in Sec. 71. The Bessel function $J_{-\nu}(x)$ ($\nu > 0$) is also well-defined when ν is replaced by $-\nu$ in equation (12):

$$J_{-\nu}(x) = \sum_{k=0}^{\infty} \frac{(-1)^k}{k!\Gamma(-\nu + k + 1)} \left(\frac{x}{2}\right)^{-\nu+2k}. \tag{13}$$

In particular, if $\nu = n = 1, 2, \ldots$, expression (13) becomes

$$J_{-n}(x) = \sum_{k=0}^{\infty} \frac{(-1)^k}{k!\Gamma(-n + k + 1)} \left(\frac{x}{2}\right)^{-n+2k} = \sum_{k=n}^{\infty} \frac{(-1)^k}{k!\Gamma(-n + k + 1)} \left(\frac{x}{2}\right)^{-n+2k},$$

since

$$\frac{1}{\Gamma(-n + k + 1)} = 0 \qquad \text{when } 0 \le k \le n - 1.$$

Replacing k by $k+n$ in this last series, so that k runs from zero to infinity again, we have

$$J_{-n}(x) = \sum_{k=0}^{\infty} \frac{(-1)^{k+n}}{(k+n)!\,\Gamma(k+1)} \left(\frac{x}{2}\right)^{n+2k} = (-1)^n \sum_{k=0}^{\infty} \frac{(-1)^k}{k!\,(n+k)!} \left(\frac{x}{2}\right)^{n+2k}.$$

That is,

$$J_{-n}(x) = (-1)^n J_n(x) \qquad (n = 1, 2, \ldots). \tag{14}$$

It is not difficult to verify by direct substitution that when $\nu > 0$ and ν is not an integer, $J_\nu(x)$ and $J_{-\nu}(x)$ are solutions of equation (6). Those solutions are arrived at by a modification, involving property (8) of the gamma function, of the procedure used in Sec. 71. The Bessel function $J_{-\nu}(x)$ is the product of $1/x^\nu$ and a power series in x whose initial term $(k=0)$ is nonzero; hence $J_{-\nu}(x)$ is unbounded as $x \to 0$. Since $J_\nu(x)$ tends to zero as $x \to 0$ for those values of ν, it is evident that $J_\nu(x)$ and $J_{-\nu}(x)$ are linearly independent functions. The general solution of Bessel's equation (6) is therefore

$$y = C_1 J_\nu(x) + C_2 J_{-\nu}(x) \qquad (\nu > 0;\ \nu \neq 1, 2, \ldots), \tag{15}$$

where C_1 and C_2 are arbitrary constants.

Because of relation (14), $J_n(x)$ and $J_{-n}(x)$ are linearly dependent; and so equation (15) does *not* provide the general solution of equation (6) when ν is a positive integer. For those values of ν the general solution is given by equation (5).

PROBLEMS

1. Let y be any solution of Bessel's equation $(xy')' + xy = 0$ of order zero, and let $\mathcal{L}$ denote the differential operator defined by means of the equation

$$\mathcal{L}[X] = (xX')' + xX.$$

(*a*) By writing $X = J_0$ and $Y = y$ in Lagrange's identity (Problem 2, Sec. 61)

$$X\mathcal{L}[Y] - Y\mathcal{L}[X] = \frac{d}{dx}\,[x(XY' - YX')]$$

for that operator and observing that $\mathcal{L}[X] = 0$ and $\mathcal{L}[Y] = 0$, show that there is a constant B such that

$$\frac{d}{dx}\left[\frac{y(x)}{J_0(x)}\right] = \frac{B}{x[J_0(x)]^2}.$$

(*b*) Assuming that the function $1/[J_0(x)]^2$ has a Maclaurin series expansion of the form†

$$\frac{1}{[J_0(x)]^2} = 1 + \sum_{k=1}^{\infty} c_k x^{2k}$$

†This valid assumption is easily justified by methods from the theory of functions of a complex variable. See the authors' book (2004, chap. 5), listed in the Bibliography.

and that the expansion obtained by multiplying each side of this by $1/x$ can be integrated term by term, use the result in part (*a*) to show formally that y can be written in the form

$$y = AJ_0(x) + B\left[J_0(x)\ln x + \sum_{k=1}^{\infty} d_k x^{2k}\right],$$

where A, B, and d_k ($k = 1, 2, \ldots$) are constants. [See equation (2), Sec. 72.]

2. According to Problem 1, Bessel's equation $xy'' + y' + xy = 0$ has, in addition to the solution $y_1 = J_0(x)$, a linearly independent solution of the form

$$y_2 = y_1 \ln x + \sum_{k=1}^{\infty} d_k x^{2k}.$$

(*a*) By substituting this expression for y_2 into Bessel's equation, noting that

$$xy_1'' + y_1' + xy_1 = 0 \quad \text{and} \quad y_1' = -J_1(x) = \sum_{k=0}^{\infty} \frac{(-1)^{k+1}}{k!(k+1)!}\left(\frac{x}{2}\right)^{2k+1},$$

and identifying the coefficients of x^{2k+1} ($k = 0, 1, 2, \ldots$) in the result, show that

$$d_1 = \frac{1}{2^2}$$

and

$$d_{k+1} = \frac{(-1)^k}{2^{2(k+1)}[(k+1)!]^2}\left[(-1)^{k+1}2^{2k}(k!)^2 d_k + \frac{1}{k+1}\right] \qquad (k = 1, 2, \ldots).$$

(*b*) Use the final result in part (*a*) and the value of d_1 found there to write expressions for d_2, d_3, and d_4, which suggest that

$$d_k = \frac{(-1)^{k+1}}{2^{2k}(k!)^2}\left(1 + \frac{1}{2} + \frac{1}{3} + \cdots + \frac{1}{k}\right) \qquad (k = 1, 2, \ldots).$$

Then use mathematical induction to show that this expression for the coefficients d_k is indeed correct. Note that it can also be written [compare with equation (2), Sec. 72]

$$d_k = \frac{(-1)^{k+1}}{2^2 4^2 6^2 \cdots (2k)^2}\left(1 + \frac{1}{2} + \frac{1}{3} + \cdots + \frac{1}{k}\right) \qquad (k = 1, 2, \ldots).$$

3. Let s_n ($n = 1, 2, \ldots$) be the sequence defined in equation (3), Sec. 72. Show that $s_n > 0$ and $s_n - s_{n+1} > 0$ for each n. Thus show that the sequence is bounded and decreasing and hence that it converges to some number γ. Also, point out why $0 \leq \gamma < 1$.

Suggestion: Observe from the graph of the function $y = 1/x$ that

$$\sum_{k=1}^{n-1} \frac{1}{k} > \int_1^n \frac{dx}{x} = \ln n \qquad (n \geq 2)$$

and

$$\frac{1}{n+1} < \int_n^{n+1} \frac{dx}{x} = \ln(n+1) - \ln n \qquad (n \geq 1).$$

4. (*a*) Derive the property $\Gamma(\nu + 1) = \nu\Gamma(\nu)$ of the gamma function, as stated in Sec. 72.

(*b*) Show that $\Gamma(1) = 1$ and, using mathematical induction, verify that $\Gamma(n + 1) = n!$ when $n = 0, 1, 2, \ldots$.

5. Verify that the function $J_\nu(x)$ $(\nu > 0)$, defined by equation (12), Sec. 72, satisfies Bessel's equation (6) in that section. Point out how it follows that $J_{-\nu}(x)$ is also a solution.

6. Refer to the result obtained in Problem 16, Sec. 50, and show that

$$\Gamma\left(\frac{1}{2}\right) = 2\int_0^\infty e^{-s^2}\,ds = \sqrt{\pi}.$$

7. With the aid of mathematical induction, verify that

$$\Gamma\left(k+\frac{1}{2}\right) = \frac{(2k)!}{k!\,2^{2k}}\sqrt{\pi} \qquad (k = 0, 1, 2, \ldots).$$

8. Use the series representations (12) and (13), Sec. 72, and the identity in Problem 7 to show that

$$(a)\ J_{1/2}(x) = \sqrt{\frac{2}{\pi x}}\sin x; \qquad (b)\ J_{-1/2}(x) = \sqrt{\frac{2}{\pi x}}\cos x.$$

9. Show that if y is a differentiable function of x and if $s = \alpha x$, where α is a nonzero constant, then

$$\frac{dy}{dx} = \alpha\frac{dy}{ds} \quad \text{and} \quad \frac{d^2y}{dx^2} = \alpha^2\frac{d^2y}{ds^2}.$$

Thus show that the substitution $s = \alpha x$ transforms the differential equation

$$x^2\frac{d^2y}{dx^2} + x\frac{dy}{dx} + (\alpha^2x^2 - n^2)y = 0 \qquad (n = 0, 1, 2, \ldots)$$

into Bessel's equation

$$s^2\frac{d^2y}{ds^2} + s\frac{dy}{ds} + (s^2 - n^2)y = 0 \qquad (n = 0, 1, 2, \ldots),$$

which is free of α. Conclude that the general solution of the first differential equation here is

$$y = C_1\,J_n(\alpha x) + C_2\,Y_n(\alpha x).$$

10. The function $I_n(x) = i^{-n}J_n(ix)$, where $i = \sqrt{-1}$, is called the *modified Bessel function* of the first kind of order n.

(*a*) Use the series representation (Sec. 71)

$$J_n(x) = \sum_{k=0}^{\infty}\frac{(-1)^k}{k!(n+k)!}\left(\frac{x}{2}\right)^{n+2k} \qquad (n = 0, 1, 2, \ldots)$$

to show that

$$I_n(x) = \sum_{k=0}^{\infty}\frac{1}{k!(n+k)!}\left(\frac{x}{2}\right)^{n+2k} \qquad (n = 0, 1, 2, \ldots).$$

Then, after establishing its convergence for all x, use this series to show that $I_n(x) > 0$ when $x > 0$ and that

$$I_n(-x) = (-1)^n I_n(x) \qquad (n = 0, 1, 2, \ldots).$$

(*b*) By referring to the final result in Problem 9, point out why $y = I_n(x)$ is a solution of the *modified Bessel equation*

$$x^2y''(x) + xy'(x) - (x^2 + n^2)y(x) = 0.$$

73. RECURRENCE RELATIONS

Starting with the expression (Sec. 71)

$$J_n(x) = \sum_{k=0}^{\infty} \frac{(-1)^k}{k!(n+k)!} \left(\frac{x}{2}\right)^{n+2k} \qquad (n = 0, 1, 2, \ldots),$$

write

$$\frac{d}{dx}[x^{-n} J_n(x)] = \frac{d}{dx}\left[\sum_{k=0}^{\infty} \frac{(-1)^k}{k!(n+k)!} \cdot \frac{x^{2k}}{2^{n+2k}}\right]$$

$$= \sum_{k=1}^{\infty} \frac{(-1)^k}{(k-1)!(n+k)!} \cdot \frac{x^{2k-1}}{2^{n+2k-1}}.$$

If we replace k by $k+1$ here, so that k runs from zero to infinity again, it follows that

$$\frac{d}{dx}[x^{-n} J_n(x)] = \sum_{k=0}^{\infty} \frac{(-1)^{k+1}}{k!(n+k+1)!} \cdot \frac{x^{2k+1}}{2^{n+2k+1}}$$

$$= -x^{-n} \sum_{k=0}^{\infty} \frac{(-1)^k}{k!(n+1+k)!} \left(\frac{x}{2}\right)^{n+1+2k},$$

or

$$\text{(1)} \qquad \frac{d}{dx}[x^{-n} J_n(x)] = -x^{-n} J_{n+1}(x) \qquad (n = 0, 1, 2, \ldots).$$

The special case

$$\text{(2)} \qquad J_0'(x) = -J_1(x)$$

was mentioned at the end of Sec. 71.

On the other hand, when $n = 1, 2, \ldots,$

$$\frac{d}{dx}[x^{n} J_n(x)] = \frac{d}{dx}\left[\sum_{k=0}^{\infty} \frac{(-1)^k}{k!(n+k)!} \cdot \frac{x^{2n+2k}}{2^{n+2k}}\right]$$

$$= \sum_{k=0}^{\infty} \frac{(-1)^k}{k!(n+k-1)!} \cdot \frac{x^{2n+2k-1}}{2^{n+2k-1}};$$

and we find that

$$\frac{d}{dx}[x^{n} J_n(x)] = x^n \sum_{k=0}^{\infty} \frac{(-1)^k}{k!(n-1+k)!} \left(\frac{x}{2}\right)^{n-1+2k},$$

or

$$\text{(3)} \qquad \frac{d}{dx}[x^{n} J_n(x)] = x^n J_{n-1}(x) \qquad (n = 1, 2, \ldots).$$

Relations (1) and (3), which are called *recurrence relations,* can sometimes be used to evaluate integrals involving Bessel functions. Relation (3) tells us, for instance, that when $n = 1, 2, \ldots,$

$$\int_0^x s^n J_{n-1}(s)\,ds = \big[s^n J_n(s)\big]_0^x .$$

That is,

$$\int_0^x s^n J_{n-1}(s)\,ds = x^n J_n(x) \qquad (n = 1, 2, \ldots). \tag{4}$$

An important special case of this is

$$\int_0^x s J_0(s)\,ds = x J_1(x). \tag{5}$$

Recurrence relations can be used (Problem 2) to derive the *reduction formula*

$$\int_0^x s^n J_0(s)\,ds = x^n J_1(x) + (n-1)x^{n-1} J_0(x) - (n-1)^2 \int_0^x s^{n-2} J_0(s)\,ds \qquad (n = 2, 3, \ldots). \tag{6}$$

Note that in view of equation (5), formula (6) can be applied successively to completely evaluate the integral on the left-hand side when the integer n is *odd.*[†]

EXAMPLE. Formula (6) can be used to show that

$$\int_0^x s^5 J_0(s)\,ds = x(x^2 - 8)^2 J_1(x) + 4x^2(x^2 - 8) J_0(x). \tag{7}$$

To be specific, one application of formula (6) yields

$$\int_0^x s^5 J_0(s)\,ds = x^5 J_1(x) + 4x^4 J_0(x) - 16 \int_0^x s^3 J_0(s)\,ds,$$

and another gives us

$$\int_0^x s^3 J_0(s)\,ds = x^3 J_1(x) + 2x^2 J_0(x) - 4 \int_0^x s J_0(s)\,ds.$$

Thus

$$\int_0^x s^5 J_0(s)\,ds = (x^5 - 16x^3) J_1(x) + (4x^4 - 32x^2) J_0(x) + 64 \int_0^x s J_0(s)\,ds.$$

Referring to the integration formula (5), we now arrive at

$$\int_0^x s^5 J_0(s)\,ds = (x^5 - 16x^3 + 64x) J_1(x) + (4x^4 - 32x^2) J_0(x),$$

which is the same as formula (7).

[†]Note, too, that when n is *even,* the reduction formula can be used to transform the problem of evaluating $\int_0^x s^n J_0(s)\,ds$ into that of evaluating $\int_0^x J_0(s)\,ds$, which is tabulated for various values of x in, for example, the book edited by Abramowitz and Stegun (1972, pp. 492–493), listed in the Bibliography. Further references are given on pp. 490–491 of that book.

Relations (1) and (3) can also be written

$$(8) \qquad xJ_n'(x) = nJ_n(x) - xJ_{n+1}(x) \qquad (n = 0, 1, 2, \ldots)$$

and

$$(9) \qquad xJ_n'(x) = -nJ_n(x) + xJ_{n-1}(x) \qquad (n = 1, 2, \ldots),$$

respectively; and eliminating $J_n'(x)$ from these equations, we find that

$$(10) \qquad xJ_{n+1}(x) = 2nJ_n(x) - xJ_{n-1}(x) \qquad (n = 1, 2, \ldots).$$

This relation, which expresses J_{n+1} in terms of the functions J_n and J_{n-1} of lower orders, is sometimes called a *pure* recurrence relation since it does not involve derivatives.

Finally, we note that relations (1), (3), and (10) remain valid when n is replaced by the unrestricted parameter ν. Modifications in the derivations simply consist of writing

$$\Gamma(\nu + k + 1) \qquad \text{or} \qquad (\nu + k)\Gamma(\nu + k)$$

in place of $(n + k)!$.

PROBLEMS

1. By differentiating each side of the recurrence relation (8) in Sec. 73 and then referring to both of the relations (8) and (9) there, show that

$$x^2 J_n''(x) = (n^2 - n - x^2)J_n(x) + xJ_{n+1}(x) \qquad (n = 0, 1, 2, \ldots).$$

2. Derive the reduction formula

$$\int_0^x s^n J_0(s)\, ds = x^n J_1(x) + (n-1)x^{n-1}J_0(x) - (n-1)^2 \int_0^x s^{n-2} J_0(s)\, ds \qquad (n = 2, 3, \ldots),$$

which was stated and illustrated in Sec. 73, by applying integration by parts twice and using the relations (Sec. 73)

$$\frac{d}{ds}[sJ_1(s)] = sJ_0(s), \qquad \frac{d}{ds}[-J_0(s)] = J_1(s)$$

in the first and second of those integrations, respectively.

Suggestion: Start by writing

$$\int_0^x s^n J_0(s)\, ds = \int_0^x s^{n-1} s J_0(s)\, ds.$$

3. Derive the differentiation formula

$$\frac{d}{dx}[x^{-\nu} J_\nu(x)] = -x^{-\nu} J_{\nu+1}(x),$$

where $\nu > 0$, and point out why it is also valid when ν is replaced by $-\nu$ $(\nu > 0)$. [Compare with relation (1), Sec. 73.]

4. Use results in Problems 3 above and 8(*a*), Sec. 72, to show that

$$J_{3/2}(x) = \sqrt{\frac{2}{\pi x}}\left(\frac{\sin x}{x} - \cos x\right).$$

74. BESSEL'S INTEGRAL FORM

We now derive a useful integral representation for $J_n(x)$. To do this, we first note that the series in the expansions

$$\text{(1)} \qquad \exp\left(\frac{xt}{2}\right) = \sum_{j=0}^{\infty} \frac{x^j}{j!2^j} t^j, \qquad \exp\left(-\frac{x}{2t}\right) = \sum_{k=0}^{\infty} \frac{(-1)^k x^k}{k!2^k} t^{-k}$$

are absolutely convergent when x is any number and $t \neq 0$. Hence the product of these exponential functions is itself represented by a series formed by multiplying each term in one series by every term in the other and then summing the resulting terms *in any order*.† Clearly, the variable t occurs in each of those resulting terms as a factor t^n $(n = 0, 1, 2, \ldots)$ or t^{-n} $(n = 1, 2, \ldots)$; and the terms involving any particular power of t may be collected as a sum.

In the case of t^n $(n = 0, 1, 2, \ldots)$, that sum is obtained by multiplying the kth term in the second series by the term in the first series whose index is $j = n + k$ and then summing from $k = 0$ to $k = \infty$. The result is

$$\sum_{k=0}^{\infty} \frac{(-1)^k}{k!(n+k)!} \left(\frac{x}{2}\right)^{n+2k} t^n = J_n(x)\, t^n.$$

Similarly, the sum of the terms involving t^{-n} $(n = 1, 2, \ldots)$ is found by multiplying the jth term in the first series by the term in the second series with index $k = n + j$ and summing from $j = 0$ to $j = \infty$. That sum may be written

$$(-1)^n \sum_{j=0}^{\infty} \frac{(-1)^j}{j!(n+j)!} \left(\frac{x}{2}\right)^{n+2j} t^{-n} = (-1)^n J_n(x)\, t^{-n}.$$

A series representation for the product of the exponential functions (1) is, therefore,

$$\text{(2)} \qquad \exp\left[\frac{x}{2}\left(t - \frac{1}{t}\right)\right] = J_0(x) + \sum_{n=1}^{\infty} [J_n(x)\, t^n + (-1)^n J_n(x)\, t^{-n}].$$

Let us write $t = e^{i\phi}$, where $i = \sqrt{-1}$, in equation (2):

$$\text{(3)} \qquad \exp\left[\frac{x}{2}(e^{i\phi} - e^{-i\phi})\right] = J_0(x) + \sum_{n=1}^{\infty} [J_n(x)\, e^{in\phi} + (-1)^n J_n(x)\, e^{-in\phi}].$$

In view of Euler's formula $e^{i\phi} = \cos\phi + i\sin\phi$,‡ we know that

$$e^{i\phi} - e^{-i\phi} = 2i\sin\phi$$

and

$$e^{in\phi} = \cos n\phi + i\sin n\phi, \qquad e^{-in\phi} = \cos n\phi - i\sin n\phi.$$

---

†For a justification of this procedure, see, for example, the book by Taylor and Mann (1983, pp. 601–602), which is listed in the Bibliography.

‡See the footnote with Problem 4, Sec. 7.

It thus follows from equation (3) that

$$\exp(ix\sin\phi) = J_0(x) + \sum_{n=1}^{\infty}[1+(-1)^n]\,J_n(x)\cos n\phi + i\sum_{n=1}^{\infty}[1-(-1)^n]\,J_n(x)\sin n\phi. \tag{4}$$

Now, again by Euler's formula,

$$\exp(ix\sin\phi) = \cos(x\sin\phi) + i\sin(x\sin\phi);$$

and if we equate the real parts on each side of equation (4), we find that

$$\cos(x\sin\phi) = J_0(x) + \sum_{n=1}^{\infty}[1+(-1)^n]\,J_n(x)\cos n\phi. \tag{5}$$

Holding x fixed and regarding this equation as a Fourier cosine series representation of the function $\cos(x\sin\phi)$ on the interval $0<\phi<\pi$, we need only recall (Sec. 2) the formula for the coefficients in such a series to write

$$[1+(-1)^n]\,J_n(x) = \frac{2}{\pi}\int_0^{\pi}\cos(x\sin\phi)\cos n\phi\,d\phi \qquad (n=0,1,2,\ldots). \tag{6}$$

If, on the other hand, we equate the imaginary parts on each side of equation (4), we obtain the Fourier sine series representation

$$\sin(x\sin\phi) = \sum_{n=1}^{\infty}[1-(-1)^n]\,J_n(x)\sin n\phi \tag{7}$$

for $\sin(x\sin\phi)$ on the same interval. Consequently (see Sec. 4),

$$[1-(-1)^n]\,J_n(x) = \frac{2}{\pi}\int_0^{\pi}\sin(x\sin\phi)\sin n\phi\,d\phi \qquad (n=1,2,\ldots). \tag{8}$$

According to expressions (6) and (8), then,

$$J_{2n}(x) = \frac{1}{\pi}\int_0^{\pi}\cos(x\sin\phi)\cos 2n\phi\,d\phi \qquad (n=0,1,2,\ldots) \tag{9}$$

and

$$J_{2n-1}(x) = \frac{1}{\pi}\int_0^{\pi}\sin(x\sin\phi)\sin(2n-1)\phi\,d\phi \qquad (n=1,2,\ldots). \tag{10}$$

A single expression for $J_n(x)$ can be obtained by adding corresponding sides of equations (6) and (8) and writing

$$2J_n(x) = \frac{2}{\pi}\int_0^{\pi}[\cos n\phi\cos(x\sin\phi) + \sin n\phi\sin(x\sin\phi)]\,d\phi.$$

That is,

$$J_n(x) = \frac{1}{\pi}\int_0^{\pi}\cos(n\phi - x\sin\phi)\,d\phi \qquad (n=0,1,2,\ldots). \tag{11}$$

This is known as *Bessel's integral form* of $J_n(x)$.

75. SOME CONSEQUENCES OF THE INTEGRAL FORMS

A number of important properties of Bessel functions follow readily from integral representations in Sec. 74. The *boundedness properties*

$$|J_n(x)| \leq 1, \qquad \left|\frac{d^k}{dx^k} J_n(x)\right| \leq 1 \qquad (k = 1, 2, \ldots) \tag{1}$$

are, for example, immediate consequences of the integral form (11) in that section. More precisely,

$$|J_n(x)| \leq \frac{1}{\pi}\int_0^{\pi} |\cos(n\phi - x\sin\phi)|\, d\phi \leq \frac{1}{\pi}\int_0^{\pi} d\phi = 1.$$

Furthermore,

$$J_n'(x) = \frac{1}{\pi}\int_0^{\pi} \sin(n\phi - x\sin\phi)\sin\phi\, d\phi;$$

and continued differentiation yields integral representations for $J_n''(x)$, etc. Since, in each case, the absolute value of the integrand does not exceed unity, the rest of inequalities (1) also hold.

Sometimes it is useful to write the integral representations (9) and (10) in Sec. 74 as

$$J_{2n}(x) = \frac{2}{\pi}\int_0^{\pi/2} \cos(x\sin\phi)\cos 2n\phi\, d\phi \qquad (n = 0, 1, 2, \ldots) \tag{2}$$

and

$$J_{2n-1}(x) = \frac{2}{\pi}\int_0^{\pi/2} \sin(x\sin\phi)\sin(2n-1)\phi\, d\phi \qquad (n = 1, 2, \ldots). \tag{3}$$

Expressions (2) and (3) follow from the fact that when x is fixed, the graphs of the integrands

$$y = g(\phi) = \cos(x\sin\phi)\cos 2n\phi,$$
$$y = h(\phi) = \sin(x\sin\phi)\sin(2n-1)\phi$$

are symmetric with respect to the line $\phi = \pi/2$:

$$g(\pi - \phi) = g(\phi), \qquad h(\pi - \phi) = h(\phi).$$

We note the special case

$$J_0(x) = \frac{2}{\pi}\int_0^{\pi/2} \cos(x\sin\phi)\, d\phi \tag{4}$$

of representation (2). It can also be written

$$J_0(x) = \frac{2}{\pi}\int_0^{\pi/2} \cos(x\cos\theta)\, d\theta \tag{5}$$

by means of the substitution

$$\theta = \frac{\pi}{2} - \phi.$$

Representations (2) and (3) may be used to verify that for each fixed n ($n = 0, 1, 2, \ldots$),

$$\lim_{x\to\infty} J_n(x) = 0. \tag{6}$$

To give the details when $n = 0$, we substitute $u = \sin\phi$ into equation (4) to write

$$\frac{\pi}{2} J_0(x) = \int_0^c \frac{\cos xu}{\sqrt{1-u^2}}\, du + \int_c^1 \frac{\cos xu}{\sqrt{1-u^2}}\, du,$$

where $0 < c < 1$. The second integral here is improper but uniformly convergent with respect to x. Hence, for any given positive number ε, the absolute value of that integral can be made less than $\varepsilon/2$, uniformly for all x, by selecting c so that the difference $1 - c$ is sufficiently small and positive. The Riemann-Lebesgue lemma (Sec. 46)

$$\lim_{r\to\infty} \int_0^c G(u)\cos ru\, du = 0 \qquad (r > 0), \tag{7}$$

involving a piecewise continuous function $G(u)$ and a cosine function, then applies to the first integral with that value of c. That is, there is a number x_ε such that the absolute value of the first integral is less than $\varepsilon/2$ whenever $x > x_\varepsilon$. Therefore,

$$\frac{\pi}{2}|J_0(x)| < \frac{\varepsilon}{2} + \frac{\varepsilon}{2} = \varepsilon \qquad \text{whenever } x > x_\varepsilon;$$

and this establishes property (6) when $n = 0$. Verification when n is a positive integer is left to the problems.

It is interesting to contrast limit (6) with the limit

$$\lim_{n\to\infty} J_n(x) = 0, \tag{8}$$

which is valid for each x ($-\infty < x < \infty$). This limit follows from the Riemann-Lebesgue lemma (7) and its version (Sec. 46)

$$\lim_{r\to\infty} \int_0^c G(u)\sin ru\, du = 0 \qquad (r > 0), \tag{9}$$

involving a sine function, when that lemma is applied to the integral representations (2) and (3) for $J_{2n}(x)$ and $J_{2n-1}(x)$, respectively.

PROBLEMS

1. Use integral representations for $J_n(x)$ to verify that
   (a) $J_0(0) = 1$; (b) $J_n(0) = 0$ ($n = 1, 2, \ldots$); (c) $J_0'(x) = -J_1(x)$.
2. Derive representation (2), Sec. 75, for $J_{2n}(x)$ by writing the Fourier cosine series (5), Sec. 74, as

$$\cos(x\sin\phi) = J_0(x) + 2\sum_{n=1}^{\infty} J_{2n}(x)\cos 2n\phi$$

   and then interpreting it as a Fourier cosine series on the interval $0 < \phi < \pi/2$.

3. Deduce from expression (2), Sec. 75, that

$$J_{2n}(x) = (-1)^n \frac{2}{\pi} \int_0^{\pi/2} \cos(x\cos\theta)\cos 2n\theta\, d\theta \qquad (n = 0, 1, 2, \ldots).$$

4. Deduce from expression (3), Sec. 75, that

$$J_{2n-1}(x) = (-1)^{n+1} \frac{2}{\pi} \int_0^{\pi/2} \sin(x\cos\theta)\cos(2n-1)\theta\, d\theta \qquad (n = 1, 2, \ldots).$$

5. Complete the verification of property (6), Sec. 75, that

$$\lim_{x\to\infty} J_n(x) = 0$$

for each fixed n $(n = 0, 1, 2, \ldots)$.

6. Apply integration by parts to representations (2) and (3) in Sec. 75 and then use the Riemann-Lebesgue lemma (Sec. 46) to show that

$$\lim_{n\to\infty} nJ_n(x) = 0$$

for each fixed x.

7. Verify directly from the representation (Sec. 75)

$$J_0(x) = \frac{2}{\pi} \int_0^{\pi/2} \cos(x\sin\phi)\, d\phi$$

that $J_0(x)$ satisfies Bessel's equation

$$xy''(x) + y'(x) + xy(x) = 0.$$

8. According to Sec. 58, if a function $f(\phi)$ and its derivative $f'(\phi)$ are continuous on the interval $-\pi \le \phi \le \pi$ and if $f(-\pi) = f(\pi)$, then Parseval's equation

$$\frac{1}{\pi} \int_{-\pi}^{\pi} [f(\phi)]^2\, d\phi = \frac{a_0^2}{2} + \sum_{n=1}^{\infty} \left(a_n^2 + b_n^2\right)$$

holds, where the numbers a_n $(n = 0, 1, 2, \ldots)$ and b_n $(n = 1, 2, \ldots)$ are the Fourier coefficients

$$a_n = \frac{1}{\pi} \int_{-\pi}^{\pi} f(\phi)\cos n\phi\, d\phi, \qquad b_n = \frac{1}{\pi} \int_{-\pi}^{\pi} f(\phi)\sin n\phi\, d\phi.$$

(*a*) By applying that result to $f(\phi) = \cos(x\sin\phi)$, which is an even function of ϕ, and referring to the Fourier (cosine) series (5) for $f(\phi)$ in Sec. 74, show that

$$\frac{1}{\pi} \int_0^{\pi} \cos^2(x\sin\phi)\, d\phi = [J_0(x)]^2 + 2\sum_{n=1}^{\infty} [J_{2n}(x)]^2 \qquad (-\infty < x < \infty).$$

(*b*) Similarly, by writing $f(\phi) = \sin(x\sin\phi)$ and referring to the Fourier (sine) series (7) for $f(\phi)$ in Sec. 74, show that

$$\frac{1}{\pi} \int_0^{\pi} \sin^2(x\sin\phi)\, d\phi = 2\sum_{n=1}^{\infty} [J_{2n-1}(x)]^2 \qquad (-\infty < x < \infty).$$

(*c*) Combine the results in parts (*a*) and (*b*) to show that

$$[J_0(x)]^2 + 2\sum_{n=1}^{\infty} [J_n(x)]^2 = 1 \qquad (-\infty < x < \infty),$$

and point out how it follows from this identity that

$$|J_0(x)| \le 1 \qquad \text{and} \qquad |J_n(x)| \le \frac{1}{\sqrt{2}} \qquad (n = 1, 2, \ldots)$$

for all x.

9. By writing $t = i$ in the series representation (2), Sec. 74, derive the expansions

$$\cos x = J_0(x) + 2\sum_{n=1}^{\infty}(-1)^n J_{2n}(x)$$

and

$$\sin x = 2\sum_{n=1}^{\infty}(-1)^{n+1} J_{2n-1}(x),$$

which are valid for all x.

10. Show that series representation (2), Sec. 74, can be written in the form

$$\exp\left[\frac{x}{2}\left(t - \frac{1}{t}\right)\right] = \lim_{N\to\infty} \sum_{n=-N}^{N} J_n(x)\, t^n \qquad (t \ne 0).$$

This exponential function is, then, a *generating function* for the Bessel functions $J_n(x)$ $(n = 0, \pm 1, \pm 2, \ldots)$.

76. THE ZEROS OF $J_n(x)$

Recall from Sec. 71 that Bessel's equation when $\nu = 0$ is

$$xy''(x) + y'(x) + xy(x) = 0. \tag{1}$$

A modified form of this equation in which the term containing the first derivative is absent will be useful to us here. That form is easily found (see Problem 1, Sec. 79) by making the substitution $y(x) = x^c u(x)$, where c is a constant, in equation (1) and observing that the coefficient of $u'(x)$ in the resulting differential equation,

$$x^2u''(x) + (1 + 2c)xu'(x) + (x^2 + c^2)u(x) = 0,$$

is zero if $c = -1/2$. The desired modified form of equation (1) is then

$$x^2u''(x) + \left(x^2 + \frac{1}{4}\right)u(x) = 0, \tag{2}$$

and the function $u(x) = \sqrt{x}J_0(x)$ is evidently a solution of equation (2).

We shall now use equation (2) to prove the following important lemma regarding the positive zeros of $J_0(x)$.[†]

Lemma. *The positive zeros of the function $J_0(x)$, or positive roots of the equation $J_0(x) = 0$, form an increasing sequence of numbers x_j $(j = 1, 2, \ldots)$ such that $x_j \to \infty$ as $j \to \infty$.*

[†]Our method is a modification of the one used by A. Czarnecki, *Amer. Math. Monthly,* vol. 71, no. 4, pp. 403–404, 1964, who considers Bessel functions $J_\nu(x)$, where $-1/2 \le \nu \le 1/2$.

To prove this, we continue with the function $u(x) = \sqrt{x}J_0(x)$. According to equation (2),

$$-u(x) - u''(x) = \frac{u(x)}{4x^2};$$

and multiplying each side of this equation by the function $v(x) = \sin x$ yields

$$u(x)v''(x) - u''(x)v(x) = u(x)\,\frac{\sin x}{4x^2}. \tag{3}$$

Also, from the identity

$$u(x)v''(x) - u''(x)v(x) = \frac{d}{dx}\,[u(x)v'(x) - u'(x)v(x)],$$

we know that

$$u(x)v''(x) - u''(x)v(x) = \frac{d}{dx}\,[u(x)\cos x - u'(x)\sin x]. \tag{4}$$

By equating the right-hand sides of equations (3) and (4), we obtain

$$u(x)\,\frac{\sin x}{4x^2} = \frac{d}{dx}\,[u(x)\cos x - u'(x)\sin x]. \tag{5}$$

Next, let k be any positive integer ($k = 1, 2, \ldots$). Since

$$\cos(2k\pi + \pi) = -1, \qquad \sin(2k\pi + \pi) = 0$$

and

$$\cos(2k\pi) = 1, \qquad \sin(2k\pi) = 0,$$

it follows from equation (5) that

$$\begin{aligned}\int_{2k\pi}^{2k\pi+\pi} u(x)\,\frac{\sin x}{4x^2}\,dx &= [u(x)\cos x - u'(x)\sin x]_{2k\pi}^{2k\pi+\pi} \\ &= -[u(2k\pi + \pi) + u(2k\pi)].\end{aligned} \tag{6}$$

It is now easy to show that our function $u(x) = \sqrt{x}J_0(x)$, and hence $J_0(x)$, has at least one zero in the interval

$$2k\pi \le x \le 2k\pi + \pi. \tag{7}$$

We do this by assuming that $u(x) \neq 0$ anywhere in that closed interval and obtaining a contradiction. According to our assumption, either $u(x) > 0$ for all x in the interval or $u(x) < 0$ for all such x, since $u(x)$ is continuous and thus cannot change sign without having a zero value at some point in the interval.

Suppose that $u(x) > 0$ everywhere in the interval (7). The integrand in integral (6) is positive when $2k\pi < x < 2k\pi + \pi$. Hence the value of the integral must be positive, while the value $-[u(2k\pi + \pi) + u(2k\pi)]$ that we obtained is evidently negative, giving a contradiction. If, on the other hand, $u(x) < 0$ throughout the interval (7), the value of the integral must be negative; but the value obtained is positive. This is again a contradiction. We thus conclude that $J_0(x)$ has at least one zero in the interval (7).

Actually, $J_0(x)$ can have *at most* a finite number of zeros in any closed bounded interval $a \le x \le b$. To see that this is so, we assume that the interval

$a \le x \le b$ *does* contain an infinite number of zeros. From advanced calculus, we know that if a given infinite set of points lies in a closed bounded interval, there is always a sequence of distinct points in that set which converges to a point in the interval.[†] In particular, then, our assumption that the interval $a \le x \le b$ contains an infinite number of zeros of $J_0(x)$ implies that there exists a sequence x_m $(m = 1, 2, \ldots)$ of distinct zeros such that $x_m \to c$ as $m \to \infty$, where c is a point which also lies in the interval. Since the function $J_0(x)$ is continuous, $J_0(c) = 0$; and by the definition of the limit of a sequence, every interval centered at c contains other zeros of $J_0(x)$. But the fact that $J_0(x)$ is not identically equal to zero and has a Maclaurin series representation which is valid for all x means that there exists some interval centered at c which contains no other zeros.[‡] Since this is contrary to what has just been shown, the number of zeros in the interval $a \le x \le b$ cannot, then, be infinite.

It is now evident that the positive zeros of $J_0(x)$ can, in fact, be arranged as an increasing sequence of numbers tending to infinity. The table below gives the values, correct to four significant figures, of the first five zeros of $J_0(x)$ and the corresponding values of $J_1(x)$. (See Fig. 62 in Sec. 71.) Extensive tables of numerical values of Bessel and related functions, together with their zeros, will be found in books listed in the Bibliography.[§]

$J_0(x_j) = 0$

j	1	2	3	4	5
x_j	2.405	5.520	8.654	11.79	14.93
$J_1(x_j)$	0.5191	−0.3403	0.2715	−0.2325	0.2065

The lemma just proved can be extended so as to apply to Bessel functions $J_n(x)$ when n is a positive integer.

Theorem. *Let n be any fixed nonnegative integer $(n = 0, 1, 2, \ldots)$. The positive zeros of the function $J_n(x)$, or positive roots of the equation*

$$J_n(x) = 0, \tag{8}$$

form an increasing sequence of numbers x_j $(j = 1, 2, \ldots)$ such that $x_j \to \infty$ as $j \to \infty$.

Our proof is by mathematical induction. First, we know from the lemma that this theorem is true when $n = 0$. Assume now that it is true when $n = m$, where m is *any* nonnegative integer, and let a and b be two distinct positive zeros of $J_m(x)$.

[†] See, for example, the book by Taylor and Mann (1983, pp. 515–519), listed in the Bibliography.

[‡] That is, the zeros of such a function are *isolated*. An argument for this is given in the authors' book (2004, p. 240), listed in the Bibliography.

[§] See especially the book edited by Abramowitz and Stegun (1972) and the ones by Jahnke, Emde, and Lösch (1960), Gray and Mathews (1966), and Watson (1995).

This means that the function $x^{-m}J_m(x)$ vanishes when $x = a$ and when $x = b$. It thus follows from Rolle's theorem that the derivative of $x^{-m}J_m(x)$ vanishes for at least one value of x between a and b; furthermore, from relation (1) in Sec. 73, we know that

$$J_{m+1}(x) = -x^m \frac{d}{dx}[x^{-m}J_m(x)].$$

Hence there is at least one zero of $J_{m+1}(x)$ between any two positive zeros of $J_m(x)$. Also, just as in the case of $J_0(x)$, the function $J_{m+1}(x)$ can have at most a finite number of zeros in any bounded interval. Inasmuch as the zeros of $J_m(x)$ form an unbounded increasing sequence of numbers, it now follows that the same is true of the zeros of $J_{m+1}(x)$. This completes the proof of the theorem.

77. ZEROS OF RELATED FUNCTIONS

We preface the theorem in this section with a needed lemma. The theorem, in addition to the one in Sec. 76, will be important to us in solving boundary value problems in which Bessel functions arise.

Lemma. *At each positive zero of $J_n(x)$ $(n = 0, 1, 2, \ldots)$, the derivative $J_n'(x)$ is nonzero. Moreover, the values of $J_n'(x)$ alternate in sign at consecutive positive zeros of $J_n(x)$.*

The proof starts with the observation that the function $y = J_n(x)$ satisfies Bessel's equation, which is of the type treated in the lemma in Sec. 62 dealing with the uniqueness of solutions of certain second-order linear differential equations. According to that lemma, there is just one continuously differentiable solution of Bessel's equation satisfying the conditions $y(c) = y'(c) = 0$, where $c > 0$, and the solution is identically equal to zero. Consequently, there is no positive number c such that $J_n(c) = J_n'(c) = 0$. That is, $J_n'(x)$ cannot vanish at a positive zero of $J_n(x)$. This means, of course, that $J_n(x)$ must change sign at such a point.

It remains to show that values of $J_n'(x)$ have different signs at consecutive positive zeros a and b $(0 < a < b)$ of $J_n(x)$. If $J_n'(a) > 0$, then $J_n(x) > 0$ $(a < x < b)$ and $J_n(x)$ is decreasing at b; hence $J_n'(b) < 0$. Similarly, if $J_n'(a) < 0$, then $J_n'(b) > 0$. [See the graphs of $y = J_0(x)$ and $y = J_1(x)$ in Fig. 62 (Sec. 71), where the slopes of these functions alternate at their zeros.] The lemma is now established.

The theorem to follow is similar to the one in Sec. 76 but involves the function $hJ_n(x) + xJ_n'(x)$, where n is a nonnegative integer and h is a nonnegative constant. Although this theorem need not exclude the possibility that h may be negative, such values will not arise in our applications.

Theorem. *Let n be any fixed nonnegative integer $(n = 0, 1, 2, \ldots)$, and suppose that h is a nonnegative constant $(h \geq 0)$. The positive zeros of the function $hJ_n(x) + xJ_n'(x)$, or positive roots of the equation*

$$hJ_n(x) + xJ_n'(x) = 0, \tag{1}$$

form an increasing sequence of numbers x_j $(j = 1, 2, \ldots)$ such that $x_j \to \infty$ as $j \to \infty$.

To prove this, we observe that if a and b are consecutive positive zeros of $J_n(x)$, it follows that $hJ_n(x) + xJ'_n(x)$ must have the values $aJ'_n(a)$ and $bJ'_n(b)$ at the points $x = a$ and $x = b$, respectively. Since one of those values is positive and the other negative, according to the above lemma, the function $hJ_n(x) + xJ'_n(x)$ vanishes at some point, or at some finite number of points, between a and b. Consequently, it has an increasing sequence of positive zeros tending to infinity.[†] So the theorem is true.

In this and the previous section, we have considered only the *positive* zeros of the functions

$$J_n(x) \quad \text{and} \quad hJ_n(x) + xJ'_n(x) \qquad (n = 0, 1, 2, \ldots) \tag{2}$$

since they are the ones that will concern us in the applications. But observe that $x = 0$ is a zero of both of these functions when $n = 1, 2, \ldots$ and that it is a zero of the second one when $n = 0$ if $h = 0$.

Furthermore, if $x = c$ is a zero of the first of the functions (2), then $x = -c$ is also a zero since $J_n(-c) = (-1)^n J_n(c)$. The same is true of the second of the functions (2). For, in view of the recurrence relation (Sec. 73)

$$xJ'_n(x) = nJ_n(x) - xJ_{n+1}(x),$$

that second function can be written

$$(h + n)J_n(x) - xJ_{n+1}(x); \tag{3}$$

and it follows that

$$(h + n)J_n(-c) - (-c)J_{n+1}(-c) = (-1)^n \left[(h + n)J_n(c) - cJ_{n+1}(c)\right].$$

78. ORTHOGONAL SETS OF BESSEL FUNCTIONS

The physical applications in this chapter will involve solutions of a singular Sturm-Liouville problem, on an interval $0 \le x \le c$, consisting of the differential equation (see Example 1, Sec. 60)

$$\frac{d}{dx}\left(x\frac{dX}{dx}\right) + \left(-\frac{n^2}{x} + \lambda x\right)X = 0 \qquad (n = 0, 1, 2, \ldots) \tag{1}$$

and a boundary condition of the type

$$b_1 X(c) + b_2 X'(c) = 0. \tag{2}$$

The constants b_1 and b_2 are real and not both zero, and *X and X' are to be continuous on the entire interval* $0 \le x \le c$.

[†]In the important special case $n = 0$, the first few zeros are tabulated for various positive values of h in, for example, the book on heat conduction by Carslaw and Jaeger (1986, p. 493), which is listed in the Bibliography.

Note how the differential equation (1) can be written

$$x^2 \frac{d^2X}{dx^2} + x\frac{dX}{dx} + (\lambda x^2 - n^2)X = 0 \qquad (n = 0, 1, 2, \ldots)$$

and that it reduces to

$$x\frac{d^2X}{dx^2} + \frac{dX}{dx} + \lambda x X = 0$$

when $n = 0$. Note, too, that when $b_2 = 0$, boundary condition (2) becomes $X(c) = 0$. When $b_2 \neq 0$, one can multiply through condition (2) by the quantity c/b_2 and write the result as $hX(c) + cX'(c) = 0$, where $h = cb_1/b_2$. In our physical applications, *h will be either positive or zero*. That constant is, of course, zero when $b_1 = 0$, in which case condition (2) becomes $X'(c) = 0$.

Theorems 1 and 2 below provide solutions of our Sturm-Liouville problem when the boundary condition (2) has one of the three forms mentioned in the paragraph just above. These theorems treat the important special case in which $n = 0$ separately from the one in which n has one of the values $n = 1, 2, \ldots$. Although the theorems could be combined, it will be more convenient to have them separate in the applications. Their proofs, appearing in Sec. 79, will, however, be combined.

Theorem 1. *For the singular Sturm-Liouville problem consisting of the differential equation*

$$x\frac{d^2X}{dx^2} + \frac{dX}{dx} + \lambda x X = 0 \qquad (0 < x < c) \tag{3}$$

and one of the boundary conditions

$$X(c) = 0, \tag{4}$$

$$hX(c) + cX'(c) = 0 \qquad (h > 0), \tag{5}$$

$$X'(c) = 0, \tag{6}$$

the eigenvalues λ_j and corresponding eigenfunctions X_j are as follows:

(*a*) *When condition* (4) *is used,*

$$\lambda_j = \alpha_j^2, \qquad X_j = J_0(\alpha_j x) \qquad (j = 1, 2, \ldots)$$

where α_j $(j = 1, 2, \ldots)$ are the positive roots of the equation

$$J_0(\alpha c) = 0;$$

(*b*) *when condition* (5) *is used,*

$$\lambda_j = \alpha_j^2, \qquad X_j = J_0(\alpha_j x) \qquad (j = 1, 2, \ldots)$$

where $\alpha_j (j = 1, 2, \ldots)$ are the positive roots of the equation

$$hJ_0(\alpha c) + (\alpha c)J_0'(\alpha c) = 0 \qquad (h > 0);$$

(*c*) *when condition* (6) *is used,*

$$\lambda_1 = 0, \quad X_1 = 1 \qquad \textit{and} \qquad \lambda_j = \alpha_j^2, \quad X_j = J_0(\alpha_j x) \qquad (j = 2, 3, \ldots)$$

where α_j $(j = 2, 3, \ldots)$ *are the positive roots of the equation*

$$J_0'(\alpha c) = 0.$$

Observe that since $J_0'(x) = -J_1(x)$ (see Sec. 73), the equations defining the numbers α_j in cases (*b*) and (*c*) can be written

$$hJ_0(\alpha c) - (\alpha c)J_1(\alpha c) = 0 \qquad \text{and} \qquad J_1(\alpha c) = 0,$$

respectively.

Theorem 2. *Let n be a positive integer* $(n = 1, 2, \ldots)$. *For the singular Sturm-Liouville problem consisting of the differential equation*

$$x^2 \frac{d^2X}{dx^2} + x\frac{dX}{dx} + (\lambda x^2 - n^2)X = 0 \qquad (0 < x < c) \tag{7}$$

and one of the boundary conditions

$$X(c) = 0, \tag{8}$$

$$hX(c) + cX'(c) = 0 \qquad (h > 0), \tag{9}$$

$$X'(c) = 0, \tag{10}$$

the eigenvalues λ_j *and corresponding eigenfunctions* X_j *are*

$$\lambda_j = \alpha_j^2, \qquad X_j = J_n(\alpha_j x) \qquad (j = 1, 2, \ldots)$$

where the numbers α_j *are defined as follows:*

(*a*) *When condition* (8) *is used,* α_j $(j = 1, 2, \ldots)$ *are the positive roots of the equation*

$$J_n(\alpha c) = 0;$$

(*b*) *when condition* (9) *is used,* α_j $(j = 1, 2, \ldots)$ *are the positive roots of the equation*

$$hJ_n(\alpha c) + (\alpha c)J_n'(\alpha c) = 0 \qquad (h > 0);$$

(*c*) *when condition* (10) *is used,* α_j $(j = 1, 2, \ldots)$ *are the positive roots of the equation*

$$J_n'(\alpha c) = 0.$$

Note that because the functions

$$hJ_n(x) + xJ_n'(x) \qquad \text{and} \qquad (h + n)J_n(x) - xJ_{n+1}(x)$$

are the same, as pointed out at the end of Sec. 77, the equation defining the numbers α_j in case (*b*) of Theorem 2 above can be written

$$(h + n)J_n(\alpha c) - (\alpha c)J_{n+1}(\alpha c) = 0. \tag{11}$$

Also, because $xJ_n'(x) = nJ_n(x) - xJ_{n+1}(x)$ (Sec. 73), the equation

$$nJ_n(\alpha c) - (\alpha c)J_{n+1}(\alpha c) = 0 \tag{12}$$

is an alternative form of the equation defining the α_j in case (*c*) of Theorem 2.

For each of the cases in the above two theorems, the orthogonality property

$$\int_0^c xJ_n(\alpha_j x)J_n(\alpha_k x)\,dx = 0 \qquad (j \neq k) \tag{13}$$

follows from case (*a*) of the theorem in Sec. 61. Observe that this orthogonality of the eigenfunctions with respect to the weight function x, on the interval $0 < x < c$, is the same as ordinary orthogonality of the functions $\sqrt{x}\,J_n(\alpha_j x)$ on the same interval. Also, many orthogonal sets are represented here, depending on the values of c, n, and h. Once we have proved the two theorems, we shall in Secs. 80 and 81 normalize these eigenfunctions and find formulas for the coefficients in generalized Fourier series expansions involving the eigenfunctions. The reader who wishes to reach such expansions and their applications more quickly may pass directly to Sec. 80 without loss of continuity.

79. PROOF OF THE THEOREMS

We turn now to the proof of the theorems in Sec. 78. We recall how we wrote the differential equation of the singular Sturm-Liouville problem there as

$$x^2\frac{d^2X}{dx^2} + x\frac{dX}{dx} + (\lambda x^2 - n^2)X = 0 \qquad (n = 0, 1, 2, \ldots) \tag{1}$$

and combined it with one of the boundary conditions

$$X(c) = 0, \tag{2}$$

$$hX(c) + cX'(c) = 0 \qquad (h > 0), \tag{3}$$

$$X'(c) = 0. \tag{4}$$

Case (*a*) of the corollary in Sec. 61, applied to the form (1)–(2), Sec. 78, of the Sturm-Liouville problem, ensures that any eigenvalue must be a real number; and, keeping in mind that X and X' must be continuous on the closed interval $0 \leq x \leq c$, we consider the possibilities that λ may be zero, positive, or negative.

When $\lambda = 0$, equation (1) is a Cauchy-Euler equation (see Problem 1, Sec. 38):

$$x^2\frac{d^2X}{dx^2} + x\frac{dX}{dx} - n^2X = 0 \qquad (n = 0, 1, 2, \ldots). \tag{5}$$

To solve it, we write $x = \exp s$ and put it in the form

$$\frac{d^2X}{ds^2} - n^2X = 0 \qquad (n = 0, 1, 2, \ldots). \tag{6}$$

If $n = 0$, equation (6) has solution $X = As + B$, where A and B are arbitrary constants; and the general solution of equation (5) is, therefore, $X(x) = A \ln x + B$. The requirement that X and X' be continuous when $0 \leq x \leq c$ forces A to be zero, and we have $X(x) = B$. When condition (2) or (3) is imposed, $B = 0$. The constant B can, however, remain arbitrary when condition (4) is used. Assigning the value $B = 1$, we then have the eigenfunction $X(x) = 1$, corresponding to $\lambda = 0$. This appears in case (c) of Theorem 1 in Sec. 78 and is the only case there in which $\lambda = 0$ is an eigenvalue. Any eigenfunction corresponding to that eigenvalue is, of course, a constant multiple of $X(x) = 1$.

If, on the other hand, n is positive, the general solution of equation (6) is $X = Ae^{ns} + Be^{-ns}$. That is,

$$X(x) = Ax^n + Bx^{-n} = Ax^n + B\frac{1}{x^n}.$$

Since our solution must be continuous and therefore bounded on $0 \le x \le c$, we require that $B = 0$. Hence $X(x) = Ax^n$. It is now easy to see that $A = 0$ if any of the conditions (2), (3), or (4) is to be satisfied, and we arrive at only the trivial solution $X(x) \equiv 0$. Thus, in Theorem 2, Sec. 78, zero is not an eigenvalue.

We consider next the possibility that $\lambda > 0$ and write $\lambda = \alpha^2$ ($\alpha > 0$). Equation (1) is then

$$x^2 \frac{d^2X}{dx^2} + x\frac{dX}{dx} + (\alpha^2x^2 - n^2)X = 0 \qquad (n = 0, 1, 2, \ldots), \tag{7}$$

and we know from Problem 9, Sec. 72, that its general solution is

$$X(x) = C_1 J_n(\alpha x) + C_2 Y_n(\alpha x).$$

Our continuity requirements imply that $C_2 = 0$, since $Y_n(\alpha x)$ is discontinuous at $x = 0$ (see Sec. 72). Hence any nontrivial solution of equation (7) that meets those requirements must be a constant multiple of the function $X(x) = J_n(\alpha x)$.

In applying one of the boundary conditions at $x = c$, we emphasize that *the symbol $J_n'(\alpha x)$ stands for the derivative of $J_n(s)$ with respect to s, evaluated at* $s = \alpha x$. Then, by the chain rule,

$$\frac{d}{dx} J_n(\alpha x) = \frac{d}{ds} J_n(s)\frac{ds}{dx} = J_n'(s)\,\alpha = \alpha J_n'(\alpha x);$$

and conditions (2), (3), and (4) require that

$$J_n(\alpha c) = 0, \tag{8}$$

$$hJ_n(\alpha c) + (\alpha c)J_n'(\alpha c) = 0 \qquad (h > 0), \tag{9}$$

and

$$J_n'(\alpha c) = 0, \tag{10}$$

respectively.

According to the theorems in Secs. 76 and 77, each of the equations (8), (9), and (10) has an infinite number of positive roots

$$\alpha_j = \frac{x_j}{c} \qquad (j = 1, 2, \ldots), \tag{11}$$

where x_j ($j = 1, 2, \ldots$) is the unbounded increasing sequence in the statement of the theorem in question. The numbers α_j here depend, of course, on the value of n and also on the value of h in the case of equation (9). Our Sturm-Liouville problem thus has eigenvalues $\lambda_j = \alpha_j^2$ $(j = 1, 2, \ldots)$, and the corresponding eigenfunctions are

$$X_j(x) = J_n(\alpha_j x) \qquad (j = 1, 2, \ldots). \tag{12}$$

The reader will note that we have not needed to consider the value $n = 0$ separately from the values $n = 1, 2, \ldots$ in the differential equation (7). Evidently, then, we have found all the positive eigenvalues and corresponding eigenfunctions listed in cases (*a*) and (*b*) of Theorem 1 in Sec. 78, as well as the ones when $j = 2, 3, \ldots$ in case (*c*) there. We have, moreover, found all the eigenvalues and eigenfunctions listed in Theorem 2 of Sec. 78. It remains only to show that there are *no negative eigenvalues* arising in either of those theorems.

To accomplish this, we assume that $\lambda < 0$ and write $\lambda = -\alpha^2$ $(\alpha > 0)$. Equation (1) then becomes

$$x^2 \frac{d^2X}{dx^2} + x\frac{dX}{dx} - (\alpha^2 x^2 + n^2)X = 0 \qquad (n = 0, 1, 2, \ldots). \tag{13}$$

The substitution $s = \alpha x$ can be used here to put equation (13) in the form (compare with Problem 9, Sec. 72)

$$s^2 \frac{d^2X}{ds^2} + s\frac{dX}{ds} - (s^2 + n^2)X = 0 \qquad (n = 0, 1, 2, \ldots). \tag{14}$$

From Problem 10, Sec. 72, we know that the modified Bessel function

$$X = I_n(s) = i^{-n} J_n(is)$$

satisfies equation (14); and since $I(s)$ has a power series representation that converges for all s, the function $X(x) = I_n(\alpha x)$ satisfies the continuity requirements in our problem. As was the case with equation (7), equation (13) has a second solution, analogous to $Y_n(\alpha x)$, that is discontinuous at $x = 0$.† Thus we know that except for an arbitrary constant factor, $X(x) = I_n(\alpha x)$ is the desired solution of equation (13).

We now show that for each positive value of α, the function $X(x) = I_n(\alpha x)$ fails to satisfy any of the boundary conditions (2), (3), or (4). In each case, our proof rests on the fact that $I_n(x) > 0$ when $x > 0$, as demonstrated in Problem 10, Sec. 72.

Since $I_n(\alpha c) > 0$ when $\alpha > 0$, it is obvious that condition (2), which requires that $I_n(\alpha c) = 0$, fails to be satisfied for any positive number α.

Condition (3), when applied to the function $X(x) = i^{-n} J_n(i\alpha x)$, becomes

$$h i^{-n} J_n(i\alpha c) + c i^{-n} i\alpha J_n'(i\alpha c) = 0 \qquad (h > 0). \tag{15}$$

But we know from the discussion at the end of Sec. 77 that the equation

$$hJ_n(x) + xJ_n'(x) = 0 \qquad (h \geq 0)$$

†For a detailed discussion of this, see, for example, the book by Tranter (1969, pp. 16ff), which is listed in the Bibliography.

has the alternative form

$$(h+n)J_n(x) - xJ_{n+1}(x) = 0 \qquad (h \geq 0).$$

Hence equation (15) can be written

$$(h+n)i^{-n}J_n(i\alpha c) + (\alpha c)i^{-(n+1)}J_{n+1}(i\alpha c) = 0 \qquad (h \geq 0).$$

Since $I_n(\alpha x) = i^{-n}J_n(i\alpha x)$, then,

$$(16) \qquad (h+n)I_n(\alpha c) + (\alpha c)I_{n+1}(\alpha c) = 0 \qquad (h \geq 0);$$

and because $\alpha > 0$, the left-hand side of this last equation is positive. So, once again, no positive values of α can occur as roots.

Finally, since condition (4) is really condition (3) if h is allowed to be zero and since relation (16) is valid when $h = 0$, we may conclude that there are no negative eigenvalues arising in either of the theorems in Sec. 78. Those theorems are now completely proved.

Except for the zero eigenvalue in case (c) of Theorem 1, Sec. 78, the eigenvalues are all represented by the numbers $\lambda_j = \alpha_j^2$ where the α_j are given by equation (11). Since the numbers x_j used in equation (11) form an unbounded increasing sequence, it is clear that the same is true of the eigenvalues λ_j. That is, $\lambda_j < \lambda_{j+1}$ and $\lambda_j \to \infty$ as $j \to \infty$.

PROBLEMS

1. By means of the substitution $y(x) = x^c u(x)$, transform Bessel's equation

$$x^2 y''(x) + xy'(x) + (x^2 - \nu^2)y(x) = 0$$

into the differential equation

$$x^2 u''(x) + (1+2c)xu'(x) + (x^2 - \nu^2 + c^2)u(x) = 0,$$

which becomes equation (2), Sec. 76, when $\nu = 0$ and $c = -1/2$.

2. Use the result in Problem 1 when $c = -1/2$ to obtain a general solution of Bessel's equation when $\nu = 1/2$. Then, using the expressions (Problem 8, Sec. 72)

$$J_{1/2}(x) = \sqrt{\frac{2}{\pi x}}\sin x \qquad \text{and} \qquad J_{-1/2}(x) = \sqrt{\frac{2}{\pi x}}\cos x,$$

point out how $J_{1/2}(x)$ and $J_{-1/2}(x)$ are special cases of that solution.

3. By referring to Theorem 1 in Sec. 78, show that the eigenvalues of the singular Sturm-Liouville problem

$$xX'' + X' + \lambda x X = 0, \qquad X(2) = 0$$

on the interval $0 \leq x \leq 2$ are the numbers $\lambda_j = \alpha_j^2$ $(j = 1, 2, \ldots)$ where α_j are the positive roots of the equation $J_0(2\alpha) = 0$ and that the corresponding eigenfunctions are $X_j = J_0(\alpha_j x)$ $(j = 1, 2, \ldots)$. With the aid of the table in Sec. 76, obtain the numerical values $\alpha_1 = 1.2$, $\alpha_2 = 2.8$, $\alpha_3 = 4.3$, valid to one decimal place.

4. Write $U(x) = \sqrt{x}J_n(\alpha x)$, where n has any one of the values $n = 0, 1, 2, \ldots$ and α is a positive constant.

(*a*) By recalling from Problem 9, Sec. 72, that $y(x) = J_n(\alpha x)$ satisfies the differential equation

$$x^2 y''(x) + xy'(x) + (\alpha^2 x^2 - n^2)y(x) = 0$$

and noting that $y(x) = x^{-1/2}U(x)$, show that $U(x)$ satisfies the differential equation

$$U''(x) + \left(\alpha^2 + \frac{1-4n^2}{4x^2}\right)U(x) = 0.$$

(*b*) Let c denote any fixed positive number and write

$$U_j(x) = \sqrt{x}\, J_n(\alpha_j x) \qquad (j=1,2,\ldots),$$

where α_j are the positive roots of the equation $J_n(\alpha c)=0$. Use the result in part (*a*) to show that

$$\left(\alpha_j^2 - \alpha_k^2\right) U_j(x)U_k(x) = U_j(x)U_k''(x) - U_j''(x)U_k(x).$$

(*c*) With the aid of the identity

$$U_j(x)U_k''(x) - U_j''(x)U_k(x) = \frac{d}{dx}\left[U_j(x)U_k'(x) - U_j'(x)U_k(x)\right],$$

show how it follows from the result in part (*b*) that the set $\{U_j(x)\}$ there is orthogonal on the interval $0 < x < c$ with weight function unity. Thus give another proof that the set $\{J_n(\alpha_j x)\}$ in case (*a*) of Theorems 1 and 2 in Sec. 78 is orthogonal on $0 < x < c$ with weight function x.

5. Let n have any one of the fixed values $n=0,1,2,\ldots$.

(*a*) Suppose that $J_n(ib)=0$ $(b \neq 0)$ and use results in Problem 10, Sec. 72, to reach a contradiction. Thus show that the function $J_n(z)$ has no pure imaginary zeros $z=ib$ $(b\neq 0)$.

(*b*) Since our series representation of $J_n(x)$ (Sec. 71) converges when x is replaced by any complex number z and since the coefficients of the powers of z in that representation are all real, it follows that $J_n(\bar{z}) = \overline{J_n(z)}$, where $\bar{z}$ denotes the complex conjugate $x - iy$ of the number $z = x + iy$.† Also, the proof of orthogonality in Problem 4 above remains valid when α is a nonzero complex number and when the set of roots α_j there is allowed to include any nonzero complex roots that may occur. Use these facts to show that if a complex number

$$\alpha_j = a + ib \qquad (a \neq 0, b \neq 0)$$

is a zero of $J_n(z)$, then so is $\overline{\alpha_j} = a - ib$. Thus

$$\int_0^1 x\left|J_n(\alpha_j x)\right|^2 dx = \int_0^1 x J_n(\alpha_j x) J_n(\overline{\alpha_j} x)\, dx = 0.$$

Point out why the first integral here is actually positive. With this contradiction, conclude that if $z = x_j$ $(j=1,2,\ldots)$ are the positive zeros of $J_n(z)$, then the only other zeros, real or complex, are $z = -x_j$ $(j=1,2,\ldots)$, and also $z=0$ when n is positive.

---

†For a discussion of power series representations in the complex plane, see the authors' book (2004, chap. 5), listed in the Bibliography.

80. THE ORTHONORMAL FUNCTIONS

The orthogonal sets of Bessel functions in Sec. 78 need to be normalized if we are to use the theory of generalized Fourier series to find expansions in terms of those sets. Theorems 1 and 2 below provide the required norms. As in Sec. 78, we state the theorems separately when $n = 0$ and when $n = 1, 2, \ldots$. Regardless of the value of n, however, the eigenfunctions of the Sturm-Liouville problem in the two theorems can be written in normalized form as

$$\phi_j(x) = \frac{J_n(\alpha_j x)}{\|J_n(\alpha_j x)\|} \qquad (j = 1, 2, \ldots). \tag{1}$$

The fact that the set (1) is orthonormal on the interval $0 < x < c$ is, of course, expressed by the equations

$$\int_0^c x\phi_j(x)\phi_k(x)\,dx = \begin{cases} 0 & \text{when } j \neq k, \\ 1 & \text{when } j = k. \end{cases} \tag{2}$$

Theorem 1. *Let c be any positive number.*

(*a*) *If* α_j $(j = 1, 2, \ldots)$ *are the positive roots of the equation*

$$J_0(\alpha c) = 0,$$

then

$$\|J_0(\alpha_j x)\|^2 = \frac{c^2}{2}[J_1(\alpha_j c)]^2 \qquad (j = 1, 2, \ldots).$$

(*b*) *If* α_j $(j = 1, 2, \ldots)$ *are the positive roots of the equation*

$$hJ_0(\alpha c) + (\alpha c)J_0'(\alpha c) = 0 \qquad (h > 0),$$

then

$$\|J_0(\alpha_j x)\|^2 = \frac{(\alpha_j c)^2 + h^2}{2\alpha_j^2}[J_0(\alpha_j c)]^2 \qquad (j = 1, 2, \ldots).$$

(*c*) *If* $\alpha_1 = 0$ *and* α_j $(j = 2, 3, \ldots)$ *are the positive roots of the equation*

$$J_0'(\alpha c) = 0,$$

then

$$\|J_0(\alpha_1 x)\|^2 = \frac{c^2}{2} \quad \textit{and} \quad \|J_0(\alpha_j x)\|^2 = \frac{c^2}{2}[J_0(\alpha_j c)]^2 \qquad (j = 2, 3, \ldots).$$

The cases (*a*), (*b*), and (*c*) in this theorem correspond to the cases (*a*), (*b*), and (*c*) in Theorem 1 of Sec. 78. Note how we are able to write the eigenfunction $X_1 = 1$ in case (*c*) of that earlier theorem as $X_1 = J_0(\alpha_1 x)$ in case (*c*) here since

$$J_0(\alpha_1 x) = J_0(0) = 1$$

when $\alpha_1 = 0$. This allows us to express its norm in a way similar to the way in which the norms occurring when $j = 2, 3, \ldots$ are expressed. The three cases in Theorem 2 just below correspond, of course, to the three cases in Theorem 2 of Sec. 78.

Theorem 2. *Let c be a positive number and n a positive integer* ($n = 1, 2, \ldots$).

(*a*) *If* α_j ($j = 1, 2, \ldots$) *are the positive roots of the equation*

$$J_n(\alpha c) = 0,$$

then

$$\|J_n(\alpha_j x)\|^2 = \frac{c^2}{2}\,[J_{n+1}(\alpha_j c)]^2 \qquad (j = 1, 2, \ldots).$$

(*b*) *If* α_j ($j = 1, 2, \ldots$) *are the positive roots of the equation*

$$hJ_n(\alpha c) + (\alpha c)J_n'(\alpha c) = 0 \qquad (h > 0),$$

then

$$\|J_n(\alpha_j x)\|^2 = \frac{(\alpha_j c)^2 - n^2 + h^2}{2\alpha_j^2}\,[J_n(\alpha_j c)]^2 \qquad (j = 1, 2, \ldots).$$

(*c*) *If* α_j ($j = 1, 2, \ldots$) *are the positive roots of the equation*

$$J_n'(\alpha c) = 0,$$

then

$$\|J_n(\alpha_j x)\|^2 = \frac{(\alpha_j c)^2 - n^2}{2\alpha_j^2}\,[J_n(\alpha_j c)]^2 \qquad (j = 1, 2, \ldots).$$

Although these two theorems are stated separately, it is convenient to prove them together. So, unless otherwise stated, n may have any one of the values $n = 0, 1, 2, \ldots$. We start the proof by recalling from Problem 9, Sec. 72, that if α is a nonzero constant, the function

$$X(x) = J_n(\alpha x) \tag{3}$$

satisfies the differential equation

$$x^2X'' + xX' + (\alpha^2x^2 - n^2)X = 0,$$

or

$$(xX')' + \left(\alpha^2 x - \frac{n^2}{x}\right)X = 0. \tag{4}$$

Next, we multiply each side of equation (4) by $2xX'$ and write

$$\frac{d}{dx}(xX')^2 + (\alpha^2x^2 - n^2)\frac{d}{dx}(X^2) = 0.$$

After integrating both terms here and using integration by parts in the second term, we find that

$$\left[(xX')^2 + (\alpha^2x^2 - n^2)X^2\right]_0^c - 2\alpha^2\int_0^c xX^2\,dx = 0,$$

where c is any positive number. When $n = 0$, the quantity inside the brackets clearly vanishes at $x = 0$; and the same is true when $n = 1, 2, \ldots$, because then $X(0) = J_n(0) = 0$. Consequently,

$$2\alpha^2\int_0^c x[X(x)]^2\,dx = [cX'(c)]^2 + (\alpha^2c^2 - n^2)[X(c)]^2.$$

Thus, in view of expression (3) and because $X'(c) = \alpha J_n'(\alpha c)$,

$$(5) \qquad 2\alpha^2 \int_0^c x[J_n(\alpha x)]^2\,dx = (\alpha c)^2[J_n'(\alpha c)]^2 + [(\alpha c)^2 - n^2][J_n(\alpha c)]^2.$$

We shall now use expression (5) to find all but one of the norms in the above theorems.

As for case (a) in these two theorems, since $J_n(\alpha_j c) = 0$, equation (5) tells us that

$$2\int_0^c x[J_n(\alpha_j x)]^2\,dx = c^2[J_n'(\alpha_j c)]^2.$$

The integral here is the square of the norm of $J_n(\alpha_j x)$ on the interval $0 < x < c$, with weight function x. Also (Sec. 73)

$$xJ_n'(x) = nJ_n(x) - xJ_{n+1}(x),$$

and it follows that $J_n'(\alpha_j c) = -J_{n+1}(\alpha_j c)$. Hence

$$2\|J_n(\alpha_j x)\|^2 = c^2[J_{n+1}(\alpha_j c)]^2.$$

This is in agreement with the expressions for the norms in case (a) of Theorems 1 and 2.

Turning to case (b) in the theorems, we see that since

$$(6) \qquad (\alpha_j c)^2[J_n'(\alpha_j c)]^2 = h^2[J_n(\alpha_j c)]^2,$$

it follows from equation (5) that

$$2\alpha_j^2 \int_0^c x[J_n(\alpha_j x)]^2\,dx = [(\alpha_j c)^2 - n^2 + h^2][J_n(\alpha_j c)]^2.$$

This gives us the norms in case (b) of the theorems.

Finally, we treat case (c) in Theorem 1 differently from case (c) in Theorem 2. Since $J_0(\alpha_1 x) = 1$ in case (c) of Theorem 1,

$$\|J_0(\alpha_1 x)\|^2 = \int_0^c x\,dx = \frac{c^2}{2}.$$

Expressions for $\|J_0(\alpha_j c)\|^2$ ($j = 2, 3, \ldots$) are obtained by writing $n = 0$ and $\alpha = \alpha_j$ in equation (5). Because $J_0'(\alpha_j c) = 0$, that gives

$$2\int_0^c x[J_0(\alpha_j x)]^2\,dx = c^2[J_0(\alpha_j c)]^2,$$

which is the same as the expression for $\|J_0(\alpha_1 x)\|^2$ ($j = 2, 3, \ldots$) in case (c) of Theorem 1.

Note that in case (b) of Theorem 2 the constant h used in the equation defining the α_j can actually be zero. By canceling out the factor (αc) in the equation

$$(\alpha c)J_n'(\alpha c) = 0$$

that results when $h = 0$, we have the defining equation for the α_j in case (c) of Theorem 2. The norm in case (c) is, moreover, the norm in case (b) with $h = 0$.

This completes the proofs of Theorems 1 and 2, and we are now prepared to find the generalized Fourier series involving the orthonormal sets (1).

81. FOURIER-BESSEL SERIES

Let f be any piecewise continuous function defined on an interval $0 < x < c$. For the normalized eigenfunctions

$$\phi_j(x) = \frac{J_n(\alpha_j x)}{\|J_n(\alpha_j x)\|} \qquad (j = 1, 2, \ldots), \tag{1}$$

stated at the beginning of Sec. 80, the Fourier constants (Sec. 54) in the correspondence

$$f(x) \sim \sum_{j=1}^{\infty} c_j \phi_j(x) \qquad (0 < x < c) \tag{2}$$

are the numbers

$$c_j = (f, \phi_j) = \int_0^c x f(x)\phi_j(x)\,dx = \frac{1}{\|J_n(\alpha_j x)\|}\int_0^c x f(x) J_n(\alpha_j x)\,dx \qquad (j = 1, 2, \ldots). \tag{3}$$

The norms found in each of the three cases (a), (b), and (c) in the theorems of Sec. 80 will now be combined with the generalized Fourier series (2) and its coefficients (3) to obtain two theorems involving what are known as *Fourier-Bessel series*. We continue to present our theorems separately when $n = 0$ and when $n = 1, 2, \ldots$. Indeed, these theorems will follow from Theorems 1 and 2, respectively, in Sec. 80.

Theorem 1. *Suppose that a function f is piecewise continuous on an interval $0 < x < c$.*

(a) *If α_j $(j = 1, 2, \ldots)$ are the positive roots of the equation*

$$J_0(\alpha c) = 0,$$

then

$$f(x) \sim \sum_{j=1}^{\infty} A_j J_0(\alpha_j x) \qquad (0 < x < c)$$

where

$$A_j = \frac{2}{c^2[J_1(\alpha_j c)]^2}\int_0^c x f(x) J_0(\alpha_j x)\,dx \qquad (j = 1, 2, \ldots).$$

(b) *If α_j $(j = 1, 2, \ldots)$ are the positive roots of the equation*

$$hJ_0(\alpha c) + (\alpha c)J_0'(\alpha c) = 0 \qquad (h > 0),$$

then

$$f(x) \sim \sum_{j=1}^{\infty} A_j J_0(\alpha_j x) \qquad (0 < x < c)$$

where

$$A_j = \frac{2\alpha_j^2}{[(\alpha_j c)^2 + h^2][J_0(\alpha_j c)]^2} \int_0^c x f(x) J_0(\alpha_j x)\, dx \qquad (j = 1, 2, \ldots).$$

(*c*) *If* α_j $(j = 2, 3, \ldots)$ *are the positive roots of the equation*

$$J_0'(\alpha c) = 0,$$

then

$$f(x) \sim A_1 + \sum_{j=2}^{\infty} A_j J_0(\alpha_j x) \qquad (0 < x < c)$$

where

$$A_1 = \frac{2}{c^2} \int_0^c x f(x)\, dx$$

and

$$A_j = \frac{2}{c^2[J_0(\alpha_j c)]^2} \int_0^c x f(x) J_0(\alpha_j x)\, dx \qquad (j = 2, 3, \ldots).$$

Note that the equations used to define the numbers α_j in parts (*b*) and (*c*) can also be written

$$hJ_0(\alpha c) - (\alpha c) J_1(\alpha c) = 0 \qquad \text{and} \qquad J_1(\alpha c) = 0,$$

respectively. (See the remarks immediately following the statement of Theorem 1 in Sec. 78.) Also, the number $\alpha_1 = 0$ in case (*c*) of Theorem 1 in Sec. 80 does not appear in case (*c*) of Theorem 1 here since $J_0(\alpha_1 x) = 1$.

Theorem 2. *Let* A_j $(j = 1, 2, \ldots)$ *be the coefficients in the correspondence*

$$f(x) \sim \sum_{j=1}^{\infty} A_j J_n(\alpha_j x) \qquad (0 < x < c),$$

where f *is piecewise continuous and* n *is a positive integer* $(n = 1, 2, \ldots)$.

(*a*) *If* α_j $(j = 1, 2, \ldots)$ *are the positive roots of the equation*

$$J_n(\alpha c) = 0,$$

then

$$A_j = \frac{2}{c^2[J_{n+1}(\alpha_j c)]^2} \int_0^c x f(x) J_n(\alpha_j x)\, dx \qquad (j = 1, 2, \ldots).$$

(*b*) *If* α_j $(j = 1, 2, \ldots)$ *are the positive roots of the equation*

$$hJ_n(\alpha c) + (\alpha c) J_n'(\alpha c) = 0 \qquad (h > 0),$$

then

$$A_j = \frac{2\alpha_j^2}{[(\alpha_j c)^2 - n^2 + h^2][J_n(\alpha_j c)]^2} \int_0^c x f(x) J_n(\alpha_j x)\, dx \qquad (j = 1, 2, \ldots).$$

(*c*) *If* α_j $(j = 1, 2, \ldots)$ *are the positive roots of the equation*

$$J_n'(\alpha c) = 0,$$

then

$$A_j = \frac{2\alpha_j^2}{[(\alpha_j c)^2 - n^2][J_n(\alpha_j c)]^2} \int_0^c x f(x) J_n(\alpha_j x)\, dx \qquad (j = 1, 2, \ldots).$$

Recalling a remark just after the statement of Theorem 2 in Sec. 78, we note that the equations used to define the α_j in cases (*b*) and (*c*) here can also be written

$$(h + n) J_n(\alpha c) - (\alpha c) J_{n+1}(\alpha c) = 0 \quad \text{and} \quad n J_n(\alpha c) - (\alpha c) J_{n+1}(\alpha c) = 0,$$

respectively.

To verify these theorems, we let n have any one of the values $n = 0, 1, 2, \ldots$ and use expression (1) to write correspondence (2) as

$$f(x) \sim \sum_{j=1}^{\infty} \frac{c_j}{\|J_n(\alpha_j x)\|} J_n(\alpha_j x) \qquad (0 < x < c). \tag{4}$$

Then if we write

$$A_j = \frac{c_j}{\|J_n(\alpha_j x)\|} \qquad (j = 1, 2, \ldots) \tag{5}$$

and refer to the last of expressions (3) for c_j, we have

$$f(x) \sim \sum_{j=1}^{\infty} A_j J_n(\alpha_j x) \qquad (0 < x < c), \tag{6}$$

where

$$A_j = \frac{1}{\|J_n(\alpha_j x)\|^2} \int_0^c x f(x) J_n(\alpha_j x)\, dx \qquad (j = 1, 2, \ldots). \tag{7}$$

The norms in the theorems in Sec. 80 can now be used in this equation to obtain the expressions for the A_j in the theorems here.

Proofs that correspondence (6) is actually an equality, under conditions similar to those used to ensure the representation of a function by its Fourier cosine or sine series, usually involve the theory of functions of a complex variable. We state, without proof, one form of such a representation theorem and refer the reader to the Bibliography.†

†This theorem is proved in the book by Watson (1995). Also, see the work by Titchmarsh (1962), as well as the books by Gray and Mathews (1966) and Bowman (1958). These are all listed in the Bibliography.

Theorem 3. *Let f denote a function that is piecewise smooth on an interval $0 < x < c$, and suppose that $f(x)$ at each point of discontinuity of f in that interval is defined as the mean value of the one-sided limits $f(x+)$ and $f(x-)$. Then all the correspondences in Theorems 1 and 2 are, in fact, equalities when the appropriate coefficients A_j are used.*

Theorem 3 and the theorems in Secs. 76 and 77 are also valid when n is replaced by an arbitrary real number ν ($\nu > -1/2$), although we have not developed properties of the function J_ν far enough to establish any such generalizations.

For functions f on the unbounded interval $x > 0$, there is an integral representation in terms of J_ν, analogous to the Fourier cosine and sine integral formulas in Sec. 48. The representation, for a fixed ν ($\nu > -1/2$), is

$$f(x) = \int_0^\infty \alpha J_\nu(\alpha x) \int_0^\infty s f(s) J_\nu(\alpha s)\, ds\, d\alpha \qquad (x > 0)$$

and is known as *Hankel's integral formula.*† It is valid if f is piecewise smooth on each bounded interval, if $\sqrt{x} f(x)$ is absolutely integrable from zero to infinity, and if $f(x)$ is defined as the mean value of $f(x+)$ and $f(x-)$ at each point of discontinuity of f.

If the interval $0 \le x \le c$ is replaced by some interval $a \le x \le b$, where $a > 0$, the Sturm-Liouville problem (1)–(2) in Sec. 78 is no longer singular when the same differential equation is used and boundary conditions of the type (2) there are applied at *each* endpoint. In general, the resulting eigenfunctions then involve both of the Bessel functions J_n and Y_n, and the coefficients in the series representations arising are considerably more involved.

82. EXAMPLES

We now illustrate, before turning to physical applications, the use of Theorems 1 and 2 in Sec. 81. The functions in this section and in the problems to follow all satisfy the conditions in Theorem 3 of Sec. 81, and so convergence will not be in doubt.

EXAMPLE 1. Let us expand the function $f(x) = 1$ ($0 < x < c$) into a series of the type

$$\sum_{j=1}^{\infty} A_j J_0(\alpha_j x),$$

where α_j ($j = 1, 2, \ldots$) are the positive roots of the equation $J_0(\alpha c) = 0$. Case (a) in Theorem 1 of Sec. 81 is evidently applicable here. It tells us that

†See the book by Sneddon (1995, chap. 2), which is listed in the Bibliography. For a summary of representations in terms of Bessel functions, see the work edited by Erdélyi (1981, vol. 2, chap. 7), which is also listed there.

$$A_j = \frac{2}{c^2[J_1(\alpha_j c)]^2} \int_0^c x J_0(\alpha_j x)\, dx \qquad (j = 1, 2, \ldots).$$

This integral is readily evaluated by substituting $s = \alpha_j x$ and using the integration formula (Sec. 73)

$$\int_0^x s J_0(s)\, ds = x J_1(x).$$

To be specific,

$$\int_0^c x J_0(\alpha_j x)\, dx = \frac{1}{\alpha_j^2} \int_0^{\alpha_j c} s J_0(s)\, ds = \frac{c}{\alpha_j} J_1(\alpha_j c).$$

Consequently,

$$A_j = \frac{2}{c^2[J_1(\alpha_j c)]^2} \cdot \frac{c J_1(\alpha_j c)}{\alpha_j} = \frac{2}{c} \cdot \frac{1}{\alpha_j J_1(\alpha_j c)};$$

and we arrive at the expansion

$$1 = \frac{2}{c} \sum_{j=1}^{\infty} \frac{J_0(\alpha_j x)}{\alpha_j J_1(\alpha_j c)} \qquad (0 < x < c), \tag{1}$$

where $J_0(\alpha_j c) = 0$ $(\alpha_j > 0)$.

EXAMPLE 2. To represent the function $f(x) = x$ $(0 < x < 1)$ in a series of the form

$$A_1 + \sum_{j=2}^{\infty} A_j J_0(\alpha_j x),$$

where α_j $(j = 2, 3, \ldots)$ are the positive roots of the equation $J_1(\alpha) = 0$, we refer to case (c) in Theorem 1 of Sec. 81. Evidently,

$$A_1 = 2 \int_0^1 x^2\, dx = \frac{2}{3}$$

and

$$A_j = \frac{2}{[J_0(\alpha_j)]^2} \int_0^1 x^2 J_0(\alpha_j x)\, dx \qquad (j = 2, 3, \ldots).$$

This last integral can be evaluated by referring to the reduction formula (see Sec. 73 and the footnote there)

$$\int_0^x s^2 J_0(s)\, ds = x^2 J_1(x) + x J_0(x) - \int_0^x J_0(s)\, ds$$

and recalling that $J_1(\alpha_j) = 0$:

$$\int_0^1 x^2 J_0(\alpha_j x)\, dx = \frac{1}{\alpha_j^3} \int_0^{\alpha_j} s^2 J_0(s)\, ds = \frac{1}{\alpha_j^3} \left[\alpha_j J_0(\alpha_j) - \int_0^{\alpha_j} J_0(s)\, ds \right].$$

Finally, then,

$$(2) \qquad x = \frac{2}{3} + 2\sum_{j=2}^{\infty}\left[\alpha_1 J_0(\alpha_j) - \int_0^{\alpha_j} J_0(s)\,ds\right]\frac{J_0(\alpha_j x)}{\alpha_j^3[J_0(\alpha_j)]^2} \qquad (0 < x < 1),$$

where $J_1(\alpha_j) = 0$ $(\alpha_j > 0)$.

EXAMPLE 3. Consider a function f that is defined on the interval $0 < x < 2$ by the equations

$$f(x) = \begin{cases} x^4 & \text{when } 0 < x < 1, \\ 0 & \text{when } 1 < x < 2, \end{cases}$$

and $f(1) = 1/2$. To show that

$$(3) \qquad f(x) = \frac{1}{2}\sum_{j=1}^{\infty}\frac{\alpha_j J_5(\alpha_j) J_4(\alpha_j x)}{(\alpha_j^2 - 4)[J_4(2\alpha_j)]^2} \qquad (0 < x < 2),$$

where α_j are the positive roots of the equation $J_4'(2\alpha) = 0$, we refer to case (c) in Theorem 2 of Sec. 81 and write

$$(4) \qquad A_j = \frac{\alpha_j^2}{2(\alpha_j^2 - 4)[J_4(2\alpha_j)]^2}\int_0^2 x f(x) J_4(\alpha_j x)\,dx \qquad (j = 1, 2, \ldots).$$

Keeping in mind that $f(x) = 0$ when $1 < x < 2$ and using the integration formula (see Sec. 73)

$$\int_0^x s^5 J_4(s)\,ds = x^5 J_5(x),$$

we find that the integral in expression (4) reduces to

$$\int_0^1 x^5 J_4(\alpha_j x)\,dx = \frac{1}{\alpha_j^6}\int_0^{\alpha_j} s^5 J_4(s)\,ds = \frac{1}{\alpha_j} J_5(\alpha_j).$$

Hence

$$A_j = \frac{1}{2}\cdot\frac{\alpha_j J_5(\alpha_j)}{(\alpha_j^2 - 4)[J_4(2\alpha_j)]^2},$$

and expansion (3) is established.

PROBLEMS

1. Show that

$$x = 2\sum_{j=1}^{\infty}\left[1 - \frac{1}{\alpha_j^2 J_1(\alpha_j)}\int_0^{\alpha_j} J_0(s)\,ds\right]\frac{J_0(\alpha_j x)}{\alpha_j J_1(\alpha_j)} \qquad (0 < x < 1),$$

where α_j $(j = 1, 2, \ldots)$ are the positive roots of the equation $J_0(\alpha) = 0$.

2. Derive the representation

$$x^2 = \frac{c^2}{2} + 4\sum_{j=2}^{\infty} \frac{J_0(\alpha_j x)}{\alpha_j^2 J_0(\alpha_j c)} \qquad (0 < x < c),$$

where α_j $(j = 2, 3, \ldots)$ are the positive roots of the equation $J_1(\alpha c) = 0$.

3. Show that if

$$f(x) = \begin{cases} 1 & \text{when } 0 < x < 1, \\ 0 & \text{when } 1 < x < 2, \end{cases}$$

and $f(1) = 1/2$, then

$$f(x) = \frac{1}{2}\sum_{j=1}^{\infty} \frac{J_1(\alpha_j)}{\alpha_j [J_1(2\alpha_j)]^2} J_0(\alpha_j x) \qquad (0 < x < 2),$$

where α_j $(j = 1, 2, \ldots)$ are the positive roots of the equation $J_0(2\alpha) = 0$.

4. Let α_j $(j = 1, 2, \ldots)$ denote the positive roots of the equation $J_0(\alpha c) = 0$, where c is a fixed positive number.

(*a*) Derive the expansion

$$x^2 = \frac{2}{c}\sum_{j=1}^{\infty} \frac{(\alpha_j c)^2 - 4}{\alpha_j^3 J_1(\alpha_j c)} J_0(\alpha_j x) \qquad (0 < x < c).$$

(*b*) Combine expansion (1) in Example 1, Sec. 82, with the one in part (*a*) here to show that

$$c^2 - x^2 = \frac{8}{c}\sum_{j=1}^{\infty} \frac{J_0(\alpha_j x)}{\alpha_j^3 J_1(\alpha_j c)} \qquad (0 < x < c).$$

5. Find the coefficients A_j $(j = 1, 2, \ldots)$ in the expansion

$$1 = A_1 + \sum_{j=2}^{\infty} A_j J_0(\alpha_j x) \qquad (0 < x < c)$$

when α_j $(j = 2, 3, \ldots)$ are the positive roots of the equation $J_0'(\alpha c) = 0$.

Answer: $A_1 = 1,\ A_j = 0\ (j = 2, 3, \ldots)$.

6. (*a*) Obtain the representation

$$1 = 2c\sum_{j=1}^{\infty} \frac{\alpha_j J_1(\alpha_j c) J_0(\alpha_j x)}{[(\alpha_j c)^2 + h^2][J_0(\alpha_j c)]^2} \qquad (0 < x < c),$$

where α_j $(j = 1, 2, \ldots)$ are the positive roots of the equation

$$hJ_0(\alpha c) + (\alpha c)J_0'(\alpha c) = 0 \qquad (h > 0).$$

[Compare this representation with representation (1) in Example 1, Sec. 82, and the one in Problem 5.]

(*b*) Show how the result in part (*a*) can be written in the form

$$1 = \frac{2}{c}\sum_{j=1}^{\infty} \frac{1}{\alpha_j} \cdot \frac{J_1(\alpha_j c) J_0(\alpha_j x)}{[J_0(\alpha_j c)]^2 + [J_1(\alpha_j c)]^2} \qquad (0 < x < c).$$

7. Show that if

$$f(x) = \begin{cases} x & \text{when } 0 < x < 1, \\ 0 & \text{when } 1 < x < 2, \end{cases}$$

and $f(1) = 1/2$, then

$$f(x) = 2\sum_{j=1}^{\infty} \frac{\alpha_j J_2(\alpha_j)}{(4\alpha_j^2 - 1)\,[J_1(2\alpha_j)]^2} J_1(\alpha_j x) \qquad (0 < x < 2),$$

where α_j $(j = 1, 2, \ldots)$ are the positive roots of the equation $J_1'(2\alpha) = 0$.

8. Let n have any one of the positive values $n = 1, 2, \ldots$. Show that

$$x^n = 2\sum_{j=1}^{\infty} \frac{\alpha_j J_{n+1}(\alpha_j)}{(\alpha_j^2 - n^2)[J_n(\alpha_j)]^2} J_n(\alpha_j x) \qquad (0 < x < 1),$$

where α_j $(j = 1, 2, \ldots)$ are the positive roots of the equation $J_n'(\alpha) = 0$.

9. Point out why the eigenvalues of the singular Sturm-Liouville problem

$$x^2 X'' + xX' + (\lambda x^2 - 1)X = 0, \qquad X(1) = 0,$$

on the interval $0 \le x \le 1$ are the numbers $\lambda_j = \alpha_j^2$ $(j = 1, 2, \ldots)$, where α_j are the positive roots of the equation $J_1(\alpha) = 0$, and why the corresponding eigenfunctions are $X_j = J_1(\alpha_j x)$ $(j = 1, 2, \ldots)$. Then obtain the representation

$$x = 2\sum_{j=1}^{\infty} \frac{J_1(\alpha_j x)}{\alpha_j J_2(\alpha_j)} \qquad (0 < x < 1)$$

in terms of those eigenfunctions.

10. As indicated in Sec. 81, there exist conditions on f under which Fourier-Bessel series representations are valid when n is replaced by ν $(\nu > -1/2)$, where ν is not necessarily an integer. In particular, suppose that

$$f(x) = \sum_{j=1}^{\infty} A_j J_{1/2}(\alpha_j x) \qquad (0 < x < c),$$

where $\sqrt{x}\, f(x)$ is piecewise smooth, where α_j are the positive roots of the equation $J_{1/2}(\alpha c) = 0$, and where [compare with case (*a*) in Theorem 2 of Sec. 81]

$$A_j = \frac{2}{c^2[J_{3/2}(\alpha_j c)]^2} \int_0^c x f(x)\, J_{1/2}(\alpha_j x)\, dx \qquad (j = 1, 2, \ldots).$$

Using the expressions [Problems 8(*a*), Sec. 72, and 4, Sec. 73]

$$J_{1/2}(x) = \sqrt{\frac{2}{\pi x}} \sin x \qquad \text{and} \qquad J_{3/2}(x) = \sqrt{\frac{2}{\pi x}} \left(\frac{\sin x}{x} - \cos x \right)$$

to substitute for the Bessel functions involved, show that this Fourier-Bessel series representation is actually the Fourier sine series representation for $\sqrt{x}\, f(x)$ on the interval $0 < x < c$.

83. TEMPERATURES IN A LONG CYLINDER

In the following examples, we shall use Bessel functions to find temperatures in an infinitely long circular cylinder $\rho \leq c$ whose lateral surface $\rho = c$ is subject to simple thermal conditions. Those conditions and others will be such that the temperatures will depend only on the space variable ρ, which is the distance from the axis of the cylinder, and time t. We assume that the material of the solid is homogeneous.

EXAMPLE 1. When the cylinder is as shown in Fig. 65 and the initial temperatures vary only with ρ, the temperatures $u = u(\rho, t)$ in the cylinder satisfy the special case (Sec. 22)

$$u_t = k\left(u_{\rho\rho} + \frac{1}{\rho}u_\rho\right) \qquad (0 < \rho < c,\ t > 0) \tag{1}$$

of the heat equation in cylindrical coordinates and the boundary conditions

$$u(c, t) = 0 \qquad (t > 0), \tag{2}$$

$$u(\rho, 0) = f(\rho) \qquad (0 < \rho < c). \tag{3}$$

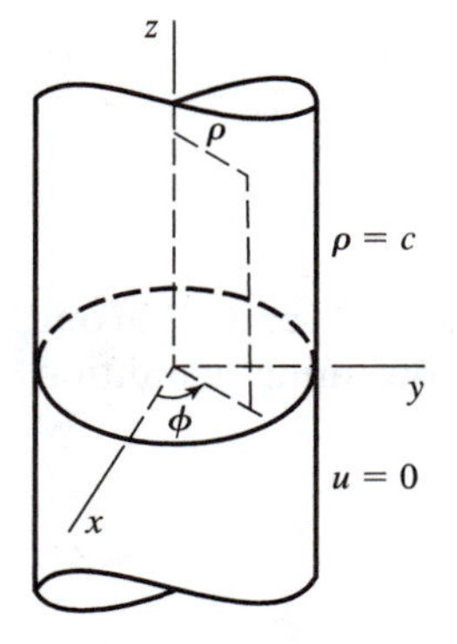

FIGURE 65

Also, when $t > 0$, the function u and its derivatives must be continuous throughout the cylinder and, in particular, on the axis $\rho = 0$. In solving this problem, we assume that f is piecewise smooth on the interval $0 < \rho < c$ and, for convenience, that f is defined as the mean value of its one-sided limits at each point in that interval where it is discontinuous.

Any solutions of the homogeneous equations (1) and (2) that are of the type $u = R(\rho)T(t)$ must satisfy the conditions

$$RT' = kT\left(R'' + \frac{1}{\rho}R'\right), \qquad R(c)T(t) = 0.$$

Separating variables in the first equation here, we have

$$\frac{T'}{kT} = \frac{1}{R}\left(R'' + \frac{1}{\rho}R'\right) = -\lambda,$$

where $-\lambda$ is a separation constant. Thus

$$\rho R''(\rho) + R'(\rho) + \lambda\rho R(\rho) = 0, \qquad R(c) = 0 \qquad (0 < \rho < c) \tag{4}$$

and

$$T'(t) + \lambda k T(t) = 0 \qquad (t > 0). \tag{5}$$

Problem (4), together with continuity conditions on R and R' on the interval $0 \le \rho \le c$, is the singular Sturm-Liouville problem in Theorem 1 of Sec. 78 when the boundary condition (4) there is taken. According to case (a) in that theorem, the eigenvalues and corresponding eigenfunctions of problem (4) here are

$$\lambda_j = \alpha_j^2, \qquad R_j = J_0(\alpha_j\rho) \qquad (j = 1, 2, \ldots),$$

where α_j are the positive roots of the equation $J_0(\alpha c) = 0$.

When $\lambda = \lambda_j$, equation (5) is satisfied by $T_j = \exp(-\alpha_j^2 kt)$. So the desired products are

$$u_j = R_j(\rho)T_j(t) = J_0(\alpha_j\rho)\exp\left(-\alpha_j^2 kt\right) \qquad (j = 1, 2, \ldots);$$

and the generalized linear combination,

$$u(\rho, t) = \sum_{j=1}^{\infty} A_j J_0(\alpha_j\rho)\exp\left(-\alpha_j^2 kt\right) \tag{6}$$

of these functions formally satisfies the homogeneous conditions (1) and (2) in our boundary value problem. It also satisfies the nonhomogeneous initial condition (3) when the coefficients A_j are such that

$$f(\rho) = \sum_{j=1}^{\infty} A_j J_0(\alpha_j\rho) \qquad (0 < \rho < c).$$

This is a valid Fourier-Bessel series representation (Theorem 3, Sec. 81) if the coefficients have the values

$$A_j = \frac{2}{c^2[J_1(\alpha_j c)]^2}\int_0^c \rho f(\rho) J_0(\alpha_j\rho)\,d\rho \qquad (j = 1, 2, \ldots), \tag{7}$$

obtained from case (a) of Theorem 1 in Sec. 81.

The formal solution of the boundary value problem is, therefore, given by equation (6) with the coefficients (7), where $J_0(\alpha_j c) = 0\,(\alpha_j > 0)$. Our temperature function can also be written

$$u(\rho, t) = \frac{2}{c^2}\sum_{j=1}^{\infty}\frac{J_0(\alpha_j\rho)}{[J_1(\alpha_j c)]^2}\exp\left(-\alpha_j^2 kt\right)\int_0^c s f(s) J_0(\alpha_j s)\,ds. \tag{8}$$

EXAMPLE 2. Let us replace the condition that the surface of the infinite cylinder in Example 1 be at temperature zero by the condition that it be insulated. The boundary value problem for the temperature function $u(\rho, t)$ is now

$$u_t = k\left(u_{\rho\rho} + \frac{1}{\rho}u_\rho\right) \qquad (0 < \rho < c,\ t > 0), \tag{9}$$

$$u_\rho(c, t) = 0 \qquad (t > 0), \tag{10}$$

$$u(\rho, 0) = f(\rho) \qquad (0 < \rho < c). \tag{11}$$

When $u = R(\rho)T(t)$, separation of variables produces the eigenvalue problem

$$\rho R''(\rho) + R'(\rho) + \lambda\rho R(\rho) = 0, \qquad R'(c) = 0 \qquad (0 < \rho < c). \tag{12}$$

Case (*c*) of Theorem 1 in Sec. 78 tells us that the eigenvalues and corresponding eigenfunctions are

$$\lambda_1 = 0, \quad R_1 = 1, \qquad \text{and} \qquad \lambda_j = \alpha_j^2, \quad R_j = J_0(\alpha_j\rho) \qquad (j = 2, 3, \ldots),$$

where α_j $(j = 2, 3, \ldots)$ are the positive roots of the equation $J_0'(\alpha c) = 0$. Since

$$T'(t) + \lambda k T(t) = 0 \qquad (t > 0),$$

the corresponding functions of t are constant multiples of

$$T_1 = 1 \qquad \text{and} \qquad T_j = \exp\left(-\alpha_j^2 kt\right) \qquad (j = 2, 3, \ldots).$$

Hence

$$u_1 = R_1(\rho)T_1(t) = 1$$

and

$$u_j = R_j(\rho)T_j(t) = J_0(\alpha_j\rho)\exp\left(-\alpha_j^2 kt\right) \qquad (j = 2, 3, \ldots).$$

The desired solution of our boundary value problem is, therefore,

$$u(\rho, t) = A_1 + \sum_{j=2}^{\infty} A_j J_0(\alpha_j\rho)\exp\left(-\alpha_j^2 kt\right) \tag{13}$$

where the coefficients A_1 and A_j $(j = 2, 3, \ldots)$ are obtained by writing $t = 0$ and referring to the nonhomogeneous condition (11). More precisely,

$$f(\rho) = A_1 + \sum_{j=2}^{\infty} A_j J_0(\alpha_j\rho) \qquad (0 < \rho < c);$$

and case (*c*) of Theorem 1 in Sec. 81 reveals that

$$A_1 = \frac{2}{c^2}\int_0^c \rho f(\rho)\,d\rho \tag{14}$$

and

$$A_j = \frac{2}{c^2[J_0(\alpha_j c)]^2}\int_0^c \rho f(\rho)J_0(\alpha_j\rho)\,d\rho \qquad (j = 2, 3, \ldots). \tag{15}$$

A number of *steady-state* temperature problems in cylindrical coordinates, giving rise to Bessel functions, appear in the problems to follow. In those problems, the temperatures will continue to be independent of ϕ. The function $u = u(\rho, z)$ will, then, be harmonic and satisfy Laplace's equation $\nabla^2 u = 0$, where (see Sec. 22)

$$\nabla^2 u = u_{\rho\rho} + \frac{1}{\rho} u_\rho + u_{zz}. \tag{16}$$

PROBLEMS

1. Let $u(\rho, t)$ denote the solution found in Example 1, Sec. 83, when $c = 1$ and $f(\rho) = u_0$, where u_0 is a constant. With the aid of the table in Sec. 76, show that the first three terms in the series for $u(\rho, t)$ are, approximately, as follows:

$$u(\rho, t) = 2u_0 [0.80\, J_0(2.4\rho) e^{-5.8kt} - 0.53\, J_0(5.5\rho) e^{-30kt} + 0.43\, J_0(8.7\rho) e^{-76kt} - \cdots].$$

2. Suppose that the surface $\rho = c$ of the infinite cylinder in Sec. 83 is such that heat transfer takes place there into surroundings at temperature zero according to Newton's law of cooling (Sec. 24):

$$K u_\rho(c, t) = -H u(c, t) \qquad (K > 0, H > 0).$$

Rewrite this condition as

$$c u_\rho(c, t) = -h u(c, t) \qquad \left(h = \frac{cH}{K} > 0\right)$$

and then use it, instead of conditions (2) and (10) in Sec. 83, to find the temperatures $u(\rho, t)$ in the cylinder.

Answer: $u(\rho, t) = \sum_{j=1}^{\infty} A_j\, J_0(\alpha_j \rho) \exp\left(-\alpha_j^2 kt\right)$,

where α_j are the positive roots of the equation $hJ_0(\alpha c) + (\alpha c) J_0'(\alpha c) = 0$ and

$$A_j = \frac{2\alpha_j^2}{[(\alpha_j c)^2 + h^2][J_0(\alpha_j c)]^2} \int_0^c \rho f(\rho) J_0(\alpha_j \rho)\, d\rho \qquad (j = 1, 2, \ldots).$$

3. Derive an expression for the steady temperatures $u(\rho, z)$ in the solid cylinder formed by the three surfaces $\rho = 1$, $z = 0$, and $z = 1$ when $u = 0$ on the side, the bottom is insulated, and $u = 1$ on the top.

Answer: $u(\rho, z) = 2 \sum_{j=1}^{\infty} \frac{J_0(\alpha_j \rho)}{\alpha_j J_1(\alpha_j)} \cdot \frac{\cosh \alpha_j z}{\cosh \alpha_j}$, where $J_0(\alpha_j) = 0$ $(\alpha_j > 0)$.

4. Find the bounded steady temperatures $u(\rho, z)$ in the semi-infinite cylinder $\rho \le 1, z \ge 0$ when $u = 1$ on the base and there is heat transfer into surroundings at temperature zero, according to Newton's law (see Problem 2), at the surface $\rho = 1, z > 0$.

Answer: $u(\rho, z) = 2h \sum_{j=1}^{\infty} \frac{J_0(\alpha_j \rho) \exp(-\alpha_j z)}{J_0(\alpha_j)\left(\alpha_j^2 + h^2\right)}$,

where $hJ_0(\alpha_j) + \alpha_j J_0'(\alpha_j) = 0$ $(\alpha_j > 0)$.

5. (*a*) A solid cylinder is formed by the three surfaces $\rho = 1$, $z = 0$, and $z = b$ $(b > 0)$. The side is insulated, the bottom is kept at temperature zero, and the top is kept at temperatures $f(\rho)$. Derive this expression for the steady temperatures $u(\rho, z)$

in the cylinder:

$$u(\rho, z) = \frac{2z}{b}\int_0^1 sf(s)\,ds + 2\sum_{j=2}^{\infty}\frac{J_0(\alpha_j\rho)}{[J_0(\alpha_j)]^2}\cdot\frac{\sinh\alpha_j z}{\sinh\alpha_j b}\int_0^1 sf(s)J_0(\alpha_j s)\,ds,$$

where $\alpha_2, \alpha_3, \ldots$ are the positive roots of the equation $J_1(\alpha) = 0$.

(*b*) Show that when $f(\rho) = 1$ $(0 < \rho < 1)$ in part (*a*), the solution there reduces to

$$u(\rho, z) = \frac{z}{b}.$$

6. A function $u(\rho, z)$ is harmonic interior to the cylinder formed by the three surfaces $\rho = c$, $z = 0$, and $z = b$ $(b > 0)$. Assuming that $u = 0$ on the first two of those surfaces and that $u(\rho, b) = f(\rho)$ $(0 < \rho < c)$, derive the expression

$$u(\rho, z) = \sum_{j=1}^{\infty} A_j J_0(\alpha_j\rho)\frac{\sinh\alpha_j z}{\sinh\alpha_j b},$$

where α_j are the positive roots of the equation $J_0(\alpha c) = 0$ and the coefficients A_j are given by equation (7), Sec. 83.

7. Solve this Dirichlet problem (Sec. 28) for $u(\rho, z)$:

$$\nabla^2 u = 0 \qquad (0 < \rho < 1, z > 0),$$

$$u(1, z) = 0, \qquad u(\rho, 0) = 1,$$

and u is to be bounded in the domain $\rho < 1$, $z > 0$.

Answer: $u(\rho, z) = 2\sum_{j=1}^{\infty}\frac{J_0(\alpha_j\rho)}{\alpha_j J_1(\alpha_j)}\exp(-\alpha_j z)$, where $J_0(\alpha_j) = 0$ $(\alpha_j > 0)$.

8. Solve the following problem for temperatures $u(\rho, t)$ in a thin circular plate with heat transfer from its faces into surroundings at temperature zero:

$$u_t = u_{\rho\rho} + \frac{1}{\rho}u_\rho - bu \qquad (0 < \rho < 1, t > 0),$$

$$u(1, t) = 0, \qquad u(\rho, 0) = 1,$$

where b is a positive constant.

Answer: $u(\rho, t) = 2\exp(-bt)\sum_{j=1}^{\infty}\frac{J_0(\alpha_j\rho)}{\alpha_j J_1(\alpha_j)}\exp\left(-\alpha_j^2 t\right)$,

where $J_0(\alpha_j) = 0$ $(\alpha_j > 0)$.

9. Solve Problem 8 after replacing the condition $u(1, t) = 0$ by this heat-transfer condition at the edge:

$$u_\rho(1, t) = -hu(1, t) \qquad (h > 0).$$

10. Solve this boundary value problem for $u(x, t)$:

$$xu_t = (xu_x)_x - \frac{n^2}{x}u \qquad (0 < x < c, t > 0),$$

$$u(c, t) = 0, \qquad \mu(x, 0) = f(x),$$

where u is continuous for $0 \le x \le c$, $t > 0$ and where n is a positive integer.

Answer: $u(x, t) = \sum_{j=1}^{\infty} A_j J_n(\alpha_j x)\exp\left(-\alpha_j^2 t\right)$, where α_j and A_j are the constants in case (*a*) of Theorem 2 in Sec. 81.

11. Let $u(\rho, z)$ denote a function that is harmonic interior to the cylinder formed by the three surfaces $\rho = c$, $z = 0$, and $z = b$ $(b > 0)$. Given that $u = 0$ on both the top and bottom of the cylinder and that $u(c, z) = f(z)$ $(0 < z < b)$, derive the expression

$$u(\rho, z) = \sum_{n=1}^{\infty} B_n \frac{I_0(n\pi\rho/b)}{I_0(n\pi c/b)} \sin \frac{n\pi z}{b},$$

where

$$B_n = \frac{2}{b} \int_0^b f(z) \sin \frac{n\pi z}{b}\, dz.$$

[See Problem 10, Sec. 72, as well as the comments immediately following equation (14), Sec. 79, regarding the solutions of the modified Bessel equation.]

12. Let the steady temperatures $u(\rho, z)$ in a semi-infinite cylinder $\rho \le 1$, $z \ge 0$, whose base is insulated, be such that $u(1, z) = f(z)$ where

$$f(z) = \begin{cases} 1 & \text{when } 0 < z < 1, \\ 0 & \text{when } \quad z > 1. \end{cases}$$

With the aid of the Fourier cosine integral formula (Sec. 48), derive the expression

$$u(\rho, z) = \frac{2}{\pi} \int_0^\infty \frac{I_0(\alpha\rho)}{\alpha I_0(\alpha)} \cos \alpha z \sin \alpha \, d\alpha$$

for those temperatures. (See the remarks at the end of Problem 11.)

13. Given a function $f(z)$ that is represented by its Fourier integral formula (Sec. 44) for all real z, derive the following expression for the harmonic function $u(\rho, z)$ inside the infinite cylinder $\rho \le c$, $-\infty < z < \infty$ such that $u(c, z) = f(z)$ $(-\infty < z < \infty)$:

$$u(\rho, z) = \frac{1}{\pi} \int_0^\infty \frac{I_0(\alpha\rho)}{I_0(\alpha c)} \int_{-\infty}^\infty f(s) \cos \alpha(s - z)\, ds\, d\alpha.$$

(See the remarks at the end of Problem 11.)

84. INTERNALLY GENERATED HEAT

When the Fourier method cannot be directly applied to obtain solutions involving Bessel functions, modifications of the method learned in earlier chapters can often be used. We illustrate here one such modification and leave others to the problems.

EXAMPLE. We suppose now that heat is generated at a constant rate per unit volume in the cylinder in Sec. 83 and that the surface and initial temperatures are both zero. The temperatures $u = u(\rho, t)$ must, therefore, satisfy the conditions

$$u_t = k\left(u_{\rho\rho} + \frac{1}{\rho} u_\rho\right) + q_0 \qquad (0 < \rho < c, t > 0), \tag{1}$$

where q_0 is a positive constant (Sec. 21), and

$$u(c, t) = 0, \qquad u(\rho, 0) = 0. \tag{2}$$

The function u is, of course, required to be continuous in the cylinder.

The differential equation (1) is nonhomogeneous because of the constant term q_0, and this suggests that we apply the method of variation of parameters, first used in Sec. 36. To be specific, we know from Example 1, Sec. 83, that without the term q_0, the eigenfunctions

$$R_j = J_0(\alpha_j \rho) \qquad (j = 1, 2, \ldots),$$

where $J_0(\alpha_j c) = 0$ $(\alpha_j > 0)$, arise. So we seek a solution of the present boundary value problem having the form

$$u(\rho, t) = \sum_{j=1}^{\infty} A_j(t)\, J_0(\alpha_j \rho), \tag{3}$$

where the α_j are as just stated.

Substituting this series into equation (1) and noting how the representation

$$q_0 = \frac{2q_0}{c} \sum_{j=1}^{\infty} \frac{J_0(\alpha_j \rho)}{\alpha_j J_1(\alpha_j c)} \qquad (0 < \rho < c)$$

follows immediately from the one obtained in Example 1, Sec. 82, we find that if series (3) is to satisfy equation (1), then

$$\sum_{j=1}^{\infty} A_j'(t)\, J_0(\alpha_j \rho) = k \sum_{j=1}^{\infty} A_j(t) \left[\frac{d^2}{d\rho^2} J_0(\alpha_j \rho) + \frac{1}{\rho} \frac{d}{d\rho} J_0(\alpha_j \rho) \right] + \sum_{j=1}^{\infty} \frac{2q_0\, J_0(\alpha_j \rho)}{c\alpha_j J_1(\alpha_j c)}.$$

But, according to Problem 9, Sec. 72,

$$\frac{d^2}{d\rho^2} J_0(\alpha_j \rho) + \frac{1}{\rho} \frac{d}{d\rho} J_0(\alpha_j \rho) = -\alpha_j^2\, J_0(\alpha_j \rho).$$

Thus

$$\sum_{j=1}^{\infty} \left[A_j'(t) + \alpha_j^2 k\, A_j(t) \right] J_0(\alpha_j \rho) = \sum_{j=1}^{\infty} \frac{2q_0}{c\alpha_j J_1(\alpha_j c)} J_0(\alpha_j \rho);$$

and by equating coefficients on each side of this equation, we arrive at the differential equation

$$A_j'(t) + \alpha_j^2 k\, A_j(t) = \frac{2q_0}{c\alpha_j J_1(\alpha_j c)} \qquad (j = 1, 2, \ldots). \tag{4}$$

Furthermore, in view of the second of conditions (2),

$$\sum_{j=1}^{\infty} A_j(0)\, J_0(\alpha_j \rho) = 0 \qquad (0 < \rho < c).$$

Consequently,

$$A_j(0) = 0 \qquad (j = 1, 2, \ldots). \tag{5}$$

To solve the linear differential equation (4), we multiply each side by the integrating factor

$$\exp \int \alpha_j^2 k\, dt = \exp(\alpha_j^2 kt).$$

This enables us to write the differential equation as

$$\frac{d}{dt}\left[e^{\alpha_j^2 kt} A_j(t)\right] = \frac{2q_0}{c\alpha_j J_1(\alpha_j c)}\, e^{\alpha_j^2 kt}.$$

After replacing t by τ here, we then integrate each side from $\tau = 0$ to $\tau = t$ and recall condition (5). The result is

$$e^{\alpha_j^2 kt} A_j(t) = \frac{2q_0}{ck\alpha_j^3 J_1(\alpha_j c)}\left(e^{\alpha_j^2 kt} - 1\right),$$

or

$$A_j(t) = \frac{2q_0}{ck} \cdot \frac{1 - \exp(-\alpha_j^2 kt)}{\alpha_j^3 J_1(\alpha_j c)} \qquad (j = 1, 2, \ldots). \tag{6}$$

Finally, by substituting this expression for the coefficients $A_j(t)$ into series (3), we arrive at the desired temperature formula:

$$u(\rho, t) = \frac{2q_0}{ck} \sum_{j=1}^{\infty} \frac{1 - \exp(-\alpha_j^2 kt)}{\alpha_j^3 J_1(\alpha_j c)} J_0(\alpha_j \rho), \tag{7}$$

where $J_0(\alpha_j c) = 0$ $(\alpha_j > 0)$.

PROBLEMS

1. Show that the solution (7) of the temperature problem in the example in Sec. 84 can be written

$$u(\rho, t) = \frac{q_0}{4k}\left[c^2 - \rho^2 - \frac{8}{c} \sum_{j=1}^{\infty} \frac{J_0(\alpha_j \rho) \exp(-\alpha_j^2 kt)}{\alpha_j^3 J_1(\alpha_j c)}\right],$$

where $J_0(\alpha_j c) = 0$ $(\alpha_j > 0)$.

Suggestion: Note that according to Problem 4(*b*), Sec. 82,

$$\sum_{j=1}^{\infty} \frac{J_0(\alpha_j \rho)}{\alpha_j^3 J_1(\alpha_j c)} = \frac{c}{8}(c^2 - \rho^2) \qquad (0 < \rho < c).$$

2. In the example in Sec. 84, suppose that the rate per unit volume at which heat is internally generated is $q(t)$, rather than simply q_0. Derive the following generalization of the solution found in that example:

$$u(\rho, t) = \frac{2}{c} \sum_{j=1}^{\infty} \frac{J_0(\alpha_j \rho)}{\alpha_j J_1(\alpha_j c)} \int_0^t q(\tau) \exp\left[-\alpha_j^2 k(t - \tau)\right] d\tau,$$

where $J_0(\alpha_j c) = 0$ $(\alpha_j > 0)$.

3. Solve the temperature problem arising when the boundary conditions (2) and (3) in Example 1, Sec. 83, are replaced by

$$u(c, t) = 1, \qquad u(\rho, 0) = 0.$$

Do this by making the substitution $u(\rho, t) = U(\rho, t) + \Phi(\rho)$ and referring to solution (6), with coefficients (7), in that earlier example and to expansion (1) in Example 1, Sec. 82.

Answer: $$u(\rho, t) = 1 - \frac{2}{c}\sum_{j=1}^{\infty} \frac{J_0(\alpha_j \rho)}{\alpha_j J_1(\alpha_j c)} \exp\left(-\alpha_j^2 kt\right),$$
where $J_0(\alpha_j c) = 0$ $(\alpha_j > 0)$.

4. Use the special case of Duhamel's theorem in Sec. 24 and the solution obtained in Problem 3 above to solve the temperature problem arising when the boundary condition $u(c, t) = 1$ in that problem is replaced by $u(c, t) = F(t)$, where $F(0) = 0$ and $F(t)$ is continuous and differentiable when $t \geq 0$.

Answer: $$u(\rho, t) = \frac{2k}{c}\sum_{j=1}^{\infty} \frac{\alpha_j J_0(\alpha_j \rho)}{J_1(\alpha_j c)} \int_0^t F(\tau) \exp\left[-\alpha_j^2 k(t-\tau)\right] d\tau,$$
where $J_0(\alpha_j c) = 0$ $(\alpha_j > 0)$.

5. Give a physical interpretation of the following boundary value problem for a function $u(\rho, t)$ (see the example in Sec. 84):

$$u_t = u_{\rho\rho} + \frac{1}{\rho} u_\rho + q_0 \qquad (0 < \rho < 1, t > 0),$$

$$u_\rho(1, t) = 0, \qquad u(\rho, 0) = a\rho^2,$$

where q_0 and a are positive constants. Then, after pointing out why it is reasonable to seek a solution of the form

$$u(\rho, t) = A_1(t) + \sum_{j=2}^{\infty} A_j(t)\, J_0(\alpha_j \rho),$$

where α_j $(j = 2, 3, \ldots)$ are the positive roots of the equation $J_1(\alpha_j) = 0$, use the method of variation of parameters to actually find that solution.

Answer: $$u(\rho, t) = \frac{a}{2} + q_0 t + 4a \sum_{j=2}^{\infty} \frac{J_0(\alpha_j \rho) \exp\left(-\alpha_j^2 t\right)}{\alpha_j^2 J_0(\alpha_j)},$$
where the α_j are as stated above.

6. Interpret this boundary value problem as a temperature problem in a cylinder (see Sec. 24):

$$u_t = u_{\rho\rho} + \frac{1}{\rho} u_\rho \qquad (0 < \rho < 1, t > 0),$$

$$u_\rho(1, t) = B, \qquad u(\rho, 0) = 0,$$

where B is a positive constant. Then, after making the substitution

$$u(\rho, t) = U(\rho, t) + \frac{B}{2}\rho^2$$

to obtain a boundary value problem for $U(\rho, t)$, refer to the solution in Problem 5 to derive the temperature formula

$$u(\rho, t) = \frac{B}{4}\left[2\rho^2 + 8t - 1 - 8\sum_{j=2}^{\infty} \frac{J_0(\alpha_j \rho) \exp\left(-\alpha_j^2 t\right)}{\alpha_j^2 J_0(\alpha_j)}\right],$$

where α_j $(j = 2, 3, \ldots)$ are the positive roots of the equation $J_1(\alpha) = 0$. [Note that the substitution for $u(\rho, t)$ made here is suggested by the fact that $U_\rho(1, t) = 0$.]

7. Over a long solid cylinder $\rho \le 1$, at uniform temperature A, there is tightly fitted a long hollow cylinder $1 \le \rho \le 2$ of the same material at temperature B. The outer surface $\rho = 2$ is then kept at temperature B. Let $u(\rho, t)$ denote the temperatures in the cylinder of radius 2 that is formed, and set up the boundary value problem for those temperatures. Then, after making the substitution

$$u(\rho, t) = U(\rho, t) + B$$

to obtain a boundary value problem for $U(\rho, t)$, refer to the solution in Example 1, Sec. 83, to derive the temperature formula

$$u(\rho, t) = B + \frac{A - B}{2} \sum_{j=1}^{\infty} \frac{J_1(\alpha_j)}{\alpha_j [J_1(2\alpha_j)]^2} J_0(\alpha_j \rho) \exp\left(-\alpha_j^2 kt\right),$$

where α_j are the positive roots of the equation $J_0(2\alpha) = 0$. (This is a temperature problem in shrunken fittings.)

8. Solve the boundary value problem

$$u_{\rho\rho} + \frac{1}{\rho} u_\rho - \frac{n^2}{\rho^2} u + u_{zz} = 0 \qquad (0 < \rho < 1, z > 0),$$

$$u(1, z) = 0, \qquad u(\rho, 0) = \rho^n,$$

where $u(\rho, z)$ is bounded and continuous for $0 \le \rho < 1$, $z > 0$ and where n is a positive integer. (When $n = 0$, this problem becomes the Dirichlet problem that was solved in Problem 7, Sec. 83.)

Answer: $u(\rho, z) = 2 \sum_{j=1}^{\infty} \dfrac{J_n(\alpha_j \rho)}{\alpha_j J_{n+1}(\alpha_j)} \exp(-\alpha_j z)$, where $J_n(\alpha_j) = 0$ $(\alpha_j > 0)$.

9. Let the function $u(\rho, \phi, z)$ satisfy Poisson's equation (Sec. 21) $\nabla^2 u + ay = 0$, where a is a constant, inside a semi-infinite half-cylinder $0 \le \rho \le 1, 0 \le \phi \le \pi, z \ge 0$, and suppose that $u = 0$ on the entire surface. The function u, which is assumed to be bounded and continuous for $0 \le \rho < 1, 0 < \phi < \pi, z > 0$, thus satisfies the boundary value problem

$$u_{\rho\rho} + \frac{1}{\rho} u_\rho + \frac{1}{\rho^2} u_{\phi\phi} + u_{zz} + a\rho \sin\phi = 0 \qquad (0 < \rho < 1, 0 < \phi < \pi, z > 0),$$

$$u(1, \phi, z) = 0, \qquad u(\rho, \phi, 0) = 0, \qquad u(\rho, 0, z) = u(\rho, \pi, z) = 0.$$

Use the following method to solve it.

(*a*) By writing $u(\rho, \phi, z) = a \sin\phi\, v(\rho, z)$, reduce the stated problem to the one

$$v_{\rho\rho} + \frac{1}{\rho} v_\rho - \frac{1}{\rho^2} v + v_{zz} + \rho = 0 \qquad (0 < \rho < 1, z > 0),$$

$$v(1, z) = 0, \qquad v(\rho, 0) = 0$$

in $v(\rho, z)$, where v is bounded and continuous for $0 \le \rho < 1, z > 0$.

(*b*) Note that when $n = 1$, the solution in Problem 8 suggests that the method of variation of parameters (see the example in Sec. 84) be used to seek a solution of the form

$$v(\rho, z) = \sum_{j=1}^{\infty} A_j(z)\, J_1(\alpha_j \rho),$$

where $J_1(\alpha_j) = 0$ $(\alpha_j > 0)$, for the problem in part (*a*). Apply that method to obtain the initial value problem

$$A_j''(z) - \alpha_j^2 A_j(z) = -\frac{2}{\alpha_j J_2(\alpha_j)}, \qquad A_j(0) = 0$$

in ordinary differential equations. Then, by adding a particular solution of this differential equation, which is a constant that is readily found by inspection, to the general solution of the complementary equation $A_j''(z) - \alpha_j^2 A_j(z) = 0$ (compare with Problem 2, Sec. 40), find $v(\rho, z)$. Thus obtain the solution

$$u(\rho, \phi, z) = 2a \sin\phi \sum_{j=1}^{\infty} \frac{1 - \exp(-\alpha_j z)}{\alpha_j^3 J_2(\alpha_j)} J_1(\alpha_j \rho),$$

where $J_1(\alpha_j) = 0$ $(\alpha_j > 0)$, of the original problem.

Suggestion: In obtaining the ordinary differential equation for $A_j(z)$ in part (*b*), one can write the needed Fourier-Bessel expansion for ρ by simply referring to the expansion already found in Problem 9, Sec. 82. Also, it is necessary to observe how the identity

$$\frac{d^2}{d\rho^2} J_1(\alpha_j \rho) + \frac{1}{\rho}\frac{d}{d\rho} J_1(\alpha_j \rho) - \frac{1}{\rho^2} J_1(\alpha_j \rho) = -\alpha_j^2 J_1(\alpha_j \rho)$$

follows immediately from the differential equation in that problem.

85. VIBRATION OF A CIRCULAR MEMBRANE

A membrane, stretched over a fixed circular frame $\rho = c$ in the plane $z = 0$, is given an initial displacement $z = f(\rho, \phi)$ and released at rest from that position. The transverse displacements $z(\rho, \phi, t)$, where ρ, ϕ, and z are cylindrical coordinates, satisfy this boundary value problem:

$$z_{tt} = a^2\left(z_{\rho\rho} + \frac{1}{\rho} z_\rho + \frac{1}{\rho^2} z_{\phi\phi}\right), \tag{1}$$

$$z(c, \phi, t) = 0 \qquad (-\pi \le \phi \le \pi, t \ge 0), \tag{2}$$

$$z(\rho, \phi, 0) = f(\rho, \phi), \qquad z_t(\rho, \phi, 0) = 0 \qquad (0 \le \rho \le c, -\pi \le \phi \le \pi), \tag{3}$$

where $z(\rho, \phi, t)$ is periodic with period 2π in the variable ϕ and is also continuous.

A function $z = R(\rho)\Phi(\phi)T(t)$ satisfies equation (1) if

$$\frac{T''}{a^2 T} = \frac{R''}{R} + \frac{R'}{\rho R} + \frac{\Phi''}{\rho^2 \Phi} = -\lambda, \tag{4}$$

where $-\lambda$ is a separation constant. We separate variables again, this time in the second of equations (4), and write

$$\frac{\Phi''}{\Phi} = -\left(\frac{\rho^2 R''}{R} + \frac{\rho R'}{R} + \lambda\rho^2\right) = -\mu,$$

where $-\mu$ is another separation constant. Evidently, then, the product $R\Phi T$ satisfies the homogeneous conditions in our boundary value problem and has the necessary periodicity with respect to ϕ if R and Φ are eigenfunctions of the

Sturm-Liouville problems

$$\rho^2 R''(\rho) + \rho R'(\rho) + (\lambda\rho^2 - \mu)R(\rho) = 0, \qquad R(c) = 0, \tag{5}$$

$$\Phi''(\phi) + \mu\Phi(\phi) = 0, \qquad \Phi(-\pi) = \Phi(\pi), \qquad \Phi'(-\pi) = \Phi'(\pi) \tag{6}$$

and T is such that

$$T''(t) + \lambda a^2 T(t) = 0, \qquad T'(0) = 0.$$

If μ has one of the values

$$\mu_0 = 0, \qquad \mu_n = n^2 \qquad (n = 1, 2, \ldots),$$

results in Sec. 78 can be applied to problem (5); and if we consider problem (6) first, we see that the constant μ must, in fact, have one of those values. According to Sec. 43, they are the eigenvalues of problem (6), the corresponding eigenfunctions being

$$\Phi_0(\phi) = 1 \quad \text{and} \quad \Phi_n(\phi) = A_n \cos n\phi + B_n \sin n\phi \qquad (n = 1, 2, \ldots),$$

where A_n and B_n are arbitrary constants. From case (a) of Theorems 1 and 2 in Sec. 78, we see that the eigenvalues of problem (5) are $\lambda_{nj} = \alpha_{nj}^2$ $(j = 1, 2, \ldots)$, where α_{nj} are the positive roots of the equation

$$J_n(\alpha c) = 0 \qquad (n = 0, 1, 2, \ldots), \tag{7}$$

and that the corresponding eigenfunctions are

$$R_{nj} = J_n(\alpha_{nj}\rho).$$

Also,

$$T_{nj}(t) = \cos \alpha_{nj} at.$$

The products $R\Phi T$ are now seen to be

$$R_{0j}\Phi_0 T_{0j} = J_0(\alpha_{0j}\rho) \cos \alpha_{0j} at \qquad (j = 1, 2, \ldots)$$

and, when $n = 1, 2, \ldots,$

$$R_{nj}\Phi_n T_{nj} = J_n(\alpha_{nj}\rho)(A_n \cos n\phi + B_n \sin n\phi) \cos \alpha_{nj} at \qquad (j = 1, 2, \ldots).$$

The generalized linear combination

$$\begin{aligned} z(\rho, \phi, t) = &\sum_{j=1}^{\infty} A_{0j} J_0(\alpha_{0j}\rho) \cos \alpha_{0j} at \\ &+ \sum_{n=1}^{\infty}\sum_{j=1}^{\infty} J_n(\alpha_{nj}\rho)(A_{nj} \cos n\phi + B_{nj} \sin n\phi) \cos \alpha_{nj} at \end{aligned} \tag{8}$$

of these products formally satisfies all the homogeneous conditions. Observe that when $n = 1, 2, \ldots$ and the products $R_{nj}\Phi_n T_{nj}$ are multiplied by arbitrary constants, those constants are absorbed into the A_n and B_n but that the resulting constants involve the index j. [Compare with the remark immediately following equation (6) in Sec. 43.] Expression (8) also satisfies the nonhomogeneous condition, which is the first of conditions (3), if the constants A_{0j}, A_{nj}, and B_{nj} are such that

$$(9) \quad f(\rho, \phi) = \frac{1}{2}\left[\sum_{j=1}^{\infty} 2A_{0j} J_0(\alpha_{0j}\rho)\right] + \sum_{n=1}^{\infty}\left\{\left[\sum_{j=1}^{\infty} A_{nj} J_n(\alpha_{nj}\rho)\right]\cos n\phi + \left[\sum_{j=1}^{\infty} B_{nj} J_n(\alpha_{nj}\rho)\right]\sin n\phi\right\}$$

when $0 \le \rho \le c, -\pi \le \phi \le \pi$.

For each fixed value of ρ, series (9) is the Fourier series for $f(\rho, \phi)$ on the interval $-\pi \le \phi \le \pi$ if

$$\sum_{j=1}^{\infty} 2\,A_{0j} J_0(\alpha_{0j}\rho) = \frac{1}{\pi}\int_{-\pi}^{\pi} f(\rho, \phi)\, d\phi$$

and

$$\sum_{j=1}^{\infty} A_{nj} J_n(\alpha_{nj}\rho) = \frac{1}{\pi}\int_{-\pi}^{\pi} f(\rho, \phi)\cos n\phi\, d\phi \qquad (n = 1, 2, \ldots),$$

$$\sum_{j=1}^{\infty} B_{nj} J_n(\alpha_{nj}\rho) = \frac{1}{\pi}\int_{-\pi}^{\pi} f(\rho, \phi)\sin n\phi\, d\phi \qquad (n = 1, 2, \ldots).$$

The series on the left-hand side of each of these three equations is the Fourier-Bessel series representation, on the interval $0 < \rho < c$, of the corresponding function of ρ on the right-hand side. Specifically, case (*a*) of Theorems 1 and 2 in Sec. 81 tells us that

$$(10) \qquad A_{0j} = \frac{1}{\pi c^2[J_1(\alpha_{0j}c)]^2}\int_0^c \rho J_0(\alpha_{0j}\rho)\int_{-\pi}^{\pi} f(\rho, \phi)\, d\phi\, d\rho$$

and

$$(11) \qquad A_{nj} = \frac{2}{\pi c^2[J_{n+1}(\alpha_{nj}c)]^2}\int_0^c \rho J_n(\alpha_{nj}\rho)\int_{-\pi}^{\pi} f(\rho, \phi)\cos n\phi\, d\phi\, d\rho,$$

$$(12) \qquad B_{nj} = \frac{2}{\pi c^2[J_{n+1}(\alpha_{nj}c)]^2}\int_0^c \rho J_n(\alpha_{nj}\rho)\int_{-\pi}^{\pi} f(\rho, \phi)\sin n\phi\, d\phi\, d\rho$$

when $n = 1, 2, \ldots$.

The displacements $z(\rho, \phi, t)$ are, then, given by equation (8) when the coefficients have the values (10), (11), and (12). We assume, of course, that the function f is such that the series in expression (8) has adequate properties of convergence and differentiability.

PROBLEMS

1. Suppose that in Sec. 85 the initial displacement function $f(\rho, \phi)$ is a linear combination of a finite number of the functions

$$J_0(\alpha_{0j}\rho) \quad \text{and} \quad J_n(\alpha_{nj}\rho)\cos n\phi, \quad J_n(\alpha_{nj}\rho)\sin n\phi \qquad (n = 1, 2, \ldots).$$

Point out why the iterated series in expression (8) of that section then contains only a finite number of terms and represents a rigorous solution of the boundary value problem.

2. Let the initial displacement of the membrane in Sec. 85 be $f(\rho)$, a function of ρ only, and derive the expression

$$z(\rho, t) = \frac{2}{c^2} \sum_{j=1}^{\infty} \frac{J_0(\alpha_j \rho) \cos \alpha_j at}{[J_1(\alpha_j c)]^2} \int_0^c s f(s) J_0(\alpha_j s)\, ds,$$

where α_j are the positive roots of the equation $J_0(\alpha c) = 0$, for the displacements when $t > 0$.

3. Show that if the initial displacement of the membrane in Sec. 85 is $A J_0(\alpha_k \rho)$, where A is a constant and α_k is some positive root of the equation $J_0(\alpha c) = 0$, then the subsequent displacements are

$$z(\rho, t) = A J_0(\alpha_k \rho) \cos \alpha_k at.$$

Observe that these displacements are all periodic in t with a common period; thus the membrane gives a musical note.

4. Replace the initial conditions (3), Sec. 85, by the conditions that $z = 0$ and $z_t = 1$ when $t = 0$. This is the case if the membrane and its frame are moving with unit velocity in the z direction and the frame is brought to rest at the instant $t = 0$. Derive the expression

$$z(\rho, t) = \frac{2}{ac} \sum_{j=1}^{\infty} \frac{\sin \alpha_j at}{\alpha_j^2 J_1(\alpha_j c)} J_0(\alpha_j \rho),$$

where α_j are the positive roots of the equation $J_0(\alpha c) = 0$, for the displacements when $t > 0$.

5. Suppose that the *damped* transverse displacements $z(\rho, t)$ in a membrane, stretched over a circular frame, satisfy the conditions

$$z_{tt} = z_{\rho\rho} + \frac{1}{\rho} z_\rho - 2b z_t \qquad (0 < \rho < 1, t > 0),$$

$$z(1, t) = 0, \qquad z(\rho, 0) = 0, \qquad z_t(\rho, 0) = v_0.$$

The constant coefficient of damping $2b$ is such that $0 < b < \alpha_1$, where α_1 is the smallest of the positive zeros of $J_0(\alpha)$. Derive the solution

$$z(\rho, t) = 2 v_0 e^{-bt} \sum_{j=1}^{\infty} \frac{J_0(\alpha_j \rho)}{\alpha_j J_1(\alpha_j)} \cdot \frac{\sin\left(t\sqrt{\alpha_j^2 - b^2}\right)}{\sqrt{\alpha_j^2 - b^2}},$$

where $J_0(\alpha_j) = 0$ ($\alpha_j > 0$), of this boundary value problem.

6. Derive the following expression for the temperatures $u(\rho, \phi, t)$ in an infinite cylinder $\rho \le c$ when $u = 0$ on the surface $\rho = c$ and $u = f(\rho, \phi)$ at time $t = 0$:

$$u(\rho, \phi, t) = \sum_{j=1}^{\infty} A_{0j} J_0(\alpha_{0j} \rho) \exp\left(-\alpha_{0j}^2 kt\right)$$
$$+ \sum_{n=1}^{\infty} \sum_{j=1}^{\infty} J_n(\alpha_{nj} \rho)(A_{nj} \cos n\phi + B_{nj} \sin n\phi) \exp\left(-\alpha_{nj}^2 kt\right),$$

where α_{nj}, A_{0j}, A_{nj}, and B_{nj} are the numbers defined in Sec. 85.

7. Derive an expression for the temperatures $u(\rho, z, t)$ in a solid cylinder $\rho \le c, 0 \le z \le \pi$ whose entire surface is kept at temperature zero and whose initial temperature is a constant A. Show that it can be written as the product

$$u(\rho, z, t) = Av(z, t)w(\rho, t)$$

of the constant A and the two functions

$$v(z, t) = \frac{4}{\pi}\sum_{n=1}^{\infty} \frac{\sin(2n-1)z}{2n-1} \exp[-(2n-1)^2 kt]$$

and

$$w(\rho, t) = \frac{2}{c}\sum_{j=1}^{\infty} \frac{J_0(\alpha_j \rho)}{\alpha_j J_1(\alpha_j c)} \exp\left(-\alpha_j^2 kt\right),$$

where α_j are the positive roots of the equation $J_0(\alpha c) = 0$. Also, show that $v(z, t)$ represents temperatures in a slab $0 \le z \le \pi$ and $w(\rho, t)$ temperatures in an infinite cylinder $\rho \le c$, both with zero boundary temperature and unit initial temperature (see Example 1, Sec. 34, and Example 1, Sec. 83).

8. Derive the following expression for temperatures $u(\rho, \phi, t)$ in the long right-angled cylindrical wedge formed by the surface $\rho = 1$ and the planes $\phi = 0$ and $\phi = \pi/2$ when $u = 0$ on its entire surface and $u = f(\rho, \phi)$ at time $t = 0$:

$$u(\rho, \phi, t) = \sum_{n=1}^{\infty}\sum_{j=1}^{\infty} B_{nj} J_{2n}(\alpha_{nj}\rho) \sin 2n\phi \exp\left(-\alpha_{nj}^2 kt\right),$$

where α_{nj} are the positive roots of the equation $J_{2n}(\alpha) = 0$ and

$$B_{nj} = \frac{8}{\pi [J_{2n+1}(\alpha_{nj})]^2} \int_0^{\pi/2} \sin 2n\phi \int_0^1 \rho f(\rho, \phi) J_{2n}(\alpha_{nj}\rho)\, d\rho\, d\phi.$$

9. Show that if the plane $\phi = \pi/2$ in Problem 8 is replaced by a plane $\phi = \phi_0$, the expression for the temperatures in the wedge will, in general, involve Bessel functions J_ν of *nonintegral* orders.

10. Solve Problem 8 when the entire surface of the wedge is insulated, instead of being kept at temperature zero.

CHAPTER 10

LEGENDRE POLYNOMIALS AND APPLICATIONS

As we shall see later in this chapter (Secs. 92 and 93), an application of the method of separation of variables to Laplace's equation in the spherical coordinates r and θ leads, after the substitution $x = \cos\theta$ is made, to *Legendre's equation*

$$[(1 - x^2)y'(x)]' + \lambda y(x) = 0, \tag{1}$$

where λ is a separation constant. The points $x = 1$ and $x = -1$ correspond to $\theta = 0$ and $\theta = \pi$, respectively, and we begin the chapter by using series to discover solutions of equation (1).

86. SOLUTIONS OF LEGENDRE'S EQUATION

To solve Legendre's equation, we write it as

$$(1 - x^2)y''(x) - 2xy'(x) + \lambda y(x) = 0 \tag{1}$$

and observe that, in standard form, it is a special case of

$$y''(x) + P(x)y'(x) + Q(x)y(x) = 0,$$

where each of the functions

$$P(x) = \frac{-2x}{1 - x^2} \qquad \text{and} \qquad Q(x) = \frac{\lambda}{1 - x^2}$$

has a Maclaurin series representation with positive radius of convergence. Thus $x = 0$ is an *ordinary point* of the differential equation (1), and we seek a solution of the form†

$$y = \sum_{j=0}^{\infty} a_j x^j, \tag{2}$$

where the series is assumed to be differentiable.

Writing

$$y' = \sum_{j=0}^{\infty} j a_j x^{j-1} \quad \text{and} \quad y'' = \sum_{j=0}^{\infty} j(j-1) a_j x^{j-2}$$

and then substituting the function (2) and these two derivatives into equation (1), we have

$$\sum_{j=0}^{\infty} j(j-1)a_j x^{j-2} - \sum_{j=0}^{\infty} j(j-1)a_j x^j - \sum_{j=0}^{\infty} 2ja_j x^j + \sum_{j=0}^{\infty} \lambda a_j x^j = 0,$$

which is the same as

$$\sum_{j=0}^{\infty} j(j-1)a_j x^{j-2} - \sum_{j=0}^{\infty} [j(j-1) + 2j - \lambda] a_j x^j = 0.$$

Since the first two terms in the first series here are actually zero and since

$$j(j-1) + 2j = j(j+1)$$

in the second series, we may write

$$\sum_{j=2}^{\infty} j(j-1)a_j x^{j-2} - \sum_{j=0}^{\infty} [j(j+1) - \lambda] a_j x^j = 0.$$

Finally, by putting the second of these series in the form

$$\sum_{j=2}^{\infty} [(j-2)(j-1) - \lambda] a_{j-2} x^{j-2},$$

we arrive at the equation

$$\sum_{j=2}^{\infty} \{ j(j-1)a_j - [(j-2)(j-1) - \lambda] a_{j-2} \} x^{j-2} = 0, \tag{3}$$

involving a single series.

Equation (3) is an identity in x if the coefficients a_j satisfy the recurrence relation

$$a_j = \frac{(j-2)(j-1) - \lambda}{j(j-1)} a_{j-2} \qquad (j = 2, 3, \ldots). \tag{4}$$

The power series (2) thus represents a solution of Legendre's equation within its interval of convergence if its coefficients satisfy relation (4). This leaves a_0 and a_1 as arbitrary constants.

---

†For a discussion of ordinary points and a justification for this, see, for example, the books referred to in the footnote in the introduction to Chap. 9.

If $a_1 = 0$, it follows from relation (4) that $a_3 = a_5 = \cdots = 0$. Thus one nontrivial solution of Legendre's equation, containing only even powers of x, is

$$y_1 = a_0 + \sum_{k=1}^{\infty} a_{2k} x^{2k} \qquad (a_0 \neq 0), \tag{5}$$

where a_0 is an arbitrary nonzero constant and where the remaining coefficients $a_2, a_4, \ldots$ are expressed in terms of a_0 by successive applications of relation (4). (See Problem 4, Sec. 87.) Another solution, containing only odd powers of x, is obtained by writing $a_0 = 0$ and letting a_1 be arbitrary. More precisely, the series

$$y_2 = a_1 x + \sum_{k=1}^{\infty} a_{2k+1} x^{2k+1} \qquad (a_1 \neq 0) \tag{6}$$

satisfies Legendre's equation for any nonzero value of a_1 when $a_3, a_5, \ldots$ are written in terms of a_1 in accordance with relation (4). These two solutions are, of course, linearly independent since they are not constant multiples of each other.

From relation (4), it is clear that the value of λ affects the values of all but the first coefficients in series (5) and (6). As we shall see in Sec. 87, there are certain values of λ that cause series (5) and (6) to terminate and become polynomials. Assuming for the moment that series (5) does not terminate, we note from relation (4), with $j = 2k$, that

$$\lim_{k\to\infty} \left| \frac{a_{2(k+1)} x^{2(k+1)}}{a_{2k} x^{2k}} \right| = \lim_{k\to\infty} \left| \frac{2k(2k+1) - \lambda}{(2k+2)(2k+1)} x^2 \right| = x^2.$$

So, according to the ratio and absolute convergence tests, series (5) converges when $x^2 < 1$ and diverges when $x^2 > 1$. Although it is somewhat more difficult to show, series (5) diverges when $x = \pm 1$.[†]

Similar arguments apply to series (6). In summary, then, *if λ is such that either of the series* (5) *or* (6) *does not terminate and become a polynomial, that series converges only when $-1 < x < 1$.*

87. LEGENDRE POLYNOMIALS

When Legendre's equation

$$(1 - x^2) y''(x) - 2xy'(x) + \lambda y(x) = 0$$

arises in the applications, it will be necessary to have a solution which, along with its derivative, is continuous on the closed interval $-1 \leq x \leq 1$. But we know from Sec. 86 that unless it terminates, neither of the series solutions

$$y_1 = a_0 + \sum_{k=1}^{\infty} a_{2k} x^{2k} \qquad (a_0 \neq 0), \tag{1}$$

[†]See, for instance, the book by Bell (2004), listed in the Bibliography.

$$(2) \qquad y_2 = a_1 x + \sum_{k=1}^{\infty} a_{2k+1} x^{2k+1} \qquad (a_1 \neq 0)$$

obtained there satisfies those continuity conditions.

Suppose now that the parameter λ in Legendre's equation has one of the integral values

$$(3) \qquad \lambda = n(n+1) \qquad (n = 0, 1, 2, \ldots),$$

and then rewrite the recurrence relation (4) in Sec. 86 as

$$(4) \qquad a_{j+2} = \frac{j(j+1) - n(n+1)}{(j+2)(j+1)} a_j \qquad (j = 0, 1, 2, \ldots).$$

Note how it follows from relation (4) here that

$$a_{n+2} = a_{n+4} = a_{n+6} = \cdots = 0.$$

Consequently, *one of the solutions* (1) *and* (2) *is actually a polynomial.*

More precisely, if $n = 0$, then

$$a_2 = a_4 = a_6 = \cdots = 0;$$

and series (1) becomes simply $y_1 = a_0$. Moreover, if n is any one of the even integers $2, 4, 6, \ldots$, that series is evidently a polynomial of degree n:

$$(5) \qquad y_1 = a_0 + a_2 x^2 + \cdots + a_n x^n \qquad (a_n \neq 0).$$

On the other hand, if $n = 1$, we see that $y_2 = a_1 x$, since

$$a_3 = a_5 = a_7 = \cdots = 0;$$

and if n is any one of the odd integers $3, 5, 7, \ldots$, series (2) becomes

$$(6) \qquad y_2 = a_1 x + a_3 x^3 + \cdots + a_n x^n \qquad (a_n \neq 0).$$

Observe that when n is even, solution (2) remains an infinite series and that when n is odd, the same is true of solution (1).

If n is even, it is customary to assign a value to a_0 such that when the coefficients $a_2, \ldots, a_n$ in expression (5) are determined by means of relation (4), the final coefficient a_n has the value

$$(7) \qquad a_n = \frac{(2n)!}{2^n (n!)^2}.$$

The reason for this requirement is that the polynomial (5) will then have the value unity when $x = 1$, as will be shown in Sec. 89. The precise value of a_0 that is needed is not important to us here. Using the convention that $0! = 1$, we note that $a_0 = 1$ if $n = 0$. In that case, $y_1 = 1$. If n is odd, we choose a_1 so that the final coefficient in expression (6) is also given by equation (7). The reason for this choice is similar to the one above regarding the value assigned to a_0. Note that $y_2 = x$ if $n = 1$, since $a_1 = 1$ for that value of n.

When $n = 2, 3, \ldots$, relation (4) can be used to write all the coefficients that precede a_n in expressions (5) and (6) in terms of a_n. To accomplish this, we first observe that the numerator on the right-hand side of relation (4) can be written

$$j(j+1) - n(n+1) = -[(n^2 - j^2) + (n - j)] = -(n - j)(n + j + 1).$$

We then solve for a_j. The result is

$$a_j = -\frac{(j+2)(j+1)}{(n-j)(n+j+1)} a_{j+2}. \tag{8}$$

To express a_{n-2k} in terms of a_n, we now use relation (8) to write the following k equations:

$$a_{n-2} = -\frac{(n)(n-1)}{(2)(2n-1)} a_n,$$

$$a_{n-4} = -\frac{(n-2)(n-3)}{(4)(2n-3)} a_{n-2},$$

$$\vdots$$

$$a_{n-2k} = -\frac{(n-2k+2)(n-2k+1)}{(2k)(2n-2k+1)} a_{n-2k+2}.$$

Equating the product of the left-hand sides of these equations to the product of their right-hand sides and then canceling the common factors

$$a_{n-2}, a_{n-4}, \ldots, a_{n-2k+2}$$

on each side of the resulting equation, we find that

$$a_{n-2k} = \frac{(-1)^k}{2^k k!} \cdot \frac{n(n-1)\cdots(n-2k+1)}{(2n-1)(2n-3)\cdots(2n-2k+1)} a_n. \tag{9}$$

Then, upon substituting expression (7) for a_n into equation (9) and combining various terms into the appropriate factorials (see Problem 1), we arrive at the desired expression:

$$a_{n-2k} = \frac{1}{2^n} \cdot \frac{(-1)^k}{k!} \cdot \frac{(2n-2k)!}{(n-2k)!(n-k)!}. \tag{10}$$

As usual, $0! = 1$.

In view of equation (10), the polynomials (5) and (6), when the nonzero constants a_0 and a_1 are such that a_n has values (7), can be written

$$P_n(x) = \frac{1}{2^n} \sum_{k=0}^{m} \frac{(-1)^k}{k!} \cdot \frac{(2n-2k)!}{(n-2k)!(n-k)!} x^{n-2k} \quad (n = 0, 1, 2, \ldots), \tag{11}$$

where

$$m = \begin{cases} n/2 & \text{if } n \text{ is even,} \\ (n-1)/2 & \text{if } n \text{ is odd.} \end{cases}$$

Another expression for $P_n(x)$ will be given in Sec. 89. Note that since $P_n(x)$ is a polynomial containing only even powers of x if n is even and only odd powers if n is odd, it is an even or an odd function, depending on whether n is even or odd; that is,

$$P_n(-x) = (-1)^n P_n(x) \quad (n = 0, 1, 2, \ldots). \tag{12}$$

The polynomial $P_n(x)$ is called the *Legendre polynomial of degree n*. For the first several values of n, expression (11) becomes (see Fig. 66)

$$P_0(x) = 1, \qquad P_1(x) = x,$$

$$P_2(x) = \frac{1}{2}(3x^2 - 1), \qquad P_3(x) = \frac{1}{2}(5x^3 - 3x),$$

$$P_4(x) = \frac{1}{8}(35x^4 - 30x^2 + 3), \qquad P_5(x) = \frac{1}{8}(63x^5 - 70x^3 + 15x).$$

Observe that the value of each of these six polynomials is unity when $x = 1$, as anticipated. Also, the polynomials $P_0(x)$, $P_2(x)$, and $P_4(x)$ contain only even powers of x, while $P_1(x)$, $P_3(x)$, and $P_5(x)$ contain only odd powers.

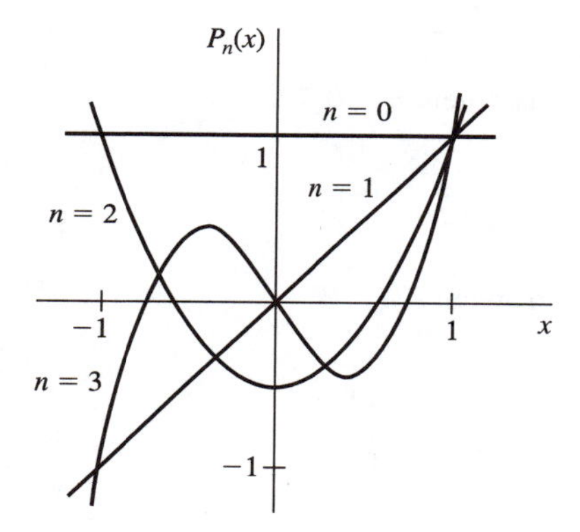

FIGURE 66

We have just seen that Legendre's equation

$$(1 - x^2)y''(x) - 2xy'(x) + n(n + 1)y(x) = 0 \qquad (n = 0, 1, 2, \ldots) \tag{13}$$

always has the polynomial solution $y = P_n(x)$, which is solution (5) (n even) or solution (6) (n odd) when appropriate values are assigned to the arbitrary constants a_0 and a_1 in those solutions. Details regarding the standard form of the accompanying series solution, which is denoted by $Q_n(x)$ and is called a *Legendre function of the second kind,* are left to the problems. We, of course, know from the statement in italics at the end of Sec. 86 that the series representing $Q_n(x)$ is convergent only when $-1 < x < 1$. It will, however, be sufficient for us to know that $Q_n(x)$ and $Q_n'(x)$ fail to be a pair of continuous functions on the *closed* interval $-1 \le x \le 1$ (Problem 9, Sec. 90). Since $P_n(x)$ and $Q_n(x)$ are linearly independent, the general solution of equation (13) is

$$y = C_1 P_n(x) + C_2 Q_n(x), \tag{14}$$

where C_1 and C_2 are arbitrary constants.

PROBLEMS

1. Give details showing how expression (10) in Sec. 87 for the coefficients a_{n-2k} in the Legendre polynomials is obtained from equations (7) and (9) there.

Suggestion: Observe that the factorials in equation (7), Sec. 87, can be written

$$(2n)! = (2n)(2n-1)(2n-2)\cdots(2n-2k+1)(2n-2k)!,$$
$$n! = n(n-1)\cdots(n-2k+1)(n-2k)!,$$
$$n! = n(n-1)\cdots(n-k+1)(n-k)!.$$

2. With the aid of expression (11), Sec. 87, for $P_n(x)$, show that when $n = 2, 3, \ldots$, the constants a_0 and a_1 in equations (5) and (6) in that section must have the following values in order for the final constant a_n to have the value specified in equation (7) there:

$$a_0 = (-1)^{n/2}\frac{(1)(3)(5)\cdots(n-1)}{(2)(4)(6)\cdots(n)} \qquad (n = 2, 4, \ldots),$$

$$a_1 = (-1)^{(n-1)/2}\frac{(1)(3)(5)\cdots(n)}{(2)(4)(6)\cdots(n-1)} \qquad (n = 3, 5, \ldots).$$

3. Establish these properties of Legendre polynomials, where $n = 0, 1, 2, \ldots$:

(*a*) $P_{2n}(0) = (-1)^n\dfrac{(2n)!}{2^{2n}(n!)^2}$; (*b*) $P'_{2n}(0) = 0$;

(*c*) $P_{2n+1}(0) = 0$; (*d*) $P'_{2n+1}(0) = (2n+1)P_{2n}(0)$.

Suggestion: For parts (*a*) and (*d*), refer to Problem 2.

4. Legendre's equation (1), Sec. 86, is often written

$$(1-x^2)y''(x) - 2xy'(x) + \nu(\nu+1)y(x) = 0,$$

where ν is an unrestricted complex number. Show that when $\lambda = \nu(\nu+1)$, recurrence relation (4), Sec. 86, can be put in the form

$$a_j = -\frac{(\nu - j + 2)(\nu + j - 1)}{j(j-1)}a_{j-2} \qquad (j = 2, 3, \ldots).$$

Then, by proceeding as we did in solving Bessel's equation (Sec. 71), use this relation to obtain the following linearly independent solutions of Legendre's equation:

$$y_1 = a_0\left\{1 + \sum_{k=1}^{\infty}(-1)^k \cdot \frac{[\nu(\nu-2)\cdots(\nu-2k+2)][(\nu+1)(\nu+3)\cdots(\nu+2k-1)]}{(2k)!}x^{2k}\right\},$$

$$y_2 = a_1\left\{x + \sum_{k=1}^{\infty}(-1)^k \cdot \frac{[(\nu-1)(\nu-3)\cdots(\nu-2k+1)][(\nu+2)(\nu+4)\cdots(\nu+2k)]}{(2k+1)!}x^{2k+1}\right\},$$

where a_0 and a_1 are arbitrary nonzero constants. (These two series converge when $-1 < x < 1$, according to Sec. 86.)

5. Show that if ν is the complex number

$$\nu = -\frac{1}{2} + i\alpha \qquad (\alpha > 0),$$

Legendre's equation in Problem 4 becomes

$$(1-x^2)y''(x) - 2xy'(x) - \left(\frac{1}{4} + \alpha^2\right)y(x) = 0.$$

Then show how it follows from the solutions obtained in Problem 4 that the functions

$$p_\alpha(x) = 1 + \sum_{k=1}^{\infty}\left[\alpha^2 + \left(\frac{1}{2}\right)^2\right]\left[\alpha^2 + \left(\frac{5}{2}\right)^2\right]\cdots\left[\alpha^2 + \left(\frac{4k-3}{2}\right)^2\right]\frac{x^{2k}}{(2k)!},$$

$$q_\alpha(x) = x + \sum_{k=1}^{\infty}\left[\alpha^2 + \left(\frac{3}{2}\right)^2\right]\left[\alpha^2 + \left(\frac{7}{2}\right)^2\right]\cdots\left[\alpha^2 + \left(\frac{4k-1}{2}\right)^2\right]\frac{x^{2k+1}}{(2k+1)!}$$

are linearly independent solutions of this differential equation, valid on the interval $-1 < x < 1$. These particular Legendre functions arise in certain boundary value problems in regions bounded by cones.

6. Note that the solutions y_1 and y_2 obtained in Problem 4 are solutions (5) and (6) in Sec. 86 when $\lambda = \nu(\nu+1)$. They remain infinite series when $\nu = n = 1, 3, 5, \ldots$ and $\nu = n = 0, 2, 4, \ldots$, respectively. When $\nu = n = 2m\,(m = 0, 1, 2, \ldots)$, the Legendre function Q_n of the second kind is defined as y_2, where

$$a_1 = \frac{(-1)^m\, 2^{2m}\,(m!)^2}{(2m)!};$$

and when $\nu = n = 2m+1\ (m = 0, 1, 2, \ldots)$, Q_n is defined as y_1, where

$$a_0 = -\frac{(-1)^m\, 2^{2m}\,(m!)^2}{(2m+1)!}.$$

Using the fact that

$$\ln\frac{1+x}{1-x} = 2\sum_{k=0}^{\infty}\frac{x^{2k+1}}{2k+1} \qquad (-1 < x < 1),$$

show that

$$Q_0(x) = \frac{1}{2}\ln\frac{1+x}{1-x} \quad \text{and} \quad Q_1(x) = \frac{x}{2}\ln\frac{1+x}{1-x} - 1 = xQ_0(x) - 1.$$

88. ORTHOGONALITY OF LEGENDRE POLYNOMIALS

Let $X(x)$ denote the dependent variable in Legendre's equation, with arbitrary λ:

$$(1-x^2)X''(x) - 2xX'(x) + \lambda X(x) = 0.$$

Writing this equation in the form

$$[(1-x^2)X'(x)]' + \lambda X(x) = 0, \tag{1}$$

we see that we have a special case of the Sturm-Liouville differential equation

$$[r(x)X'(x)]' + [q(x) + \lambda p(x)]X(x) = 0, \tag{2}$$

where $p(x) = 1$, $q(x) = 0$, and $r(x) = 1 - x^2$. The function $r(x)$ vanishes at $x = \pm 1$; thus, as already pointed out in Example 2, Sec. 60, equation (1) here, without boundary conditions, is a singular Sturm-Liouville problem on the closed interval $-1 \leq x \leq 1$, where X and X' are required to be continuous on that interval.

The following theorem provides us with all the solutions of problem (1).

Theorem. *The eigenvalues and corresponding eigenfunctions of the singular Sturm-Liouville problem*

$$[(1-x^2)X'(x)]' + \lambda X(x) = 0 \qquad (-1 < x < 1) \tag{3}$$

are

$$\lambda_n = n(n+1), \qquad X_n = P_n(x) \qquad (n = 0, 1, 2, \ldots), \tag{4}$$

where $P_n(x)$ are Legendre polynomials. The set $\{P_n(x)\}$ $(n = 0, 1, 2, \ldots)$ is, moreover, orthogonal on the interval $-1 < x < 1$, with weight function unity.

We start the proof by recalling from Sec. 87 that $P_n(x)$ and $Q_n(x)$ are linearly independent solutions of equation (3) when λ has any one of the values

$$\lambda_n = n(n+1) \qquad (n = 0, 1, 2, \ldots).$$

Since the polynomial $P_n(x)$ and its derivative are continuous on the entire interval $-1 \leq x \leq 1$ and since this is not true of the Legendre function $Q_n(x)$, it is clear that the continuity requirements on X and X' are met only when X is a constant multiple of $P_n(x)$. Hence the λ_n and X_n in the statement of the theorem are, in fact, eigenvalues and eigenfunctions. It remains to show that there are no other eigenvalues.

To accomplish this, we first observe that since the eigenfunctions just noted all correspond to different eigenvalues, the set $\{P_n(x)\}$ $(n = 0, 1, 2, \ldots)$ is, in fact, orthogonal on the interval $-1 < x < 1$, with weight function $p(x) = 1$. (See the theorem in Sec. 61.) That is,

$$\int_{-1}^{1} P_m(x) P_n(x)\, dx = 0 \qquad (m \neq n). \tag{5}$$

In the notation used for inner products, property (5) reads $(P_m, P_n) = 0$ $(m \neq n)$. Later (Sec. 90) there will be a theorem telling us that if a function f is piecewise smooth on the interval $-1 < x < 1$, then the generalized Fourier series for f with respect to the orthonormal set of functions

$$\phi_n(x) = \frac{P_n(x)}{\|P_n\|} \qquad (n = 0, 1, 2, \ldots) \tag{6}$$

converges to $f(x)$ at all but possibly a finite number of points in the interval $-1 < x < 1$. The set $\{\phi_n(x)\}$ is, therefore, closed (Sec. 54) in the function space $C_p'(-1, 1)$, defined in Sec. 9. That is, there is no function in $C_p'(-1, 1)$, with positive norm, that is orthogonal to each of the functions (6).

Suppose now that λ is another eigenvalue, different from those listed in the statement of the theorem, and let X denote an eigenfunction corresponding to λ. Because of the orthogonality of eigenfunctions corresponding to distinct eigenvalues, $(X, \phi_n) = 0$ $(n = 0, 1, 2, \ldots)$ where the functions ϕ_n are those in equation (6). But the fact that $\{\phi_n(x)\}$ is closed requires that X, which is continuous on the entire interval $-1 \leq x \leq 1$, have value zero for each x in that interval. Consequently, since an eigenfunction cannot be identically equal to zero, X is not an eigenfunction. In view of this contradiction, there are no other eigenvalues; and the proof of the theorem is finished.

If the interval $0 \le x \le 1$, rather than $-1 \le x \le 1$, is used, the differential equation (1) along with either one of the boundary conditions $X'(0) = 0$, $X(0) = 0$ is also a singular Sturm-Liouville problem (Sec. 60).

Corollary. *The eigenvalues and corresponding eigenfunctions of the singular Sturm-Liouville problem consisting of the differential equation*

$$[(1 - x^2)X'(x)]' + \lambda X(x) = 0 \qquad (0 < x < 1) \tag{7}$$

and the boundary condition $X'(0) = 0$ are

$$\lambda_n = 2n(2n + 1), \qquad X_n = P_{2n}(x) \qquad (n = 0, 1, 2, \ldots). \tag{8}$$

If the condition $X(0) = 0$ is used instead, the eigenvalues and eigenfunctions are

$$\lambda_n = (2n + 1)(2n + 2), \qquad X_n = P_{2n+1}(x) \qquad (n = 0, 1, 2, \ldots). \tag{9}$$

The sets $\{P_{2n}(x)\}$ $(n = 0, 1, 2, \ldots)$ and $\{P_{2n+1}(x)\}$ $(n = 0, 1, 2, \ldots)$ are, in addition, orthogonal on the interval $0 < x < 1$, with weight function unity.

To see how these solutions arise, we consider first the solutions in the theorem when the condition $X'(0) = 0$ is imposed on them. Since $P_n'(0) = 0$ only when n is even (Problem 3, Sec. 87), it follows that the polynomials $P_{2n+1}(x)$ $(n = 0, 1, 2, \ldots)$ must be eliminated. This leaves the eigenvalues and eigenfunctions (8). If, on the other hand, the condition $X(0) = 0$ is imposed, the fact that $P_n(0) = 0$ only when n is an odd integer leads us to the eigenvalues and eigenfunctions (9).

The theorem in Sec. 61, regarding the orthogonality of eigenfunctions, ensures the orthogonality stated in the above corollary:

$$\int_0^1 P_{2m}(x) P_{2n}(x)\, dx = 0 \qquad (m \ne n) \tag{10}$$

and

$$\int_0^1 P_{2m+1}(x) P_{2n+1}(x)\, dx = 0 \qquad (m \ne n), \tag{11}$$

where $m = 0, 1, 2, \ldots$ and $n = 0, 1, 2, \ldots$. Valid representations of piecewise smooth functions on the interval $0 < x < 1$ will follow (Sec. 90) from representations on the interval $-1 < x < 1$ in terms of the set $\{P_n(x)\}$ $(n = 0, 1, 2, \ldots)$, just as Fourier cosine and sine series follow from Fourier series involving both cosines and sines. Hence the same argument, involving closed sets, that was used in the proof of the theorem above can be used to show that there are no other eigenvalues of the Sturm-Liouville problems in the corollary.

89. RODRIGUES' FORMULA AND NORMS

According to expression (11), Sec. 87,

$$P_n(x) = \frac{1}{2^n n!} \sum_{k=0}^{m} (-1)^k \frac{n!}{k!(n-k)!} \cdot \frac{(2n-2k)!}{(n-2k)!} x^{n-2k} \tag{1}$$

where

$$m = \begin{cases} n/2 & \text{if } n \text{ is even,} \\ (n-1)/2 & \text{if } n \text{ is odd.} \end{cases}$$

Since

$$\frac{d^n}{dx^n} x^{2n-2k} = \frac{(2n-2k)!}{(n-2k)!} x^{n-2k} \qquad (0 \le k \le m)$$

and because of the linearity of the differential operator d^n/dx^n, expression (1) can be written

$$(2) \qquad P_n(x) = \frac{1}{2^n n!} \frac{d^n}{dx^n} \sum_{k=0}^{m} (-1)^k \frac{n!}{k!(n-k)!} x^{2n-2k}.$$

The powers of x in the sum here evidently decrease in steps of 2 as the index k increases; and the lowest power is $2n - 2m$, which is n if n is even and $n + 1$ if n is odd. The sum can actually be extended so that k ranges from 0 to n. This is because the additional polynomial that is introduced is of degree less than n, and its nth derivative is, therefore, zero. Since the resulting sum is the binomial expansion of $(x^2 - 1)^n$, it follows from equation (2) that

$$(3) \qquad P_n(x) = \frac{1}{2^n n!} \frac{d^n}{dx^n} (x^2 - 1)^n \qquad (n = 0, 1, 2, \ldots).$$

This is *Rodrigues' formula* for the Legendre polynomials.

Various useful properties of Legendre polynomials are readily obtained from Rodrigues' formula with the aid of *Leibnitz' rule* for the nth derivative of the product of two functions (Problem 5):

$$(4) \qquad D^n(fg) = \sum_{k=0}^{n} \frac{n!}{k!(n-k)!} (D^k f)(D^{n-k} g),$$

where it is understood that all the required derivatives exist and that the zero-order derivative of a function is the function itself.

EXAMPLE 1. If we write $u = x^2 - 1$, so that

$$u^n = (x^2 - 1)^n = (x + 1)^n (x - 1)^n,$$

it follows from Leibnitz' rule that

$$D^n u^n = \sum_{k=0}^{n} \frac{n!}{k!(n-k)!} [D^k (x+1)^n][D^{n-k}(x-1)^n].$$

Now the first term in this sum is

$$[D^0 (x+1)^n][D^n (x-1)^n] = (x+1)^n \, n!,$$

and the remaining terms all contain the factor $(x - 1)$ to some positive power. Hence the value of the sum when $x = 1$ is $2^n n!$, and it follows from Rodrigues' formula (3) that

$$(5) \qquad P_n(1) = 1 \qquad (n = 0, 1, 2, \ldots).$$

Observe how it follows from this and the relation

$$P_n(-x) = (-1)^n P_n(x) \qquad (n = 0, 1, 2, \ldots),$$

obtained in Sec. 87, that

$$P_n(-1) = (-1)^n \qquad (n = 0, 1, 2, \ldots). \tag{6}$$

EXAMPLE 2. We turn now to the derivation of an important recurrence relation involving Legendre polynomials. We shall need the following two identities, where $u = x^2 - 1$:

$$Du^{n+1} = D(x^2 - 1)^{n+1} = 2(n+1)xu^n \qquad (n = 0, 1, 2, \ldots), \tag{7}$$

$$D^2u^{n+1} = 2(n+1)[(2n+1)u^n + 2nu^{n-1}] \qquad (n = 0, 1, 2, \ldots). \tag{8}$$

Identity (7) is obvious and enables us to write

$$\begin{aligned} D(Du^{n+1}) &= 2(n+1)(u^n + xDu^n) \\ &= 2(n+1)(u^n + 2nx^2u^{n-1}) \\ &= 2(n+1)[u^n + 2n(x^2-1)u^{n-1} + 2nu^{n-1}], \end{aligned}$$

which is the same as identity (8).

We begin our derivation of the recurrence relation by using Rodrigues' formula (3) to write

$$P_{n+1}(x) = \frac{D^{n-1}(D^2u^{n+1})}{2^{n+1}(n+1)!}.$$

Hence, by expression (8),

$$P_{n+1}(x) = \frac{(2n+1)D^{n-1}u^n + 2nD^{n-1}u^{n-1}}{2^n n!}.$$

Referring to Rodrigues' formula once again, we have

$$D^{n-1}u^{n-1} = 2^{n-1}(n-1)!P_{n-1}(x);$$

and so it follows that

$$P_{n+1}(x) = \frac{(2n+1)D^{n-1}u^n + 2^n n!P_{n-1}(x)}{2^n n!}.$$

That is,

$$\frac{P_{n+1}(x) - P_{n-1}(x)}{2n+1} = \frac{D^{n-1}u^n}{2^n n!}. \tag{9}$$

On the other hand, we see from Rodrigues' formula (3) and expression (7) that

$$P_{n+1}(x) = \frac{D^n(Du^{n+1})}{2^{n+1}(n+1)!} = \frac{D^n(xu^n)}{2^n n!}.$$

Inasmuch as the first two terms of Leibnitz' rule (4) are

$$D^n(fg) = (f)(D^ng) + n(Df)(D^{n-1}g) + \cdots,$$

this yields

$$P_{n+1}(x) = \frac{xD^n u^n + nD^{n-1}u^n}{2^n n!}.$$

Because

$$D^n u^n = 2^n n! P_n(x), \tag{10}$$

then,

$$\frac{P_{n+1}(x) - xP_n(x)}{n} = \frac{D^{n-1}u^n}{2^n n!}. \tag{11}$$

Finally, by equating the left-hand sides of equations (9) and (11), we arrive at the desired recurrence relation:

$$(n+1)P_{n+1}(x) + nP_{n-1}(x) = (2n+1)xP_n(x) \qquad (n = 1, 2, \ldots). \tag{12}$$

Note, too, that the relation

$$P'_{n+1}(x) - P'_{n-1}(x) = (2n+1)P_n(x) \qquad (n = 1, 2, \ldots) \tag{13}$$

is an immediate consequence of equations (9) and (10).

We now show how relation (12) and its form

$$nP_n(x) + (n-1)P_{n-2}(x) = (2n-1)xP_{n-1}(x) \qquad (n = 2, 3, \ldots), \tag{14}$$

obtained by replacing n by $n-1$ in that relation, can be used to find the norms

$$\|P_n\| = (P_n, P_n)^{1/2} = \left\{ \int_{-1}^{1} [P_n(x)]^2 \, dx \right\}^{1/2}$$

of the orthogonal (see Sec. 88) polynomials P_n. Keeping in mind that

$$(P_{n+1}, P_{n-1}) = 0 \quad \text{and} \quad (P_{n-2}, P_n) = 0,$$

we find from relations (12) and (14), respectively, that

$$n(P_{n-1}, P_{n-1}) = (2n+1)(xP_n, P_{n-1}) \tag{15}$$

and

$$n(P_n, P_n) = (2n-1)(xP_{n-1}, P_n). \tag{16}$$

The integrals representing (xP_n, P_{n-1}) and (xP_{n-1}, P_n) are identical, and we need only eliminate those quantities from equations (15) and (16) to see that

$$(2n+1)(P_n, P_n) = (2n-1)(P_{n-1}, P_{n-1}),$$

or

$$(2n+1)\|P_n\|^2 = (2n-1)\|P_{n-1}\|^2 \qquad (n = 2, 3, \ldots). \tag{17}$$

It is easy to verify directly that equation (17) is also valid in the case $n = 1$, involving $P_1(x) = x$ and $P_0(x) = 1$.

Next, we let n be any fixed positive integer and use equation (17) to write the following n equations:

$$\begin{aligned}
(2n+1)\|P_n\|^2 &= (2n-1)\|P_{n-1}\|^2,\\
(2n-1)\|P_{n-1}\|^2 &= (2n-3)\|P_{n-2}\|^2,\\
&\vdots\\
(5)\|P_2\|^2 &= (3)\|P_1\|^2,\\
(3)\|P_1\|^2 &= (1)\|P_0\|^2.
\end{aligned}$$

Setting the product of the left-hand sides of these equations equal to the product of their right-hand sides and then canceling appropriately, we arrive at the result

$$(2n+1)\|P_n\|^2 = \|P_0\|^2 \qquad (n = 1, 2, \ldots).$$

Since $\|P_0\|^2 = 2$, this means that

$$\|P_n\| = \sqrt{\frac{2}{2n+1}} \qquad (n = 0, 1, 2, \ldots). \tag{18}$$

The set of polynomials

$$\phi_n(x) = \sqrt{\frac{2n+1}{2}}\, P_n(x) \qquad (n = 0, 1, 2, \ldots) \tag{19}$$

is, therefore, *orthonormal* (Sec. 52) on the interval $-1 < x < 1$.

PROBLEMS

1. From the orthogonality of the set $\{P_n(x)\}$, state why

(*a*) $\int_{-1}^{1} P_n(x)\,dx = 0$ $(n = 1, 2, \ldots)$;

(*b*) $\int_{-1}^{1} (Ax + B) P_n(x)\,dx = 0$ $(n = 2, 3, \ldots)$, where A and B are constants.

2. Verify directly that the Legendre polynomials

$$P_0(x) = 1, \qquad P_1(x) = x, \qquad P_2(x) = \frac{1}{2}(3x^2 - 1), \qquad P_3(x) = \frac{1}{2}(5x^3 - 3x)$$

form an orthogonal set on the interval $-1 < x < 1$. Show that their graphs are as indicated in Fig. 66 (Sec. 87).

3. Use the fact that the set $\{\phi_n(x)\}$ defined by equation (19), Sec. 89, is orthonormal on the interval $-1 < x < 1$ to show that the following sets are orthonormal on the interval $0 < x < 1$:

(*a*) $\{\sqrt{4n+1}\, P_{2n}(x)\}$ $(n = 0, 1, 2, \ldots)$; (*b*) $\{\sqrt{4n+3}\, P_{2n+1}(x)\}$ $(n = 0, 1, 2, \ldots)$.

Suggestion: Note that

$$2\int_0^1 f(x)\,dx = \int_{-1}^{1} f(x)\,dx$$

when f is an even function.

4. From recurrence relation (13), Sec. 89, obtain the integration formula

$$\int_a^1 P_n(x)\,dx = \frac{1}{2n+1}\,[P_{n-1}(a) - P_{n+1}(a)] \qquad (n = 1, 2, \ldots).$$

5. Use mathematical induction on the integer n to verify Leibnitz' rule (4), Sec. 89.

6. Write

$$F(x,t) = (1 - 2xt + t^2)^{-1/2},$$

where $|x| \le 1$ and t is as yet unrestricted.

(*a*) Note that $x = \cos\theta$ for some uniquely determined value of θ $(0 \le \theta \le \pi)$, and show that

$$F(x,t) = (1 - e^{i\theta}t)^{-1/2}\,(1 - e^{-i\theta}t)^{-1/2}.$$

Then, using the fact that $(1-z)^{-1/2}$ has a valid Maclaurin series expansion when $|z| < 1$, point out why the factors $(1 - e^{i\theta}t)^{-1/2}$ and $(1 - e^{-i\theta}t)^{-1/2}$, considered as functions of t, can be represented by Maclaurin series which are valid when $|t| < 1$. It follows that the product $F(x,t)$ also has such a representation when $|t| < 1$.[†] That is, there are functions $f_n(x)$ $(n = 0, 1, 2, \ldots)$ such that

$$F(x,t) = \sum_{n=0}^{\infty} f_n(x)\,t^n \qquad (|t| < 1).$$

(*b*) Show that the function $F(x,t)$ satisfies the identity

$$(1 - 2xt + t^2)\,\frac{\partial F}{\partial t} = (x - t)F,$$

and use this result to show that the functions $f_n(x)$ in part (*a*) satisfy the recurrence relation

$$(n+1)\,f_{n+1}(x) + nf_{n-1}(x) = (2n+1)xf_n(x) \qquad (n = 1, 2, \ldots).$$

(*c*) Show that the first two functions $f_0(x)$ and $f_1(x)$ in part (*a*) are 1 and x, respectively, and notice that the recurrence relation obtained in part (*b*) can then be used to determine $f_n(x)$ when $n = 2, 3, \ldots$. Compare that relation with relation (12), Sec. 89, and conclude that the functions $f_n(x)$ are, in fact, the Legendre polynomials $P_n(x)$; that is, show that

$$(1 - 2xt + t^2)^{-1/2} = \sum_{n=0}^{\infty} P_n(x)\,t^n \qquad (|x| \le 1, |t| < 1).$$

The function F is, therefore, a *generating function* for the Legendre polynomials.

7. Give an alternative proof of the property (Sec. 89) $P_n(1) = 1$ $(n = 0, 1, 2, \ldots)$, using
(*a*) recurrence relation (12), Sec. 89, and mathematical induction;
(*b*) the generating function obtained in Problem 6(*c*).

[†]For a discussion of this point, see, for example, the authors' book (2004, pp. 215–216), listed in the Bibliography.

90. LEGENDRE SERIES

In Sec. 89, we saw that the set of polynomials

$$\phi_n(x) = \sqrt{\frac{2n+1}{2}}\,P_n(x) \qquad (n = 0, 1, 2, \ldots) \tag{1}$$

is orthonormal, with weight function unity, on the interval $-1 < x < 1$. If f denotes a function in $C_p(-1, 1)$, the Fourier constants (Sec. 54) in the correspondence

$$f(x) \sim \sum_{n=0}^{\infty} c_n\,\phi_n(x) \qquad (-1 < x < 1) \tag{2}$$

are

$$c_n = (f, \phi_n) = \sqrt{\frac{2n+1}{2}} \int_{-1}^{1} f(x) P_n(x)\,dx \qquad (n = 0, 1, 2, \ldots). \tag{3}$$

This observation enables us to prove the following theorem regarding *Legendre series.*

Theorem 1. *Let f be piecewise continuous on the interval stated in each part below.*

(a) If A_n $(n = 0, 1, 2, \ldots)$ are the coefficients in the correspondence

$$f(x) \sim \sum_{n=0}^{\infty} A_n P_n(x) \qquad (-1 < x < 1),$$

then

$$A_n = \frac{2n+1}{2} \int_{-1}^{1} f(x) P_n(x)\,dx \qquad (n = 0, 1, 2, \ldots).$$

(b) If A_{2n} $(n = 0, 1, 2, \ldots)$ are the coefficients in the correspondence

$$f(x) \sim \sum_{n=0}^{\infty} A_{2n} P_{2n}(x) \qquad (0 < x < 1),$$

then

$$A_{2n} = (4n + 1) \int_{0}^{1} f(x) P_{2n}(x)\,dx \qquad (n = 0, 1, 2, \ldots).$$

(c) If A_{2n+1} $(n = 0, 1, 2, \ldots)$ are the coefficients in the correspondence

$$f(x) \sim \sum_{n=0}^{\infty} A_{2n+1} P_{2n+1}(x) \qquad (0 < x < 1),$$

then

$$A_{2n+1} = (4n + 3) \int_{0}^{1} f(x) P_{2n+1}(x)\,dx \qquad (n = 0, 1, 2, \ldots).$$

Part (a) of the theorem is verified by writing correspondence (2) as

$$f(x) \sim \sum_{n=0}^{\infty} c_n \sqrt{\frac{2n+1}{2}}\, P_n(x) \qquad (-1 < x < 1)$$

and then putting

$$A_n = c_n \sqrt{\frac{2n+1}{2}} \qquad (n = 0, 1, 2, \ldots).$$

According to Sec. 87, each $P_{2n}(x)$ is even and each $P_{2n+1}(x)$ is odd. That is,

$$P_{2n}(-x) = P_{2n}(x) \qquad \text{and} \qquad P_{2n+1}(-x) = -P_{2n+1}(x) \qquad (n = 0, 1, 2, \ldots).$$

Evidently, then, if the function f in part (a) of the theorem is *even,* the product $f(x)P_{2n+1}(x)$ is odd and the graph of $y = f(x)P_{2n+1}(x)$ is symmetric with respect to the origin. On the other hand, $f(x)P_{2n}(x)$ is even, and the graph of $y = f(x)P_{2n}(x)$ is symmetric with respect to the y axis. Consequently,

$$\int_{-1}^{1} f(x)P_{2n+1}(x)\,dx = 0 \qquad \text{and} \qquad \int_{-1}^{1} f(x)P_{2n}(x)\,dx = 2\int_{0}^{1} f(x)P_{2n}(x)\,dx.$$

Hence it follows from the expression for A_n in part (a) that

$$A_{2n+1} = 0 \qquad (n = 0, 1, 2, \ldots).$$

Also, the coefficients A_{2n} $(n = 0, 1, 2, \ldots)$ are as shown in part (b). Thus if we apply part (a) of the theorem to the even extension of the function in part (b), we have the correspondence in part (b), with the coefficients shown there. Part (b) of the theorem is now verified.

Similarly, if f is an *odd* function, the products $f(x)P_{2n}(x)$ and $f(x)P_{2n+1}(x)$ are odd and even, respectively, and so

$$\int_{-1}^{1} f(x)P_{2n}(x)\,dx = 0 \quad \text{and} \quad \int_{-1}^{1} f(x)P_{2n+1}(x)\,dx = 2\int_{0}^{1} f(x)P_{2n+1}(x)\,dx.$$

Consequently,

$$A_{2n} = 0 \qquad (n = 0, 1, 2, \ldots);$$

and the coefficients A_{2n+1} $(n = 0, 1, 2, \ldots)$ are those shown in part (c). Application of part (a) of the theorem to the odd extension of the function in part (c) now completes the proof of the theorem.

We state here, without proof, a representation theorem involving Legendre series that is applicable to piecewise smooth functions.[†]

Theorem 2. *Let f denote a function that is piecewise smooth on the interval $-1 < x < 1$, and suppose that $f(x)$ at each point of discontinuity of f in that*

[†]The proof, which is rather lengthy, can be found in, for example, the book by Kreider, Kuller, Ostberg, and Perkins (1966, pp. 425–432), listed in the Bibliography. A simplified proof of a special case of the theorem appears in the book by Rainville (1972, pp. 177–179), also listed there.

interval is defined as the mean value of the one-sided limits $f(x+)$ and $f(x-)$. Then

$$f(x) = \sum_{n=0}^{\infty} A_n P_n(x) \qquad (-1 < x < 1), \tag{4}$$

where the coefficients A_n are given in part (a) of Theorem 1.

Adaptations of this theorem to even and odd extensions of a function that is piecewise smooth on the interval $0 < x < 1$ are obvious.

EXAMPLE. Let us expand the function $f(x) = 1$ $(0 < x < 1)$ in a series of Legendre polynomials of odd degree. According to part (c) of Theorem 1,

$$A_{2n+1} = (4n+3)\int_0^1 P_{2n+1}(x)\,dx \qquad (n = 0, 1, 2, \ldots).$$

The integral here is readily evaluated with the aid of the integration formula (Problem 4, Sec. 89)

$$\int_a^1 P_n(x)\,dx = \frac{1}{2n+1}\,[P_{n-1}(a) - P_{n+1}(a)] \qquad (n = 1, 2, \ldots),$$

which tells us that

$$A_{2n+1} = P_{2n}(0) - P_{2n+2}(0). \tag{5}$$

Thus

$$1 = \sum_{n=0}^{\infty} [P_{2n}(0) - P_{2n+2}(0)]\, P_{2n+1}(x) \qquad (0 < x < 1). \tag{6}$$

Since [Problem 3(a), Sec. 87]

$$P_{2n}(0) = (-1)^n \frac{(2n)!}{2^{2n}(n!)^2} \qquad (n = 0, 1, 2, \ldots),$$

the coefficients (5) can also be written

$$A_{2n+1} = (-1)^n \left(\frac{4n+3}{2n+2}\right) \frac{(2n)!}{2^{2n}(n!)^2}. \tag{7}$$

This alternative form of representation (6) is then obtained:

$$1 = \sum_{n=0}^{\infty} (-1)^n \left(\frac{4n+3}{2n+2}\right) \frac{(2n)!}{2^{2n}(n!)^2} P_{2n+1}(x) \qquad (0 < x < 1). \tag{8}$$

PROBLEMS

1. Let F denote the odd extension of the function $f(x) = 1$ $(0 < x < 1)$ to the interval $-1 < x < 1$, where $F(0) = 0$. Also, let g be the function defined by the equations

$$g(x) = \begin{cases} 0 & \text{when } -1 < x < 0, \\ 1 & \text{when } \quad 0 < x < 1, \end{cases}$$

and $g(0) = 1/2$. Then, by observing that

$$g(x) = \frac{1}{2} + \frac{1}{2}F(x) \qquad (-1 < x < 1)$$

and referring to expansion (6), Sec. 90, show that

$$g(x) = \frac{1}{2}P_0(x) + \frac{1}{2}\sum_{n=0}^{\infty}[P_{2n}(0) - P_{2n+2}(0)]\,P_{2n+1}(x) \qquad (-1 < x < 1).$$

2. Let f denote the function defined by the equations

$$f(x) = \begin{cases} 0 & \text{when } -1 < x \le 0, \\ x & \text{when } \quad 0 < x < 1. \end{cases}$$

(*a*) State why $f(x)$ is represented by its Legendre series (4), Sec. 90, at each point of the interval $-1 < x < 1$.
(*b*) Show that $A_{2n+1} = 0$ $(n = 1, 2, \ldots)$ in the series in part (*a*).
(*c*) Find the first four nonzero terms of the series in part (*a*) to show that

$$f(x) = \frac{1}{4}P_0(x) + \frac{1}{2}P_1(x) + \frac{5}{16}P_2(x) - \frac{3}{32}P_4(x) + \cdots \qquad (-1 < x < 1).$$

3. Show that for all x,

(*a*) $x^2 = \frac{1}{3}P_0(x) + \frac{2}{3}P_2(x)$; (*b*) $x^3 = \frac{3}{5}P_1(x) + \frac{2}{5}P_3(x)$.

4. Obtain the first three nonzero terms in the series of Legendre polynomials of even degree representing the function $f(x) = x$ $(0 < x < 1)$ to show that

$$x = \frac{1}{2}P_0(x) + \frac{5}{8}P_2(x) - \frac{3}{16}P_4(x) + \cdots \qquad (0 < x < 1).$$

Point out why this expansion remains valid when $x = 0$, and state what function the series represents on the interval $-1 < x < 1$.

5. By applying Theorem 1 in Sec. 57 to the Fourier constants (3) in Sec. 90, state why

$$\lim_{n\to\infty}\sqrt{2n+1}\int_{-1}^{1} f(x)P_n(x)\,dx = 0$$

when f is piecewise continuous on the interval $-1 < x < 1$.

6. Let f denote a function that is piecewise smooth on the interval $0 < x < 1$, and suppose that $f(x)$ at each point of discontinuity there is the mean value of the one-sided limits $f(x+)$ and $f(x-)$.

(*a*) By finding the Fourier constants for f with respect to the orthonormal set $\{\sqrt{4n+1}\,P_{2n}(x)\}$ $(n = 0, 1, 2, \ldots)$ in Problem 3(*a*), Sec. 89, derive the coefficients A_{2n} appearing in case (*b*) of Theorem 1 in Sec. 90.
(*b*) Apply Theorem 1 in Sec. 57 to the Fourier constants in part (*a*) to show that (compare with Problem 5)

$$\lim_{n\to\infty}\sqrt{4n+1}\int_{0}^{1} f(x)P_{2n}(x)\,dx = 0.$$

7. (*a*) By recalling that $P_m(x)$ is a polynomial of degree m containing only the powers $x^m, x^{m-2}, x^{m-4}, \ldots$ of x (Sec. 87), state why

$$x^m = c\,P_m(x) + c_{m-2}x^{m-2} + c_{m-4}x^{m-4} + \cdots,$$

where the coefficients are constants. Apply the same argument to x^{m-2}, etc., to conclude that x^m is a finite linear combination of the polynomials

$$P_m(x),\ P_{m-2}(x),\ P_{m-4}(x),\ \ldots.$$

(*b*) With the aid of the result in part (*a*), point out why

$$\int_{-1}^{1} P_n(x)p(x)\,dx = 0,$$

where $P_n(x)$ is a Legendre polynomial of degree n ($n = 1, 2, \ldots$) and $p(x)$ is any polynomial whose degree is less than n.

8. Let n have any one of the values $n = 1, 2, \ldots$.

(*a*) By recalling the result in Problem 1(*a*), Sec. 89, state why $P_n(x)$ must change sign at least once in the open interval $-1 < x < 1$. Then let $x_1, x_2, \ldots, x_k$ denote the totality of distinct points in that interval at which $P_n(x)$ changes sign. Since any polynomial of degree n has at most n distinct zeros, we know that $1 \le k \le n$.

(*b*) Assume that the number of points $x_1, x_2, \ldots, x_k$ in part (*a*) is such that $k < n$, and consider the polynomial

$$p(x) = (x - x_1)(x - x_2)\cdots(x - x_k).$$

Use the result in Problem 7(*b*) to show that the integral

$$\int_{-1}^{1} P_n(x)p(x)\,dx$$

has value zero; and, after noting that $P_n(x)$ and $p(x)$ change sign at precisely the same points in the interval $-1 < x < 1$, state why the value of the integral cannot be zero. Having reached this contradiction, conclude that $k = n$ and hence that *the zeros of a Legendre polynomial $P_n(x)$ are all real and distinct and lie in the open interval $-1 < x < 1$.*

9. Show in the following way that for each value of n ($n = 0, 1, 2, \ldots$), the Legendre function of the second kind $Q_n(x)$ (Sec. 87) and its derivative $Q_n'(x)$ fail to be a pair of continuous functions on the closed interval $-1 \le x \le 1$. Suppose that there is an integer N such that $Q_N(x)$ and $Q_N'(x)$ are continuous on that interval. The functions $Q_N(x)$ and $P_n(x)$ ($n \ne N$) are then eigenfunctions corresponding to different eigenvalues of the singular Sturm-Liouville problem (1), Sec. 88. Point out how it follows that

$$\int_{-1}^{1} Q_N(x)P_n(x)\,dx = 0 \qquad (n \ne N),$$

and then use Theorem 2 in Sec. 90 to show that $Q_N(x) = A_N P_N(x)$, where A_N is some constant. This is, however, impossible since $P_N(x)$ and $Q_N(x)$ are linearly independent.

91. THE EIGENFUNCTIONS $P_n(\cos\theta)$

The boundary value problems to be treated in Secs. 92 and 93 will involve singular Sturm-Liouville problems whose eigenfunctions are

$$P_n(\cos\theta) \qquad (n = 0, 1, 2, \ldots).$$

In this section, we give some modifications of earlier results involving

$$P_n(x) \qquad (n = 0, 1, 2, \ldots)$$

that will apply to those trigonometric eigenfunctions.

Theorem 1. *Let*

$$\frac{d}{d\theta}\left(\sin\theta\,\frac{d\Theta}{d\theta}\right) + \lambda\sin\theta\,\Theta = 0 \tag{1}$$

be the differential equation in the singular Sturm-Liouville problem described in one of the three cases below. The eigenvalues λ_n *and corresponding eigenfunctions* Θ_n *are as follows:*

(*a*) *When the interval is* $0 \le \theta \le \pi$ *and no boundary condition is used,*

$$\lambda_n = n(n+1), \qquad \Theta_n = P_n(\cos\theta) \qquad (n = 0, 1, 2, \ldots).$$

(*b*) *When the interval is* $0 \le \theta \le \pi/2$ *and the boundary condition* $\Theta'(\pi/2) = 0$ *is used,*

$$\lambda_n = 2n(2n+1), \qquad \Theta_n = P_{2n}(\cos\theta) \qquad (n = 0, 1, 2, \ldots).$$

(*c*) *When the interval is* $0 \le \theta \le \pi/2$ *and the boundary condition* $\Theta(\pi/2) = 0$ *is used,*

$$\lambda_n = (2n+1)(2n+2), \qquad \Theta_n = P_{2n+1}(\cos\theta) \qquad (n = 0, 1, 2, \ldots).$$

The fact that the problem in each of the three cases in this theorem is actually a singular Sturm-Liouville problem, where Θ and Θ' are required to be continuous on the stated interval, is readily established by referring to Sec. 60.

As for the proof of the theorem, the substitution $x = \cos\theta$ enables us to write

$$\frac{d\Theta}{d\theta} = \frac{d\Theta}{dx}\frac{dx}{d\theta} = -\frac{d\Theta}{dx}\sin\theta,$$

or

$$\frac{1}{\sin\theta}\frac{d\Theta}{d\theta} = -\frac{d\Theta}{dx}.$$

Consequently,

$$\sin\theta\,\frac{d\Theta}{d\theta} = (1-\cos^2\theta)\left(\frac{1}{\sin\theta}\frac{d\Theta}{d\theta}\right) = -(1-x^2)\frac{d\Theta}{dx};$$

and so

$$\frac{1}{\sin\theta}\frac{d}{d\theta}\left(\sin\theta\,\frac{d\Theta}{d\theta}\right) = \frac{1}{\sin\theta}\frac{d}{dx}\left[-(1-x^2)\frac{d\Theta}{dx}\right]\frac{dx}{d\theta} = \frac{d}{dx}\left[(1-x^2)\frac{d\Theta}{dx}\right].$$

Equation (1) thus becomes

$$\frac{d}{dx}\left[(1-x^2)\frac{d\Theta}{dx}\right] + \lambda\Theta = 0, \tag{2}$$

which, except for notation, is Legendre's equation (1) in Sec. 88. It is now evident that under the transformation $x = \cos\theta$, the above theorem is simply an alternative form of the theorem and its corollary in Sec. 88.

The next theorem is just a restatement of the first theorem in Sec. 90 on Legendre series.

Theorem 2. *Let F be piecewise continuous on the interval stated in each part below.*

(*a*) *If* A_n $(n = 0, 1, 2, \ldots)$ *are the coefficients in the correspondence*

$$F(\theta) \sim \sum_{n=0}^{\infty} A_n P_n(\cos\theta) \qquad (0 < \theta < \pi),$$

then

$$A_n = \frac{2n+1}{2} \int_0^{\pi} F(\theta) P_n(\cos\theta) \sin\theta \, d\theta \qquad (n = 0, 1, 2, \ldots).$$

(*b*) *If* A_{2n} $(n = 0, 1, 2, \ldots)$ *are the coefficients in the correspondence*

$$F(\theta) \sim \sum_{n=0}^{\infty} A_{2n} P_{2n}(\cos\theta) \qquad \left(0 < \theta < \frac{\pi}{2}\right),$$

then

$$A_{2n} = (4n+1) \int_0^{\pi/2} F(\theta) P_{2n}(\cos\theta) \sin\theta \, d\theta \qquad (n = 0, 1, 2, \ldots).$$

(*c*) *If* A_{2n+1} $(n = 0, 1, 2, \ldots)$ *are the coefficients in the correspondence*

$$F(\theta) \sim \sum_{n=0}^{\infty} A_{2n+1} P_{2n+1}(\cos\theta) \qquad \left(0 < \theta < \frac{\pi}{2}\right),$$

then

$$A_{2n+1} = (4n+3) \int_0^{\pi/2} F(\theta) P_{2n+1}(\cos\theta) \sin\theta \, d\theta \qquad (n = 0, 1, 2, \ldots).$$

To verify this theorem, we need only write $x = \cos\theta$ and define the function

$$f(x) = F(\cos^{-1} x), \tag{3}$$

where the principal values of the inverse cosine function are to be taken. When this $f(x)$ is used in Theorem 1 of Sec. 90, that theorem becomes Theorem 2 here. If F is piecewise smooth on the stated interval, the correspondence is, of course, an equality for each point θ at which F is continuous.

92. DIRICHLET PROBLEMS IN SPHERICAL REGIONS

For our first application of Legendre series, we shall determine the harmonic function u in the region $r < c$ such that u assumes prescribed values $F(\theta)$ on the spherical surface $r = c$. Here r, ϕ, and θ are spherical coordinates, and u is independent of ϕ. Thus $u = u(r, \theta)$ satisfies Laplace's equation (Sec. 22)

$$r \frac{\partial^2}{\partial r^2}(ru) + \frac{1}{\sin\theta}\frac{\partial}{\partial\theta}\left(\sin\theta \frac{\partial u}{\partial\theta}\right) = 0 \qquad (r < c, 0 < \theta < \pi) \tag{1}$$

and the condition (Fig. 67)

$$u(c, \theta) = F(\theta) \qquad (0 < \theta < \pi). \tag{2}$$

The function u and its partial derivatives of the first and second orders are to be continuous throughout the interior $0 \le r < c, 0 \le \theta \le \pi$ of the sphere.

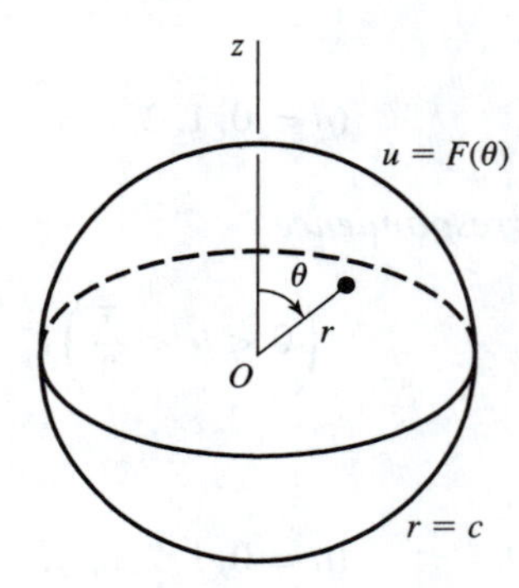

FIGURE 67

Physically, the function u may denote steady temperatures in a solid sphere $r \le c$ whose surface temperatures depend only on θ; that is, the surface temperatures are uniform over each circle $r = c, \theta = \theta_0$. Also, u represents electrostatic potential in the space $r < c$, which is free of charges, when $u = F(\theta)$ on the boundary $r = c$.

We start by seeking a product solution $u = R(r)\Theta(\theta)$ of equation (1) that satisfies the stated continuity requirements. Separation of variables shows that for some constant λ,

$$\frac{1}{\sin\theta\,\Theta}\frac{d}{d\theta}\left(\sin\theta \frac{d\Theta}{d\theta}\right) = -\frac{r}{R}\frac{d^2}{dr^2}(rR) = -\lambda.$$

Consequently, R must satisfy the ordinary differential equation

$$r \frac{d^2}{dr^2}(rR) - \lambda R = 0 \qquad (r < c) \tag{3}$$

and be continuous when $0 \le r < c$. Also, for the same constant λ, the function Θ satisfies the equation

$$\frac{d}{d\theta}\left(\sin\theta \frac{d\Theta}{d\theta}\right) + \lambda \sin\theta\,\Theta = 0 \qquad (0 < \theta < \pi), \tag{4}$$

where Θ and Θ' are to be continuous on the closed interval $0 \le \theta \le \pi$.

Case (*a*) of Theorem 1 in Sec. 91 tells us that equation (4) is a singular Sturm-Liouville problem whose eigenvalues and corresponding eigenfunctions are

$$\lambda_n = n(n+1), \qquad \Theta_n = P_n(\cos\theta) \qquad (n = 0, 1, 2, \ldots). \tag{5}$$

Writing equation (3) in the form

$$r^2R'' + 2rR' - \lambda R = 0,$$

we see that it is a Cauchy-Euler equation, which reduces to a differential equation with constant coefficients after the substitution $r = \exp s$ is made (see Problem 1, Sec. 38). When $\lambda = n(n+1)$, its general solution is

$$R(r) = C_1 r^n + C_2 r^{-n-1} = C_1 r^n + \frac{C_2}{r^{n+1}} \qquad (0 < r < c), \tag{6}$$

as is easily verified. The continuity of R at $r = 0$ requires that $C_2 = 0$, and so the desired functions of r are $R_n(r) = r^n$ $(n = 0, 1, 2, \ldots)$.

The functions $u_n = r^n P_n(\cos\theta)$ $(n = 0, 1, 2, \ldots)$, therefore, satisfy Laplace's equation (1) and the continuity conditions accompanying it. Formally, their generalized linear combination

$$u(r, \theta) = \sum_{n=0}^{\infty} B_n r^n P_n(\cos\theta) \tag{7}$$

is a solution of our boundary value problem if the constants B_n are such that $u(c, \theta) = F(\theta)$, or

$$F(\theta) = \sum_{n=0}^{\infty} B_n c^n P_n(\cos\theta) \qquad (0 < \theta < \pi). \tag{8}$$

To find these constants, we need only refer to case (*a*) of Theorem 2 in Sec. 91. Evidently, $B_n c^n = A_n$ where

$$A_n = \frac{2n+1}{2} \int_0^{\pi} F(\theta) P_n(\cos\theta) \sin\theta \, d\theta \qquad (n = 0, 1, 2, \ldots); \tag{9}$$

and the formal solution of our Dirichlet problem can be written in terms of the constants (9) as

$$u(r, \theta) = \sum_{n=0}^{\infty} A_n \left(\frac{r}{c}\right)^n P_n(\cos\theta) \qquad (r \le c). \tag{10}$$

We note that *the harmonic function v in the unbounded region $r > c$, exterior to the spherical surface $r = c$, which assumes the values $F(\theta)$ on that surface and is bounded as $r \to \infty$ can be found in like manner.* Here $C_1 = 0$ in our solution (6) of equation (3) if R is to remain bounded as $r \to \infty$; and the solutions of equation (1) are

$$v_n = \frac{1}{r^{n+1}} P_n(\cos\theta) \qquad (n = 0, 1, 2, \ldots).$$

Thus

$$v(r, \theta) = \sum_{n=0}^{\infty} \frac{B_n}{r^{n+1}} P_n(\cos\theta) \qquad (r \geq c), \tag{11}$$

where the B_n are this time related to the constants (9) by means of the equation $A_n = B_n/c^{n+1}$. That is,

$$v(r, \theta) = \sum_{n=0}^{\infty} A_n \left(\frac{c}{r}\right)^{n+1} P_n(\cos\theta) \qquad (r \geq c). \tag{12}$$

PROBLEMS

1. Suppose that u is harmonic throughout the regions $r < c$ and $r > c$, that $u \to 0$ as $r \to \infty$, and that $u = 1$ on the spherical surface $r = c$. Show from results found in Sec. 92 that $u = 1$ when $r \leq c$ and $u = c/r$ when $r > c$.
2. Suppose that for all ϕ, the steady temperatures $u(r, \theta)$ in a solid sphere $r \leq 1$ are such that $u(1, \theta) = F(\theta)$ where

$$F(\theta) = \begin{cases} 1 & \text{when } 0 < \theta < \dfrac{\pi}{2}, \\ 0 & \text{when } \dfrac{\pi}{2} < \theta < \pi. \end{cases}$$

Derive the expression

$$u(r, \theta) = \frac{1}{2} + \frac{1}{2}\sum_{n=0}^{\infty} [P_{2n}(0) - P_{2n+2}(0)]\, r^{2n+1} P_{2n+1}(\cos\theta)$$

for those temperatures.

3. Let $u(r, \theta)$ denote steady temperatures in a hollow sphere $a \leq r \leq b$ when

$$u(a, \theta) = F(\theta) \quad \text{and} \quad u(b, \theta) = 0 \qquad (0 < \theta < \pi).$$

Derive the expression

$$u(r, \theta) = \sum_{n=0}^{\infty} A_n \frac{b^{2n+1} - r^{2n+1}}{b^{2n+1} - a^{2n+1}} \left(\frac{a}{r}\right)^{n+1} P_n(\cos\theta),$$

where

$$A_n = \frac{2n+1}{2} \int_0^{\pi} F(\theta) P_n(\cos\,\theta)\, \sin\theta\, d\theta \qquad (n = 0, 1, 2 \ldots).$$

4. Let $u(x, t)$ represent the temperatures in a nonhomogeneous insulated bar $-1 \leq x \leq 1$ alng the x axis, and suppose that the thermal conductivity is proportional to $1 - x^2$. The heat equation takes the form

$$\frac{\partial u}{\partial t} = b \frac{\partial}{\partial x}\left[(1 - x^2)\frac{\partial u}{\partial x}\right] \qquad (b > 0).$$

Here b is constant since we assume that the product of the physical constants σ and δ used in Sec. 20 and in Problem 2, Sec. 21, is constant. Note that the ends $x = \pm 1$ are insulated because the conductivity vanishes there. Assuming that

$$u(x, 0) = f(x) \qquad (-1 < x < 1),$$

derive the expression

$$u(x,t) = \sum_{n=0}^{\infty} A_n \exp[-n(n+1)bt] P_n(x),$$

where

$$A_n = \frac{2n+1}{2} \int_{-1}^{1} f(x) P_n(x)\, dx \qquad (n = 0, 1, 2 \ldots).$$

5. Show that if $f(x) = x^2$ $(-1 < x < 1)$ in Problem 4, then

$$u(x,t) = \frac{1}{3} + \left(x^2 - \frac{1}{3}\right) \exp(-6bt).$$

6. Give a physical interpretation of the following boundary value problem in spherical coordinates for a harmonic function $u(r,\theta)$:

$$r \frac{\partial^2}{\partial r^2}(ru) + \frac{1}{\sin\theta} \frac{\partial}{\partial\theta}\left(\sin\theta \frac{\partial u}{\partial\theta}\right) = 0 \qquad (1 < r < b, \theta_1 < \theta < \theta_2),$$

$$u(1,\theta) = 0, \qquad u(b,\theta) = 0,$$
$$u(r,\theta_1) = f(r), \qquad u(r,\theta_2) = 0,$$

where $0 < \theta_1 < \theta_2 < \pi$. Then, using the normalized eigenfunctions found in Problem 11, Sec. 63, and the functions p_α and q_α in Problem 5, Sec. 87, derive the expression

$$u(r,\theta) = \frac{1}{\sqrt{r}} \sum_{n=1}^{\infty} B_n \frac{F_n(\theta)}{F_n(\theta_1)} \sin(\alpha_n \ln r),$$

where

$$\alpha_n = \frac{n\pi}{\ln b}, \qquad B_n = \frac{2}{\ln b} \int_1^b \frac{f(r)}{\sqrt{r}} \sin(\alpha_n \ln r)\, dr,$$

and

$$F_n(\theta) = p_{\alpha_n}(\cos\theta)\, q_{\alpha_n}(\cos\theta_2) - p_{\alpha_n}(\cos\theta_2)\, q_{\alpha_n}(\cos\theta).$$

93. STEADY TEMPERATURES IN A HEMISPHERE

The base $r < 1, \theta = \pi/2$ of a solid hemisphere $r \le 1, 0 \le \theta \le \pi/2$ is insulated; and the flux of heat inward through the hemispherical surface is kept at prescribed values $F(\theta)$. The boundary value problem for steady temperatures $u(r,\theta)$ in the hemisphere consists of Laplace's equation

$$r \frac{\partial^2}{\partial r^2}(ru) + \frac{1}{\sin\theta} \frac{\partial}{\partial\theta}\left(\sin\theta \frac{\partial u}{\partial\theta}\right) = 0 \qquad \left(r < 1, 0 < \theta < \frac{\pi}{2}\right), \tag{1}$$

the condition of insulation (see Problem 9, Sec. 24)

$$u_\theta\left(r, \frac{\pi}{2}\right) = 0 \qquad (0 < r < 1) \tag{2}$$

at the base, part of which is shown in Fig. 68, and the flux condition (see Sec. 24)

$$Ku_r(1, \theta) = F(\theta) \qquad \left(0 < \theta < \frac{\pi}{2}\right), \tag{3}$$

where K is thermal conductivity. In order that temperatures be steady, we assume that the values $F(\theta)$ are such that the resultant rate of flow through the hemispherical surface is zero. That is,

$$\int_0^{\pi/2} F(\theta)\, 2\pi \sin\theta \, d\theta = 0. \tag{4}$$

Also, we assume that F is piecewise smooth on the interval $0 < \theta < \pi/2$ and that u satisfies the usual continuity conditions when $0 \le r < 1$ and $0 \le \theta \le \pi/2$.

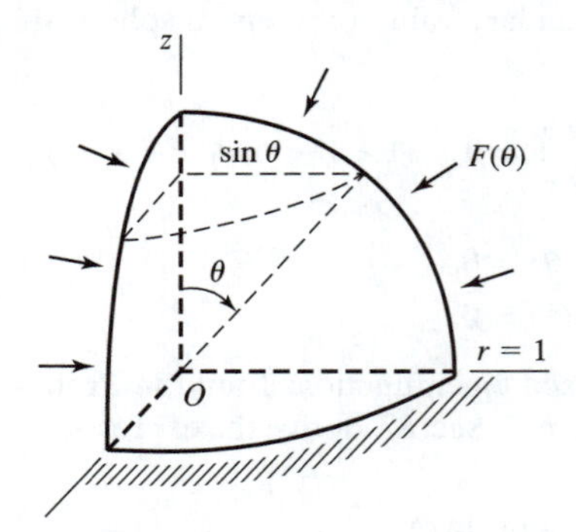

FIGURE 68

Writing $u = R(r)\Theta(\theta)$ and separating variables in equations (1) and (2), we obtain the conditions

$$r(rR)'' - \lambda R = 0 \qquad (r < 1), \tag{5}$$

where R must be continuous when $0 \le r < 1$, and

$$\frac{d}{d\theta}\left(\sin\theta \frac{d\Theta}{d\theta}\right) + \lambda \sin\theta \, \Theta = 0 \qquad \left(0 < \theta < \frac{\pi}{2}\right), \tag{6}$$

$$\Theta'\left(\frac{\pi}{2}\right) = 0, \tag{7}$$

where Θ and Θ' are to be continuous when $0 \le \theta \le \pi/2$.

From case (*b*) in Theorem 1, Sec. 91, we know that the singular Sturm-Liouville problem (6)–(7) has eigenvalues and corresponding eigenfunctions

$$\lambda_n = 2n(2n+1), \qquad \Theta_n = P_{2n}(\cos\theta) \qquad (n = 0, 1, 2, \ldots).$$

When $\lambda = \lambda_n$, equation (5) is the Cauchy-Euler equation

$$r^2 R'' + 2rR' - 2n(2n+1)R = 0,$$

whose bounded solution is $R_n = r^{2n}$ when $0 < r < 1$.

Formally, then,

$$u(r, \theta) = \sum_{n=0}^{\infty} B_n r^{2n} P_{2n}(\cos\theta)$$

if the constants B_n are such that condition (3) is satisfied. That condition requires that

$$F(\theta) = \sum_{n=1}^{\infty} (2KnB_n) P_{2n}(\cos\theta) \qquad \left(0 < \theta < \frac{\pi}{2}\right). \tag{8}$$

Case (*b*) of Theorem 2 in Sec. 91 tells us that this is a valid Legendre series representation if $2KnB_n = A_{2n}$ where

$$A_{2n} = (4n+1) \int_0^{\pi/2} F(\theta) P_{2n}(\cos\theta) \sin\theta \, d\theta \qquad (n = 1, 2, \ldots), \tag{9}$$

and if F is such that the condition $A_0 = 0$, which is precisely condition (4), is satisfied. Thus B_0 *is left arbitrary,* and

$$u(r, \theta) = B_0 + \frac{1}{2K} \sum_{n=1}^{\infty} \frac{1}{n} A_{2n} r^{2n} P_{2n}(\cos\theta) \qquad \left(r \le 1, 0 \le \theta \le \frac{\pi}{2}\right) \tag{10}$$

where the coefficients A_{2n} have the values (9).

The constant B_0 is the temperature at the origin $r = 0$. Solutions of such problems with just Neumann conditions (Sec. 28) are determined only up to such an arbitrary additive constant because all the boundary conditions prescribe only values of derivatives of the harmonic functions.

PROBLEMS

1. The base $r < c, \theta = \pi/2$ of a solid hemisphere $r \le c, 0 \le \theta \le \pi/2$ is insulated. The temperature distribution on the hemispherical surface is $u = F(\theta)$. Derive the expression

$$u(r, \theta) = \sum_{n=0}^{\infty} A_{2n} \left(\frac{r}{c}\right)^{2n} P_{2n}(\cos\theta),$$

where

$$A_{2n} = (4n+1) \int_0^{\pi/2} F(\theta) P_{2n}(\cos\theta) \sin\theta \, d\theta \qquad (n = 0, 1, 2, \ldots),$$

for the steady temperatures in the solid. Also, show that $u(r, \theta) = 1$ when $F(\theta) = 1$.

2. A function u is harmonic and bounded in the unbounded region $r > c, 0 \le \theta < \pi/2$. Also, $u = 0$ everywhere on the flat boundary surface $r > c, \theta = \pi/2$, and $u = F(\theta)$ on the hemispherical boundary surface $r = c, 0 < \theta < \pi/2$. Derive the expression

$$u(r, \theta) = \sum_{n=0}^{\infty} A_{2n+1} \left(\frac{c}{r}\right)^{2n+2} P_{2n+1}(\cos\theta),$$

where

$$A_{2n+1} = (4n+3) \int_0^{\pi/2} F(\theta) P_{2n+1}(\cos\theta) \sin\theta \, d\theta \qquad (n = 0, 1, 2, \ldots).$$

3. The flux of heat $Ku_r(1, \theta)$ into a solid sphere at its surface $r = 1$ is a prescribed function $F(\theta)$, where F is such that the net time rate of flow of heat into the solid is zero. Thus (see Sec. 93)

$$\int_0^{\pi} F(\theta)\, 2\pi \sin\theta \, d\theta = 0.$$

Assuming that $u = 0$ at the center $r = 0$, derive the expression

$$u(r, \theta) = \frac{1}{K} \sum_{n=1}^{\infty} \frac{1}{n} A_n r^n P_n(\cos\theta),$$

where

$$A_n = \frac{2n+1}{2} \int_0^{\pi} F(\theta) P_n(\cos\theta) \sin\theta \, d\theta \qquad (n = 0, 1, 2, \ldots),$$

for the steady temperatures throughout the entire sphere $0 \le r \le 1$.

4. The base $r < 1, \theta = \pi/2$ of a solid hemisphere $r \le 1, 0 \le \theta \le \pi/2$ is kept at temperature $u = 0$, while $u = 1$ on the hemispherical surface $r = 1, 0 < \theta < \pi/2$. Derive the expression

$$u(r, \theta) = \sum_{n=0}^{\infty} (-1)^n \left(\frac{4n+3}{2n+2} \right) \frac{(2n)!}{2^{2n}(n!)^2} r^{2n+1} P_{2n+1}(\cos\theta)$$

for the steady temperatures in that solid.

5. Heat is generated at a steady and uniform rate throughout the interior of a solid hemisphere $0 \le r \le 1, 0 \le \theta \le \pi/2$, and the entire surface is kept at temperature zero. Thus the steady temperatures $u = u(r, \theta)$ satisfy the nonhomogeneous differential equation

$$\frac{1}{r} \frac{\partial^2}{\partial r^2}(ru) + \frac{1}{r^2 \sin\theta} \frac{\partial}{\partial\theta} \left(\sin\theta \frac{\partial u}{\partial\theta} \right) + q_0 = 0 \qquad \left(0 < r < 1, 0 < \theta < \frac{\pi}{2} \right)$$

and the boundary conditions

$$u(1, \theta) = 0, \qquad u\left(r, \frac{\pi}{2}\right) = 0.$$

Also, $u(r, \theta)$ is continuous at $r = 0$. Point out how Problem 4 suggests seeking a solution of the form

$$u(r, \theta) = \sum_{n=0}^{\infty} B_n(r)\, P_{2n+1}(\cos\theta)$$

and applying the method of variation of parameters, which was first used in Sec. 36. Follow the steps below to find the solution by that method.

(*a*) Observe how it follows immediately from case (*c*) in Theorem 1, Sec. 91, that

$$\frac{1}{\sin\theta} \frac{d}{d\theta} \left[\sin\theta \frac{d}{d\theta} P_{2n+1}(\cos\theta) \right] = -(2n+1)(2n+2)\, P_{2n+1}(\cos\theta)$$

$$(n = 0, 1, 2, \ldots).$$

Then, with the aid of this identity and expansion (6), Sec. 90, obtain the initial value problem

$$r^2 B_n''(r) + 2rB_n'(r) - (2n+1)(2n+2)B_n(r) = -q_0 A_{2n+1} r^2, \qquad B_n(1) = 0$$

$$(n = 0, 1, 2, \ldots),$$

where $A_{2n+1} = P_{2n}(0) - P_{2n+2}(0)$ and where $B_n(r)$ is to be continuous on the interval $0 \le r \le 1$.

(*b*) Solve the differential equation in part (*a*) by adding a particular solution of it to the general solution of the complementary equation (compare with Problem 2, Sec. 40). Then apply the required conditions on $B_n(r)$, stated in part (*a*), to complete the solution of the initial value problem in ordinary differential equations there. Thus arrive at the desired temperature function:

$$u(r, \theta) = \frac{q_0}{2} \sum_{n=0}^{\infty} \frac{P_{2n}(0) - P_{2n+2}(0)}{(2n-1)(n+2)} (r^2 - r^{2n+1}) P_{2n+1}(\cos\theta).$$

Suggestion: Observe that the differential equation in part (*a*) has a particular solution of the form $B_n(r) = ar^2$, where a is a constant. Also, note that the complementary equation in part (*b*) is of Cauchy-Euler type, and solve it by the method described in Problem 1, Sec. 38.

CHAPTER 11

VERIFICATION OF SOLUTIONS AND UNIQUENESS

In this chapter, we examine in some detail the question of verifying solutions of boundary value problems. Careful verifications of the solutions of the two boundary value problems solved in Chap. 4, where the Fourier method was introduced, will be made. More precisely, the solutions found for the temperature problem in Sec. 31 and the vibrating string problem in Sec. 32 will be verified.

We shall also consider the question of establishing that a solution of a given problem is the only possible solution. A multiplicity of solutions may actually arise when the statement of the problem does not demand adequate continuity or boundedness of a solution and its derivatives. This was illustrated in Problem 14, Sec. 50.

We begin the chapter with an important theorem that enables us to establish uniform convergence of solutions obtained in the form of series and is often useful in both verifying such a solution and proving that it is unique. The remaining theorems in the chapter give conditions under which a solution is unique. They apply only to specific types of problems, and their applications are further limited because they require a rather high degree of regularity of the functions involved.

94. ABEL'S TEST FOR UNIFORM CONVERGENCE

We start here with some needed background on uniform convergence. Let $s_n(x)$ denote the sum of the first n terms of a series

$$\sum_{i=1}^{\infty} X_i(x) \tag{1}$$

of functions $X_i(x)$ that converges to a sum $s(x)$:

$$(2) \qquad s_n(x) = \sum_{i=1}^{n} X_i(x), \qquad s(x) = \lim_{n\to\infty} s_n(x).$$

Suppose that the series converges *uniformly* with respect to x for all x in some set. Then (see Sec. 16), for each positive number ε, there exists a positive integer n_ε, independent of x, such that

$$(3) \qquad |s(x) - s_n(x)| < \varepsilon \qquad \text{whenever } n > n_\varepsilon$$

for every x in the set. The following lemma, known as the *Cauchy criterion,* provides us with an alternative characterization of uniform convergence.

Lemma. *A necessary and sufficient condition for the uniform convergence of series* (1) *on a given set is that for each positive number ε, there exists a positive integer n_ε, independent of x, such that for all points x in the set and all positive integers j,*

$$(4) \qquad |s_{n+j}(x) - s_n(x)| < \varepsilon \qquad \textit{whenever } n > n_\varepsilon.$$

To verify the necessity of condition (4), we assume that condition (3) is satisfied and let j denote any positive integer. Since

$$|s_{n+j} - s_n| = |(s_{n+j} - s) + (s - s_n)| \le |s - s_{n+j}| + |s - s_n|,$$

we find that

$$|s_{n+j} - s_n| < 2\varepsilon \qquad \text{whenever } n > n_\varepsilon;$$

and since ε is an arbitrary positive number, this is the same as statement (4).

To show that condition (4) is also sufficient for uniform convergence, we assume that condition (4) holds and recall from calculus that it is sufficient for the pointwise convergence of series (1). Hence it implies that the sum $s(x)$, defined by means of the second of equations (2), exists. Keeping n fixed in inequality (4) and letting j tend to infinity, we now have the inequality

$$|s(x) - s_n(x)| \le \varepsilon \qquad \text{whenever } n > n_\varepsilon$$

if x is in the set. That is, since ε is arbitrary, series (1) converges uniformly.

Note that x here may equally well denote elements $(x_1, x_2, \ldots, x_N)$ of some set in N-dimensional space. The uniform convergence is then with respect to all N variables $x_1, x_2, \ldots, x_N$ together.

We now derive a test for the uniform convergence of infinite series whose terms are products of certain types of functions. Its application in verifying formal solutions of boundary value problems will be illustrated in Sec. 95. The test, known as *Abel's test,*[†] involves functions in a sequence $T_i(t)$ $(i = 1, 2, \ldots)$ which is *uniformly bounded* for all points t in an interval. That is, there exists a positive constant M, independent of i, such that

$$(5) \qquad |T_i(t)| < M \qquad (i = 1, 2, \ldots)$$

[†]Niels Henrik Abel, Norwegian, 1802–1829.

for all t in the interval. The sequence is, moreover, *monotonic with respect to* i. Thus, for every t in the interval, either

$$T_{i+1}(t) \leq T_i(t) \qquad (i = 1, 2, \ldots) \tag{6}$$

or

$$T_{i+1}(t) \geq T_i(t) \qquad (i = 1, 2, \ldots). \tag{7}$$

We state the test as a theorem which shows that when the terms of a uniformly convergent series are multiplied by functions $T_i(t)$ of the type just described, the new series is also uniformly convergent.

Theorem. *The series*

$$\sum_{i=1}^{\infty} X_i(x) T_i(t) \tag{8}$$

converges uniformly with respect to the two variables x and t together in a region R of the xt plane if the following two conditions are satisfied:

(*i*) *The series*

$$\sum_{i=1}^{\infty} X_i(x)$$

converges uniformly with respect to x for all x such that (x, t) *is in R.*

(*ii*) *The functions* $T_i(t)$ *are uniformly bounded and monotonic with respect to i* $(i = 1, 2, \ldots)$ *for all t such that* (x, t) *is in R.*

To start the proof, we let S_n denote partial sums of series (8):

$$S_n(x, t) = \sum_{i=1}^{n} X_i(x) T_i(t).$$

In view of the lemma above, the uniform convergence of that series will be established if we prove that to each positive number ε there corresponds a positive integer n_ε, independent of x and t, such that

$$|S_m(x, t) - S_n(x, t)| < \varepsilon \qquad \text{whenever } n > n_\varepsilon,$$

for all integers $m = n + 1, n + 2, \ldots$ and for all points (x, t) in R.

We write the partial sum

$$s_n(x) = X_1(x) + X_2(x) + \cdots + X_n(x).$$

Then, for each pair of integers m and n $(m > n)$,

$$\begin{aligned}
S_m - S_n &= X_{n+1} T_{n+1} + X_{n+2} T_{n+2} + \cdots + X_m T_m \\
&= (s_{n+1} - s_n) T_{n+1} + (s_{n+2} - s_{n+1}) T_{n+2} + \cdots + (s_m - s_{m-1}) T_m \\
&= (s_{n+1} - s_n) T_{n+1} + (s_{n+2} - s_n) T_{n+2} - (s_{n+1} - s_n) T_{n+2} \\
&\quad + \cdots + (s_m - s_n) T_m - (s_{m-1} - s_n) T_m.
\end{aligned}$$

By pairing alternate terms here, we find that

$$S_m - S_n = (s_{n+1} - s_n)(T_{n+1} - T_{n+2}) + (s_{n+2} - s_n)(T_{n+2} - T_{n+3}) + \cdots + (s_{m-1} - s_n)(T_{m-1} - T_m) + (s_m - s_n)T_m. \tag{9}$$

Suppose now that the functions T_i are nonincreasing with respect to i, so that they satisfy condition (6), and that they also satisfy the uniform boundedness condition (5). Then the factors $T_{n+1} - T_{n+2}$, $T_{n+2} - T_{n+3}$, etc., in equation (9) are nonnegative, and $|T_i(t)| < M$. Since the series with terms $X_i(x)$ converges uniformly, an integer n_ε exists such that

$$|s_{n+j}(x) - s_n(x)| < \frac{\varepsilon}{3M} \qquad \text{whenever } n > n_\varepsilon,$$

for all positive integers j, where ε is any given positive number and n_ε is independent of x. Then if $n > n_\varepsilon$ and $m > n$, it follows from equation (9) that

$$|S_m - S_n| < \frac{\varepsilon}{3M}[(T_{n+1} - T_{n+2}) + (T_{n+2} - T_{n+3}) + \cdots + |T_m|]$$
$$= \frac{\varepsilon}{3M}(T_{n+1} - T_m + |T_m|) \le \frac{\varepsilon}{3M}(|T_{n+1}| + 2|T_m|).$$

Therefore,

$$|S_m(x, t) - S_n(x, t)| < \varepsilon \qquad \text{whenever } m > n > n_\varepsilon;$$

and the uniform convergence of series (8) is established.

The proof is similar when the functions T_i are nondecreasing with respect to i.

When x is kept fixed, the series with terms X_i is a series of constants; and the only requirement placed on it is that it be convergent. Then the theorem shows that when T_i are bounded and monotonic, the series of terms $X_i T_i(t)$ is uniformly convergent with respect to t.

Extensions of the theorem to cases in which X_i are functions of x and t, or both X_i and T_i are functions of several variables, become evident when it is observed that our proof rests on the uniform convergence of the series of terms X_i and the bounded monotonic nature of the functions T_i.

95. VERIFICATION OF SOLUTION OF TEMPERATURE PROBLEM

We turn now to the full verification of the solution of the temperature problem

$$u_t(x, t) = k u_{xx}(x, t) \qquad (0 < x < c, t > 0), \tag{1}$$

$$u_x(0, t) = 0, \qquad u_x(c, t) = 0 \qquad (t > 0), \tag{2}$$

$$u(x, 0) = f(x) \qquad (0 < x < c) \tag{3}$$

that was obtained in Sec. 31. We recall that the continuous functions

$$u_0 = 1, \qquad u_n = \exp\left(-\frac{n^2\pi^2 k}{c^2}t\right)\cos\frac{n\pi x}{c} \qquad (n = 1, 2, \ldots) \tag{4}$$

were found to satisfy the homogeneous conditions (1) and (2). In view of Example 1, Sec. 30, which was based on the superposition theorem in that section, the generalized linear combination

$$u = A_0\, u_0 + \sum_{n=1}^{\infty} A_n\, u_n \tag{5}$$

formally satisfied conditions (1) and (2). Expression (5), when written as

$$u(x,t) = A_0 + \sum_{n=1}^{\infty} A_n \exp\left(-\frac{n^2\pi^2 k}{c^2}\, t\right) \cos\frac{n\pi x}{c}, \tag{6}$$

then gave us the formal solution of the boundary value problem when its coefficients were assigned the values

$$A_0 = \frac{1}{c}\int_0^c f(x)\,dx, \qquad A_n = \frac{2}{c}\int_0^c f(x)\cos\frac{n\pi x}{c}\,dx \qquad (n = 1, 2, \ldots). \tag{7}$$

We assume here that f is piecewise smooth (Sec. 9) on $0 < x < c$. Also, at a point of discontinuity of f in that interval, we define $f(x)$ as the mean value of the one-sided limits $f(x+)$ and $f(x-)$. Note how it follows from expressions (7) that

$$|A_0| \le \frac{1}{c}\int_0^c |f(x)|\,dx, \qquad |A_n| \le \frac{2}{c}\int_0^c |f(x)|\,dx \qquad (n = 1, 2, \ldots)$$

and hence that there is a positive constant M, independent of n, such that

$$|A_n| \le M \qquad (n = 0, 1, 2, \ldots). \tag{8}$$

We begin our verification by showing that series (5), with coefficients (7), actually converges in the region $0 \le x \le c$, $t > 0$ of the xt plane and that it satisfies the homogeneous conditions (1) and (2). To accomplish this, we first note from expressions (4) and inequalities (8) that if t_0 is a fixed positive number,

$$|A_n u_n| \le M \exp\left(-\frac{n^2\pi^2 k}{c^2}\, t_0\right) \qquad (n = 0, 1, 2, \ldots) \tag{9}$$

whenever $0 \le x \le c$ and $t \ge t_0$ (Fig. 69). An application of the ratio test shows that the series

$$\sum_{n=0}^{\infty} n^i \exp\left(-\frac{n^2\pi^2 k}{c^2}\, t_0\right) \tag{10}$$

of constants converges when i is any nonnegative integer and, in particular, when $i = 0$. So we know from the comparison and absolute convergence tests that the series (5) converges when $0 \le x \le c$, $t \ge t_0$. One can use series (10) and the

Weierstrass M-test (Sec. 16) to show that the series

$$(11) \qquad \sum_{n=0}^{\infty}(A_n u_n)_x, \qquad \sum_{n=0}^{\infty}(A_n u_n)_{xx}$$

of derivatives converge uniformly on the interval $0 \le x \le c$ for any fixed t $(t \ge t_0)$. Likewise, the series

$$(12) \qquad \sum_{n=0}^{\infty}(A_n u_n)_t$$

converges uniformly on the semi-infinite interval $t \ge t_0$ for any fixed x $(0 \le x \le c)$.

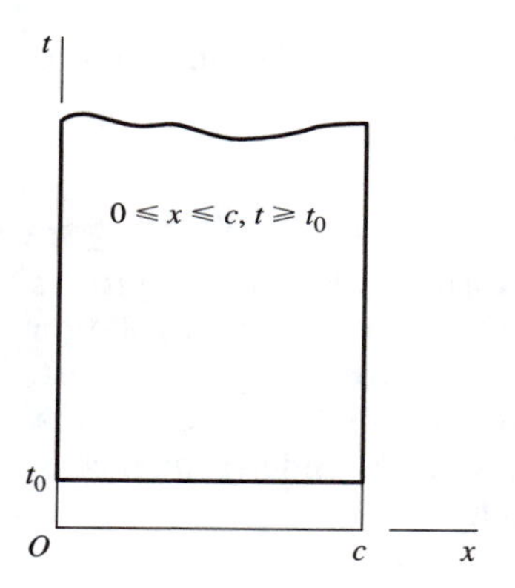

FIGURE 69

The uniformity of the convergence of these series ensures that series (5) is differentiable twice with respect to x and once with respect to t, provided that $0 \le x \le c, t \ge t_0$. Consequently, if we write

$$L = k\frac{\partial^2}{\partial x^2} - \frac{\partial}{\partial t}$$

and note that $Lu_n = 0$ $(n = 0, 1, 2, \ldots)$, it follows from the superposition theorem in Sec. 30 that $Lu = 0$ when $0 \le x \le c, t \ge t_0$. Thus series (5) converges and satisfies the heat equation (1) in the domain $0 < x < c, t > 0$ since the positive number t_0 can be chosen arbitrarily small.

Writing $L = \partial/\partial x$ and again using the theorem in Sec. 30, we see that series (5) also satisfies boundary conditions (2). Observe that since the first of series (11) is uniformly convergent on the interval $0 \le x \le c$ for any fixed t $(t \ge t_0)$, the derivative $u_x(x,t)$ of series (5) is *continuous* in x on that interval. Hence the one-sided limits

$$u_x(0+,t) = \lim_{\substack{x \to 0 \\ x > 0}} u_x(x,t), \qquad u_x(c-,t) = \lim_{\substack{x \to c \\ x < c}} u_x(x,t)$$

at the endpoints of the interval $0 \le x \le c$ $(t \ge t_0)$ exist and have the values $u_x(0,t)$ and $u_x(c,t)$, respectively. Since conditions (2) are satisfied and since t_0 can be chosen arbitrarily small, then,

$$u_x(0+, t) = 0, \qquad u_x(c-, t) = 0 \qquad (t > 0). \tag{13}$$

In seeking solutions of boundary value problems, we tacitly require that those solutions satisfy such continuity conditions at boundary points. Thus, when conditions (2) are part of a boundary value problem, it is understood that conditions (13) must also be satisfied. As we have just seen, series (5) has that property.

The nonhomogeneous condition (3) is clearly satisfied by our solution since series (6) reduces to the Fourier cosine series

$$A_0 + \sum_{n=1}^{\infty} A_n \cos \frac{n\pi x}{c} \qquad (0 < x < c) \tag{14}$$

for f when $t = 0$; and Theorem 2 in Sec. 14 ensures that series (14) converges to $f(x)$ when $0 < x < c$.

It remains to show that

$$u(x, 0+) = f(x) \qquad (0 < x < c). \tag{15}$$

This is a continuity requirement that must be satisfied when $t = 0$, just as conditions (13) must hold when $x = 0$ and $x = c$. One can show that solution (6) has this property by appealing to Abel's test in Sec. 94. According to that test, the series formed by multiplying the terms of a convergent series of constants, such as series (14) with x fixed, by corresponding terms of a bounded sequence of functions of t whose values never increase with n, such as

$$\exp\left(-\frac{n^2\pi^2 k}{c^2}t\right) \qquad (n = 0, 1, 2, \ldots),$$

is uniformly convergent with respect to t. So, for any fixed x $(0 < x < c)$, the series in expression (6) converges uniformly with respect to t when $t \geq 0$ and thus represents a function that is continuous in t $(t \geq 0)$. This shows that our solution $u(x, t)$ is continuous in t when $t \geq 0$, in particular when $t = 0$. That is,

$$\lim_{\substack{t \to 0 \\ t > 0}} u(x, t) = u(x, 0),$$

or $u(x, 0+) = u(x, 0)$, for each fixed x $(0 < x < c)$. Property (15) now follows from the fact that $u(x, 0) = f(x)$ $(0 < x < c)$. This completes the verification that the function (6), with coefficients (7), is a solution of the boundary value problem (1)–(3).

96. UNIQUENESS OF SOLUTIONS OF THE HEAT EQUATION

Let D denote the domain consisting of all points interior to a closed surface S; and let $\overline{D}$ be the *closure* of that domain, consisting of all points in D and all points on S. We assume always that the closed surface S is *piecewise smooth*. That is, it is a continuous surface consisting of a finite number of parts over each of which the outward unit normal vector exists and varies continuously from point to point.

Then if U is a function of x, y, and z which is continuous in $\overline{D}$, together with its partial derivatives of the first and second orders, a special case of Green's identity that we shall need here states that

$$\iint_S U \frac{dU}{dn}\, dA = \iiint_D (U\nabla^2 U + U_x^2 + U_y^2 + U_z^2)\, dV. \tag{1}$$

Here dA is the area element on S, the symbol dV represents $dx\, dy\, dz$, and dU/dn is the derivative in the direction of the outward unit vector normal to S.[†]

Consider a homogeneous solid whose interior is the domain D and whose temperatures at time t are denoted by $u(x, y, z, t)$. A fairly general problem in heat conduction is the following:

$$u_t = k\nabla^2 u + q(x, y, z, t) \qquad [(x, y, z) \text{ in } D, t > 0], \tag{2}$$

$$u(x, y, z, 0) = f(x, y, z) \qquad [(x, y, z) \text{ in } \overline{D}], \tag{3}$$

$$u = g(x, y, z, t) \qquad [(x, y, z) \text{ on } S, t \geq 0]. \tag{4}$$

This is the problem of determining temperatures in a body, with prescribed initial temperatures $f(x, y, z)$ and surface temperatures $g(x, y, z, t)$, interior to which heat may be generated continuously at a rate per unit volume proportional to $q(x, y, z, t)$.

Suppose that the problem has two solutions

$$u = u_1(x, y, z, t), \qquad u = u_2(x, y, z, t)$$

where both u_1 and u_2 are continuous functions in the closed region $\overline{D}$ when $t \geq 0$, while their derivatives of the first order with respect to t and of the first and second orders with respect to x, y, and z are continuous in $\overline{D}$ when $t > 0$. Since u_1 and u_2 satisfy the linear equations (2), (3), and (4), their difference

$$U(x, y, z, t) = u_1(x, y, z, t) - u_2(x, y, z, t)$$

satisfies the homogeneous problem

$$U_t = k\nabla^2 U \qquad [(x, y, z) \text{ in } D, t > 0], \tag{5}$$

$$U(x, y, z, 0) = 0 \qquad [(x, y, z) \text{ in } \overline{D}], \tag{6}$$

$$U = 0 \qquad [(x, y, z) \text{ on } S, t \geq 0]. \tag{7}$$

Moreover, U and its derivatives have the continuity properties of u_1 and u_2 assumed above.

We shall now show that $U = 0$ in D when $t > 0$, so that the two solutions u_1 and u_2 are identical. That is, not more than one solution of the boundary value problem in u can exist if the solution is required to satisfy the stated continuity conditions.

[†]Identity (1) is found by applying Gauss's divergence theorem to the vector field $U \operatorname{grad} U$. See the book by Taylor and Mann (1983, pp. 492–493), listed in the Bibliography.

The continuity of U with respect to x, y, z, and t together in the closed region $\overline{D}$ when $t \geq 0$ implies that the integral

$$I(t) = \frac{1}{2} \iiint_D [U(x, y, z, t)]^2 \, dV \tag{8}$$

is a continuous function of t when $t \geq 0$; and, according to equation (6), $I(0) = 0$. Also, in view of the continuity of U_t when $t > 0$, we may use equation (5) to write

$$I'(t) = \iiint_D U U_t \, dV = k \iiint_D U \nabla^2 U \, dV \qquad (t > 0).$$

Identity (1) applies to the last integral here because of the continuity of the derivatives of U when $t > 0$. Thus

$$\iiint_D U \nabla^2 U \, dV = \iint_S U \frac{du}{dn} \, dA - \iiint_D \left(U_x^2 + U_y^2 + U_z^2\right) dV \tag{9}$$

when $t > 0$. But $U = 0$ on S, and $k > 0$; consequently,

$$I'(t) = -k \iiint_D \left(U_x^2 + U_y^2 + U_z^2\right) dV \leq 0.$$

The mean value theorem for derivatives applies to $I(t)$. That is, for each positive t, a number t_1 $(0 < t_1 < t)$ exists such that

$$I(t) - I(0) = tI'(t_1);$$

and since $I(0) = 0$ and $I'(t_1) \leq 0$, it follows that $I(t) \leq 0$. However, definition (8) of the integral shows that $I(t) \geq 0$. Therefore,

$$I(t) = 0 \qquad (t \geq 0);$$

and so the nonnegative integrand U^2 cannot have a positive value at any point in D. For if it did, the continuity of U^2 would require that U^2 be positive throughout some neighborhood of the point, and that would mean $I(t) > 0$. Consequently,

$$U(x, y, z, t) = 0 \qquad [(x, y, z) \text{ in } \overline{D}, t \geq 0];$$

and we arrive at the following theorem on uniqueness.

Theorem 1. *Let a function $u = u(x, y, z, t)$ satisfy these conditions of regularity:*

(*i*) *It is a continuous function of the variables x, y, z, and t together when the point (x, y, z) is in the closure $\overline{D}$ of a given domain D and $t \geq 0$.*

(*ii*) *The derivatives of u appearing in the heat equation* (2) *are continuous in the same sense when $t > 0$.*

If u is a solution of the boundary value problem (2)–(4), *it is the only possible solution satisfying conditions* (*i*) *and* (*ii*).

When conditions (*i*) and (*ii*) in Theorem 1 are added to the requirement that u is to satisfy the heat equation and the boundary conditions, our boundary value problem is completely stated; and u will be the only possible solution if it exists.

The condition that u be continuous in $\overline{D}$ when $t = 0$ restricts the usefulness of our theorem. It is clearly not satisfied if the initial temperature function f in condition (3) fails to be continuous throughout $\overline{D}$, or if at some point on S the initial value $g(x, y, z, 0)$ of the prescribed surface temperature differs from the value $f(x, y, z)$. The continuity requirement at $t = 0$ can be relaxed in some cases.†

The proof of Theorem 1 required that the integral

$$\iint_S U \frac{dU}{dn}\, dA$$

in equation (9) either vanish or have a negative value. It vanished because $U = 0$ on S. Note that it is never positive if condition (4) is replaced by the boundary condition

$$\frac{du}{dn} + hu = g(x, y, z, t) \qquad [(x, y, z) \text{ on } S, t > 0] \tag{10}$$

where $h \geq 0$, since in that case $dU/dn = -hU$ and $U dU/dn \leq 0$ on S. Thus our theorem can be modified as follows.

Theorem 2. *The conclusion in Theorem* 1 *is true if boundary condition* (4) *is replaced by condition* (10), *or if condition* (4) *is satisfied on part of the surface S and condition* (10) *is satisfied on the rest.*

EXAMPLE. In the problem of temperature distribution in a slab with insulated faces $x = 0$ and $x = c$ and initial temperatures $f(x)$ (Sec. 95), write $c = \pi$ and assume that f is continuous and f' is piecewise continuous on the interval $0 \leq x \leq \pi$. Then the Fourier cosine series for f converges uniformly to $f(x)$ on that interval (Sec. 16). Let $u(x, t)$ denote the sum of the series

$$A_0 + \sum_{n=1}^{\infty} A_n e^{-n^2kt} \cos nx \qquad (0 \leq x \leq \pi, t \geq 0), \tag{11}$$

which is the formal solution (6), Sec. 95, of the boundary value problem, the constants $2A_0$ and A_n $(n = 1, 2, \ldots)$ being the coefficients in the Fourier cosine series for f on $0 < x < \pi$.

We can see from Abel's test (Sec. 94) that series (11) converges uniformly with respect to x and t together in the region $0 \leq x \leq \pi, t \geq 0$ of the xt plane; thus u is continuous there. When $t \geq t_0$, where t_0 is any positive number, the series obtained by differentiating series (11) term by term any number of times with respect to x to t is uniformly convergent, according to the Weierstrass M-test (Sec. 16). Consequently, we now know that u satisfies all the equations in the boundary value problem (compare with Sec. 95) and also that u_t, u_x, and u_{xx} are continuous

†Integral transforms can sometimes be used to prove uniqueness of solutions of certain types of boundary value problems. This is illustrated in the book by Churchill (1972, sec. 79), listed in the Bibliography.

functions in the region $0 \le x \le \pi$, $t > 0$. Thus u satisfies the regularity conditions (*i*) and (*ii*) stated in Theorem 1, and Theorem 2 applies to show that the sum $u(x, t)$ of series (11) is the only solution that satisfies those conditions.

97. VERIFICATION OF SOLUTION OF VIBRATING STRING PROBLEM

In this section, we shall verify the formal solution that we found in Sec. 32 for the boundary value problem

$$y_{tt}(x,t) = a^2 y_{xx}(x,t) \qquad (0 < x < c, t > 0), \tag{1}$$

$$y(0,t) = 0, \qquad y(c,t) = 0, \tag{2}$$

$$y(x,0) = f(x), \qquad y_t(x,0) = 0. \tag{3}$$

The given function f was assumed to be continuous on the interval $0 \le x \le c$; also, $f(0) = f(c) = 0$. Assuming further that f' *is at least piecewise continuous,* we know (Sec. 14) that $f(x)$ is represented by its Fourier sine series when $0 \le x \le c$. The coefficients

$$B_n = \frac{2}{c}\int_0^c f(x)\sin\frac{n\pi x}{c}\,dx \qquad (n = 1, 2, \ldots) \tag{4}$$

in that series are the ones in the series solution

$$y(x,t) = \sum_{n=1}^{\infty} B_n \sin\frac{n\pi x}{c}\cos\frac{n\pi at}{c} \tag{5}$$

that we obtained. So when $t = 0$, the series in expression (5) converges to $f(x)$; that is, $y(x, 0) = f(x)$ when $0 \le x \le c$.

The nature of the problem calls for a solution $y(x, t)$ that is continuous in x and t when $0 \le x \le c$ and $t \ge 0$ and is such that $y_t(x, t)$ is continuous in t at $t = 0$. Hence the prescribed boundary values in conditions (2) and (3) are also limiting values on the boundary of the domain $0 < x < c$, $t > 0$:

$$y(0+,t) = 0, \qquad y(c-,t) = 0 \qquad (t \ge 0),$$

$$y(x,0+) = f(x), \qquad y_t(x,0+) = 0 \qquad (0 \le x \le c).$$

To verify that expression (5) represents a solution, we must prove that the series there converges to a continuous function $y(x, t)$ which satisfies the wave equation (1) and all the boundary conditions. But series (5), with coefficients (4), can fail to be twice differentiable with respect to x and t even when it has a sum that satisfies the wave equation. This is, in fact, the case with the solution in the example in Sec. 32, where the coefficients B_n were found to be

$$B_n = \frac{8h}{n^2\pi^2}\sin\frac{n\pi}{2} \qquad (n = 1, 2, \ldots).$$

After series (5) is differentiated twice with respect to x or t when those values of B_n are used, it is apparent that the resulting series cannot converge since its nth term does not tend to zero. The closed form of series (5) that was obtained in

Sec. 39 will, however, enable us to verify our solution. That closed form was

$$y(x,t) = \frac{1}{2}[F(x+at) + F(x-at)], \tag{6}$$

where F is the odd periodic extension of f, with period $2c$:

$$F(x) = f(x) \qquad \text{when } 0 \le x \le c \tag{7}$$

and

$$F(-x) = -F(x), \qquad F(x+2c) = F(x) \qquad \text{for all } x. \tag{8}$$

We turn now to the verification of our solution in the form (6). From our assumption that f is continuous when $0 \le x \le c$ and that $f(0) = f(c) = 0$, we see that the odd periodic extension F is continuous for all x (Fig. 70). Let us also *assume that f' and f'' are continuous when $0 \le x \le c$ and that*

$$f''(0) = f''(c) = 0.$$

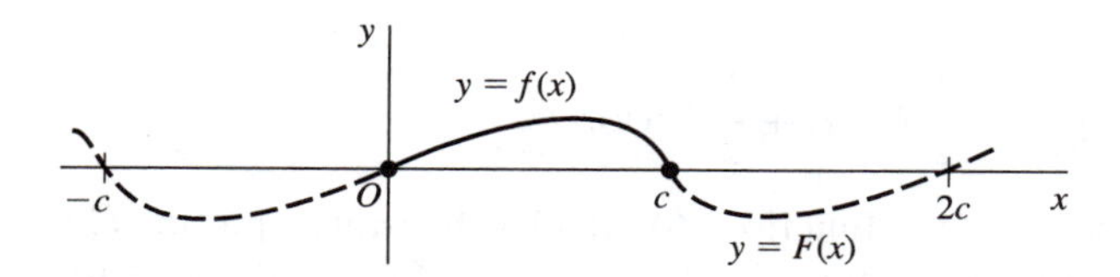

FIGURE 70

It is then easy to show that the derivatives F' and F'' are continuous for all x. For, by recalling that $F(x) = -F(-x)$ and then applying the chain rule, one can write

$$F'(x) = -\frac{d}{dx}F(-x) = F'(-x),$$

where $F'(-x)$ denotes the derivative of F evaluated at $-x$. Thus F' is an *even* periodic function; likewise, F'' is an *odd* periodic function. Consequently, F' and F'' are continuous, as indicated in Fig. 71.

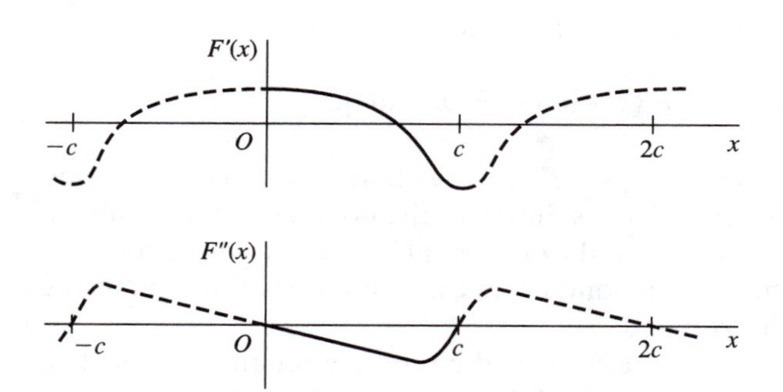

FIGURE 71

To show that the function (6) satisfies the wave equation, we write it as

$$y = \frac{1}{2}F(u) + \frac{1}{2}F(v),$$

where $u = x + at$ and $v = x - at$. The chain rule for differentiating composite functions reveals that

$$\frac{\partial y}{\partial t} = \frac{\partial y}{\partial u}\frac{\partial u}{\partial t} + \frac{\partial y}{\partial v}\frac{\partial v}{\partial t},$$

or

$$\frac{\partial y}{\partial t} = \frac{a}{2}F'(u) - \frac{a}{2}F'(v);$$

and by letting $\partial y/\partial t$ play the role of y in this last expression, we find that

$$\frac{\partial^2 y}{\partial t^2} = \frac{\partial}{\partial t}\left(\frac{\partial y}{\partial t}\right) = \frac{a^2}{2}F''(u) + \frac{a^2}{2}F''(v). \tag{9}$$

Similarly,

$$\frac{\partial^2 y}{\partial x^2} = \frac{1}{2}F''(u) + \frac{1}{2}F''(v). \tag{10}$$

In view of expressions (9) and (10), the function (6) satisfies the wave equation (1). Furthermore, because F is continuous for all x, the function (6) is continuous for all x and t, in particular when $0 \le x \le c$ and $t \ge 0$.

While it is evident from series (5) that our solution $y(x, t)$ satisfies the conditions $y(0, t) = y(c, t) = 0$ and $y(x, 0) = f(x)$, expression (6) can also be used to verify this. For example, when $x = c$ in expression (6), one can write

$$F(c - at) = -F(-c + at) = -F(-c + at + 2c) = -F(c + at).$$

Therefore,

$$y(c, t) = \frac{1}{2}[F(c + at) - F(c + at)] = 0.$$

As for the final boundary condition $y_t(x, 0) = 0$, we observe that

$$y_t(x, t) = \frac{a}{2}[F'(x + at) - F'(x - at)].$$

Hence $y_t(x, 0) = 0$, and the continuity of F' ensures that $y_t(x, t)$ is continuous. The function (6) is now fully verified as a solution of the boundary value problem (1)–(3). In Sec. 98, we shall show why it is the only possible solution which, together with its derivatives of the first and second orders, is continuous throughout the region $0 \le x \le c, t \ge 0$ of the xt plane.

If the conditions on f' and f'' are relaxed by merely requiring those two functions to be *piecewise continuous,* we find that at each instant t there may be a finite number of points x $(0 \le x \le c)$ where the partial derivatives of y fail to exist.

Except at those points, our function satisfies the wave equation and the condition $y_t(x, 0) = 0$. The other boundary conditions are satisfied as before, but we have a solution of our boundary value problem in a broader sense.

Finally, we note that except for the nonhomogeneous condition, our boundary value problem is satisfied by any partial sum

$$y_N(x, t) = \sum_{n=1}^{N} B_n \sin\frac{n\pi x}{c} \cos\frac{n\pi at}{c} \tag{11}$$

of the series solution (5). Instead of satisfying the given nonhomogeneous condition, it satisfies the condition

$$y(x, 0) = \sum_{n=1}^{N} B_n \sin\frac{n\pi x}{c}. \tag{12}$$

The sum on the right-hand side of equation (12) is, of course, a partial sum of the Fourier sine series for f on the interval $0 < x < c$. Since the odd periodic extension of f is clearly continuous and f' is piecewise continuous, that series converges uniformly to $f(x)$ on the interval $0 \le x \le c$ (Sec. 16). Hence if N is taken sufficiently large, the sum $y_N(x, 0)$ can be made to approximate $f(x)$ arbitrarily closely for all values of x in that interval.

The function $y_N(x, t)$, which is everywhere continuous together with all its partial derivatives, is therefore established as a solution of the approximating problem obtained by replacing the nonhomogeneous condition in the original problem by condition (12).

Corresponding approximations can be made to other problems. But a remarkable feature in the present case is that $y_N(x, t)$ never deviates from the actual displacement $y(x, t)$ by more than the greatest deviation of $y_N(x, 0)$ from $f(x)$. To see this, we need only recall the trigonometric identity

$$2\sin A\cos B = \sin(A + B) + \sin(A - B)$$

and write

$$2\sin\frac{n\pi x}{c}\cos\frac{n\pi at}{c} = \sin\frac{n\pi(x + at)}{c} + \sin\frac{n\pi(x - at)}{c}. \tag{13}$$

Expression (11) then becomes (compare with Sec. 39)

$$y_N(x, t) = \frac{1}{2}\left[\sum_{n=1}^{N} B_n \sin\frac{n\pi(x + at)}{c} + \sum_{n=1}^{N} B_n \sin\frac{n\pi(x - at)}{c}\right];$$

and the two sums here are those of the first N terms of the sine series for the odd periodic extension F of the function f, with arguments $x + at$ and $x - at$. But the greatest deviation of the first sum from $F(x + at)$, or of the second from $F(x - at)$, is the same as the greatest deviation of $y_N(x, 0)$ from $f(x)$.

98. UNIQUENESS OF SOLUTIONS OF THE WAVE EQUATION

Consider the following generalization of the problem verified in Sec. 97 for the transverse displacements in a stretched string:

$$y_{tt}(x,t) = a^2 y_{xx}(x,t) + \phi(x,t) \qquad (0 < x < c, t > 0), \tag{1}$$

$$y(0,t) = p(t), \qquad y(c,t) = q(t) \qquad (t \geq 0), \tag{2}$$

$$y(x,0) = f(x), \qquad y_t(x,0) = g(x) \qquad (0 \leq x \leq c). \tag{3}$$

Here we require y to be of class C^2 in the region R: $0 \leq x \leq c, t \geq 0$, by which we shall mean that y and its derivatives of the first and second orders, including y_{xt} and y_{tx}, are to be continuous functions in R. The prescribed functions ϕ, p, q, f, and g must be restricted if the problem is to have a solution of class C^2.

Suppose that there are two solutions $y_1(x,t)$ and $y_2(x,t)$ in that class. Then the difference

$$Y(x,t) = y_1(x,t) - y_2(x,t)$$

is of class C^2 in R and satisfies the homogeneous problem

$$Y_{tt}(x,t) = a^2 Y_{xx}(x,t) \qquad (0 < x < c, t > 0), \tag{4}$$

$$Y(0,t) = 0, \qquad Y(c,t) = 0 \qquad (t \geq 0), \tag{5}$$

$$Y(x,0) = 0, \qquad Y_t(x,0) = 0 \qquad (0 \leq x \leq c). \tag{6}$$

We shall prove that $Y = 0$ throughout R; thus $y_1 = y_2$, as stated in the following theorem.

Theorem. *The boundary value problem* (1)–(3) *cannot have more than one solution of class C^2 in R.*

To start the proof, we note that the integrand of the integral

$$I(t) = \frac{1}{2}\int_0^c \left(Y_x^2 + \frac{1}{a^2}Y_t^2\right) dx \qquad (t \geq 0) \tag{7}$$

satisfies conditions such that

$$I'(t) = \int_0^c \left(Y_x Y_{xt} + \frac{1}{a^2} Y_t Y_{tt}\right) dx. \tag{8}$$

Since $Y_{tt} = a^2 Y_{xx}$, the integrand here can be written

$$Y_x Y_{tx} + Y_t Y_{xx} = \frac{\partial}{\partial x}(Y_x Y_t).$$

So in view of equations (5), from which it follows that

$$Y_t(0,t) = 0, \qquad Y_t(c,t) = 0,$$

one can write

$$I'(t) = Y_x(c,t)Y_t(c,t) - Y_x(0,t)Y_t(0,t) = 0. \tag{9}$$

Hence $I(t)$ is a constant. But equation (7) shows that $I(0) = 0$ because $Y(x, 0) = 0$, and so $Y_x(x, 0) = 0$; also, $Y_t(x, 0) = 0$. Thus $I(t) = 0$. The nonnegative continuous integrand of that integral must, therefore, vanish; that is,

$$Y_x(x, t) = Y_t(x, t) = 0 \qquad (0 \le x \le c, t \ge 0).$$

So Y is constant. In fact, $Y(x, t) = 0$ since $Y(x, 0) = 0$; and the proof of the theorem is complete.

If y_x, instead of y, is prescribed at the endpoint in either or both of conditions (2), the proof of uniqueness is still valid because condition (9) is again satisfied.

The requirement of continuity on derivatives of y is severe. Solutions of many simple problems in a wave equation have discontinuities in their derivatives.

APPENDIX 1

BIBLIOGRAPHY

The following list of books and articles for supplementary study of the various topics that have been introduced is far from exhaustive. Further references are given in the ones mentioned here.

Abramowitz, M., and I. A. Stegun (Eds.): *Handbook of Mathematical Functions with Formulas, Graphs, and Mathematical Tables,* Dover Publications, Inc., New York, 1972.

Apostol, T. M.: *Mathematical Analysis,* 2d ed., Addison-Wesley Publishing Company, Inc., Reading, MA, 1974.

Articolo, G. A.: *Partial Differential Equations and Boundary Value Problems with Maple V,* Academic Press, San Diego, CA, 1998.

Bell, W. W.: *Special Functions for Scientists and Engineers,* Dover Publications, Inc., New York, 2004.

Berg, P. W., and J. L. McGregor: *Elementary Partial Differential Equations,* The McGraw-Hill Companies, New York, 1988.

Birkhoff, G., and G.-C. Rota: *Introduction to Ordinary Differential Equations,* 4th ed., John Wiley & Sons, Inc., New York, 1989.

Bowman, F.: *Introduction to Bessel Functions,* Dover Publications, Inc., New York, 1958.

Boyce, W. E., and R. C. DiPrima: *Elementary Differential Equations,* 8th ed., John Wiley & Sons, Inc., New York, 2005.

Broman, A.: *Introduction to Partial Differential Equations,* Dover Publications, Inc., New York, 1989.

Brown, J. W., and R. V. Churchill: *Complex Variables and Applications,* 7th ed., McGraw-Hill Higher Education, Boston, 2004.

Brown, J. W., and F. Farris: On the Laplacian, *Math. Mag.,* vol. 59, no. 4, pp. 227–229, 1986.

Buck, R. C.: *Advanced Calculus,* 3d ed., Waveland Press, Inc., Long Grove, IL, 2004.

Budak, B. M., A. A. Samarskii, and A. N. Tikhonov: *A Collection of Problems on Mathematical Physics,* Dover Publications, Inc., New York, 1988.

Byerly, W. E.: *Fourier's Series and Spherical Harmonics,* Dover Publications, Inc., New York, 2003.

Cannon, J. T., and S. Dostrovsky: *The Evolution of Dynamics, Vibration Theory from 1687 to 1742,* Springer-Verlag, New York, 1981.

Carslaw, H. S.: *Introduction to the Theory of Fourier's Series and Integrals,* 3d ed., Dover Publications, Inc., New York, 1952.

______, and J. C. Jaeger: *Conduction of Heat in Solids,* 2d ed., Oxford University Press, New York, 1986.

Churchill, R. V.: *Operational Mathematics,* 3d ed., The McGraw-Hill Companies, New York, 1972.

Coddington, E. A.: *An Introduction to Ordinary Differential Equations,* Dover Publications, Inc., New York, 1989.

______, and N. Levinson: *Theory of Ordinary Differential Equations,* Krieger Publishing Co., Inc., Melbourne, FL, 1984.

Courant, R., and D. Hilbert: *Methods of Mathematical Physics,* vols. 1 and 2, John Wiley & Sons, Inc., New York, 1996.

Davis, H. F.: *Fourier Series and Orthogonal Functions,* Dover Publications, Inc., New York, 1990.

Derrick, W. R., and S. I. Grossman: *Elementary Differential Equations,* 4th ed., Addison-Wesley Educational Publishers, Inc., Reading, MA, 1997.

Erdélyi, A. (Ed.): *Higher Transcendental Functions,* vols. 1–3, Krieger Publishing Co., Inc., Melbourne, FL, 1981.

Farrell, O. J., and B. Ross: *Solved Problems in Analysis as Applied to Gamma, Beta, Legendre, and Bessel Functions,* Dover Publications, Inc., New York, 1971.

Folland, G. B.: *Fourier Analysis and Its Applications,* Wadsworth, Inc., Belmont, CA, 1992.

Fourier, J.: *The Analytical Theory of Heat,* translated by A. Freeman, Dover Publications, Inc., New York, 2003.

Franklin, P.: *A Treatise on Advanced Calculus,* Dover Publications, Inc., New York, 1964.

Grattan-Guinness, I.: *Joseph Fourier, 1768–1830,* The MIT Press, Cambridge, MA, and London, 1972.

Gray, Alfred, and M. A. Pinsky: Gibbs' Phenomenon for Fourier-Bessel Series, *Expo. Math.,* vol. 11, pp. 123–135, 1993.

Gray, Andrew, and G. B. Mathews: *A Treatise on Bessel Functions and Their Applications to Physics,* 2d ed., with T. M. MacRobert, Dover Publications, Inc., New York, 1966.

Haberman, R.: *Elementary Applied Partial Differential Equations: With Fourier Series and Boundary Value Problems,* 4th ed., Prentice-Hall, Inc., Upper Saddle River, NJ, 2004.

Hanna, J. R., and J. H. Rowland: *Fourier Series, Transforms, and Boundary Value Problems,* 2d ed., John Wiley & Sons, Inc., New York, 1990.

Hayt, W. H., Jr., and J. A. Buck: *Engineering Electromagnetics,* 7th ed., McGraw-Hill Higher Education, Burr Ridge, IL, 2005.

Herivel, J.: *Joseph Fourier: The Man and the Physicist,* Oxford University Press, London, 1975.

Hewitt, E., and R. E. Hewitt: The Gibbs-Wilbraham Phenomenon: An Episode in Fourier Analysis, *Arch. Hist. Exact Sci.,* vol. 21, pp. 129–160, 1979.

Hobson, E. W.: *The Theory of Spherical and Ellipsoidal Harmonics,* 2d printing, Chelsea Publishing Company, Inc., New York, 1965.

Hochstadt, H.: *The Functions of Mathematical Physics,* Dover Publications, Inc., New York, 1987.

Ince, E. L.: *Ordinary Differential Equations,* Dover Publications, Inc., New York, 1956.

Jackson, D.: *Fourier Series and Orthogonal Polynomials,* Dover Publications, Inc., New York, 2004.

Jahnke, E., F. Emde, and F. Lösch: *Tables of Higher Functions,* 6th ed., The McGraw-Hill Companies, New York, 1960.

Kaplan, W.: *Advanced Calculus,* 5th ed., Addison-Wesley Publishing Company, Inc., Reading, MA, 2003.

______: *Advanced Mathematics for Engineers,* Addison-Wesley Publishing Company, Inc., Reading, MA, 1981.

Kellogg, O. D.: *Foundations of Potential Theory,* Dover Publications, Inc., New York, 1974.

Kevorkian, J.: *Partial Differential Equations: Analytical Solution Techniques,* 2d ed., Springer-Verlag, New York, 1999.

Körner, T. W.: *Fourier Analysis,* Cambridge University Press, Cambridge, England, 1990.

Kreider, D. L., R. G. Kuller, D. R. Ostberg, and F. W. Perkins: *An Introduction to Linear Analysis,* Addison-Wesley Publishing Company, Inc., Reading, MA, 1966.

Lanczos, C.: *Discourse on Fourier Series,* Hafner Publishing Company, New York, 1966.

Langer, R. E.: Fourier's Series: The Genesis and Evolution of a Theory, Slaught Memorial Papers, no. 1, *Am. Math. Monthly,* vol. 54, no. 7, part 2, pp. 1–86, 1947.

Lebedev, N. N.: *Special Functions and Their Applications,* Dover Publications, Inc., New York, 1972.

______, I. P. Skalskaya, and Y. S. Uflyand: *Worked Problems in Applied Mathematics,* Dover Publications, Inc., New York, 1979.

Luke, Y. L.: *Integrals of Bessel Functions,* The McGraw-Hill Companies, New York, 1962.

MacRobert, T. M.: *Spherical Harmonics,* 3d ed., with I. N. Sneddon, Pergamon Press, Ltd., Oxford, England, 1967.

Magnus, W., F. Oberhettinger, and R. P. Soni: *Formulas and Theorems for the Special Functions of Mathematical Physics,* 3d ed., Springer-Verlag, New York, 1966.

McLachlan, N. W.: *Bessel Functions for Engineers,* 2d ed., Oxford University Press, London, 1955.

Oberhettinger, F.: *Fourier Expansions: A Collection of Formulas,* Academic Press, New York and London, 1973.

Özişik, M. N.: *Boundary Value Problems of Heat Conduction,* Dover Publications, Inc., New York, 2003.

Papp, F. J.: Two Equivalent Properties for Orthonormal Sets of Functions in Complete Spaces, *Int. J. Math. Educ. Sci. Technol.,* vol. 22, pp. 147–149, 1991.

Pinsky, M. A.: *Partial Differential Equations and Boundary-Value Problems with Applications,* 3d ed., Waveland Press, Inc., Long Grove, IL, 2003.

Powers, D. L.: *Boundary Value Problems,* 5th ed., Elsevier Science & Technology Books, San Diego, CA, 2006.

Rainville, E. D.: *Special Functions,* Chelsea Publishing Company, Inc., New York, 1972.

______, P. E. Bedient, and R. E. Bedient: *Elementary Differential Equations,* 8th ed., Prentice-Hall, Inc., Upper Saddle River, NJ, 1997.

Rogosinski, W.: *Fourier Series,* 2d ed., Chelsea Publishing Company, Inc., New York, 1959.

Sagan, H.: *Boundary and Eigenvalue Problems in Mathematical Physics,* Dover Publications, Inc., New York, 1989.

Sansone, G., and A. H. Diamond: *Orthogonal Functions,* rev. ed., Dover Publications, Inc., New York, 1991.

Seeley, R. T.: *An Introduction to Fourier Series and Integrals,* W. A. Benjamin, Inc., New York, 1966.

Smirnov, V. I.: *Complex Variables—Special Functions,* Addison-Wesley Publishing Company, Inc., Reading, MA, 1964.

Sneddon, I. N.: *Elements of Partial Differential Equations,* The McGraw-Hill Companies, New York, 1957.

______: *Fourier Transforms,* Dover Publications, Inc., New York, 1995.

______: *Special Functions of Mathematical Physics and Chemistry,* 3d ed., Longman, New York, 1980.

______: *The Use of Integral Transforms,* The McGraw-Hill Companies, New York, 1972.

Snider, A. D.: *Partial Differential Equations,* Prentice-Hall, Inc., Upper Saddle River, NJ, 1999.

Stakgold, I.: *Boundary Value Problems of Mathematical Physics,* Society for Industrial and Applied Mathematics, Philadelphia, 2000.

______: *Green's Functions and Boundary Value Problems,* 2d ed., John Wiley & Sons, Inc., New York, 1998.

Streeter, V. L., E. B. Wylie, and K. W. Bedford: *Fluid Mechanics,* 9th ed., The McGraw-Hill Companies, New York, 1998.

Stubhaug, A.: *Niels Henrik Abel and His Times,* Springer-Verlag, Berlin, 2000.

Taylor, A. E., and W. R. Mann: *Advanced Calculus,* 3d ed., John Wiley & Sons, Inc., New York, 1983.

Timoshenko, S. P., and J. N. Goodier: *Theory of Elasticity,* 3d ed., The McGraw-Hill Companies, New York, 1970.

Titchmarsh, E. C.: *Eigenfunction Expansions Associated with Second-Order Differential Equations,* Oxford University Press, London, part I, 2d ed., 1962; part II, 1958.

______: *Introduction to the Theory of Fourier Integrals,* 3d ed., Chelsea Publishing Company, Inc., New York, 1986.

Tolstov, G. P.: *Fourier Series,* Dover Publications, Inc., New York, 1976.

Tranter, C. J.: *Bessel Functions with Some Physical Applications,* Hart Publishing Company, Inc., New York, 1969.

Trim, D. W.: *Applied Partial Differential Equations,* PWS-KENT Publishing Company, Boston, 1990.

Van Vleck, E. B.: The Influence of Fourier's Series upon the Development of Mathematics, *Science,* vol. 39, pp. 113–124, 1914.

Watson, G. N.: *A Treatise on the Theory of Bessel Functions,* 2d ed., Cambridge University Press, Cambridge, England, 1995.

Weinberger, H. F.: *A First Course in Partial Differential Equations with Complex Variables and Transform Methods,* Dover Publications, Inc., New York, 1995.

Whittaker, E. T., and G. N. Watson: *A Course of Modern Analysis,* 4th ed., Cambridge University Press, Cambridge, England, 1996.

Zygmund, A.: *Trigonometric Series,* 2d ed., vols. I and II, Cambridge University Press, Cambridge, England, 1988.

APPENDIX 2

SOME FOURIER SERIES EXPANSIONS

Some of the Fourier series expansions found in this book are listed on this and the following page.

Fourier Cosine Series

$$x = \frac{\pi}{2} - \frac{4}{\pi}\sum_{n=1}^{\infty}\frac{\cos(2n-1)x}{(2n-1)^2} \qquad (0 < x < \pi)$$

$$\pi - x = \frac{\pi}{2} + \frac{4}{\pi}\sum_{n=1}^{\infty}\frac{\cos(2n-1)x}{(2n-1)^2} \qquad (0 < x < \pi)$$

$$x^2 = \frac{c^2}{3} + \frac{4c^2}{\pi^2}\sum_{n=1}^{\infty}\frac{(-1)^n}{n^2}\cos\frac{n\pi x}{c} \qquad (0 < x < c)$$

$$x^4 = \frac{\pi^4}{5} + 8\sum_{n=1}^{\infty}(-1)^n\frac{(n\pi)^2-6}{n^4}\cos nx \qquad (0 < x < \pi)$$

$$\cos ax = \frac{2a\sin a\pi}{\pi}\left[\frac{1}{2a^2} + \sum_{n=1}^{\infty}\frac{(-1)^{n+1}}{n^2-a^2}\cos nx\right] \qquad (0 < x < \pi),$$

where $a \neq 0, \pm 1, \pm 2, \ldots$

$$\cosh ax = \frac{\sinh a\pi}{a\pi}\left[1 + 2a^2\sum_{n=1}^{\infty}\frac{(-1)^n}{a^2+n^2}\cos nx\right] \qquad (0 < x < \pi),$$

where $a \neq 0$

$$\sin x = \frac{2}{\pi} - \frac{4}{\pi}\sum_{n=1}^{\infty}\frac{\cos 2nx}{4n^2-1} \qquad (0 < x < \pi)$$

Fourier Sine Series

$$1 = \frac{4}{\pi}\sum_{n=1}^{\infty}\frac{\sin(2n-1)x}{2n-1} \qquad (0 < x < \pi)$$

$$x = 2\sum_{n=1}^{\infty}\frac{(-1)^{n+1}}{n}\sin nx \qquad (0 < x < \pi)$$

$$\pi - x = 2\sum_{n=1}^{\infty}\frac{\sin nx}{n} \qquad (0 < x < \pi)$$

$$x^2 = 2c^2\sum_{n=1}^{\infty}\left[\frac{(-1)^{n+1}}{n\pi} - 2\,\frac{1-(-1)^n}{(n\pi)^3}\right]\sin\frac{n\pi x}{c} \qquad (0 < x < c)$$

$$x(2c - x) = \frac{32\,c^2}{\pi^3}\sum_{n=1}^{\infty}\frac{1}{(2n-1)^3}\sin\frac{(2n-1)\pi x}{2c} \qquad (0 < x < 2c)$$

$$x^3 = 2\sum_{n=1}^{\infty}(-1)^{n+1}\frac{(n\pi)^2 - 6}{n^3}\sin nx \qquad (0 < x < \pi)$$

$$x(\pi^2 - x^2) = 12\sum_{n=1}^{\infty}\frac{(-1)^{n+1}}{n^3}\sin nx \qquad (0 < x < \pi)$$

$$x(1 - x^2) = \frac{12}{\pi^3}\sum_{n=1}^{\infty}\frac{(-1)^{n+1}}{n^3}\sin n\pi x \qquad (0 < x < 1)$$

$$x(x-1)(x-2) = \frac{12}{\pi^3}\sum_{n=1}^{\infty}\frac{\sin n\pi x}{n^3} \qquad (0 < x < 1)$$

$$\cos\pi x = \frac{8}{\pi}\sum_{n=1}^{\infty}\frac{n}{4n^2-1}\sin 2n\pi x \qquad (0 < x < 1)$$

$$\sinh ax = \frac{2\sinh a\pi}{\pi}\sum_{n=1}^{\infty}(-1)^{n+1}\frac{n}{a^2+n^2}\sin nx \qquad (0 < x < \pi)$$

Fourier Series

$$e^{ax} = \frac{\sinh a\pi}{a\pi} + \frac{2\sinh a\pi}{\pi}\sum_{n=1}^{\infty}\frac{(-1)^n}{a^2+n^2}(a\cos nx - n\sin nx) \qquad (-\pi < x < \pi),$$

where $a \neq 0$

$$e^{x} = \frac{\sinh c}{c} + 2\sinh c\sum_{n=1}^{\infty}\frac{(-1)^n}{c^2+(n\pi)^2}\left(c\cos\frac{n\pi x}{c} - n\pi\sin\frac{n\pi x}{c}\right) \qquad (-c < x < c)$$

APPENDIX 3

SOLUTIONS OF SOME REGULAR STURM-LIOUVILLE PROBLEMS

The eigenvalues and normalized eigenfunctions, together with weight functions, of some of the regular Sturm-Liouville problems solved in this book are listed on this and the following pages.

1. $X'' + \lambda X = 0, \qquad X'(0) = 0, \qquad hX(c) + X'(c) = 0 \qquad (h > 0).$

Solutions: Weight function $p(x) = 1$;

$$\lambda_n = \alpha_n^2, \qquad \phi_n(x) = \sqrt{\frac{2h}{hc + \sin^2 \alpha_n c}} \cos \alpha_n x \qquad (n = 1, 2, \ldots),$$

where
$$\tan \alpha_n c = \frac{h}{\alpha_n} \qquad (\alpha_n > 0).$$

2. $X'' + \lambda X = 0, \qquad X(0) = 0, \qquad hX(1) + X'(1) = 0 \qquad (h > 0).$

Solutions: Weight function $p(x) = 1$;

$$\lambda_n = \alpha_n^2, \qquad \phi_n(x) = \sqrt{\frac{2h}{h + \cos^2 \alpha_n}} \sin \alpha_n x \qquad (n = 1, 2, \ldots),$$

where
$$\tan \alpha_n = -\frac{\alpha_n}{h} \qquad (\alpha_n > 0).$$

3. $(xX')' + \dfrac{\lambda}{x} X = 0, \qquad X'(1) = 0, \qquad X(b) = 0.$

Solutions: Weight function $p(x) = \dfrac{1}{x}$;

$$\lambda_n = \alpha_n^2, \qquad \phi_n(x) = \sqrt{\frac{2}{\ln b}} \cos(\alpha_n \ln x) \qquad (n = 1, 2, \ldots),$$

where
$$\alpha_n = \frac{(2n-1)\pi}{2 \ln b}.$$

4. $(xX')' + \dfrac{\lambda}{x}X = 0, \qquad X(1) = 0, \qquad X(b) = 0.$

Solutions: Weight function $p(x) = \dfrac{1}{x}$;

$$\lambda_n = \alpha_n^2, \qquad \phi_n(x) = \sqrt{\frac{2}{\ln b}}\sin(\alpha_n \ln x) \qquad (n = 1, 2, \ldots),$$

where
$$\alpha_n = \frac{n\pi}{\ln b}.$$

5. $(xX')' + \dfrac{\lambda}{x}X = 0, \qquad X'(1) = 0, \qquad hX(b) + X'(b) = 0 \qquad (h > 0).$

Solutions: Weight function $p(x) = \dfrac{1}{x}$;

$$\lambda_n = \alpha_n^2, \qquad \phi_n(x) = \sqrt{\frac{2hb}{hb\ln b + \sin^2(\alpha_n \ln b)}}\cos(\alpha_n \ln x) \qquad (n = 1, 2, \ldots),$$

where
$$\tan(\alpha_n \ln b) = \frac{hb}{\alpha_n} \qquad (\alpha_n > 0).$$

6. $X'' + \lambda X = 0, \qquad X'(0) = 0, \qquad X(c) = 0.$

Solutions: Weight function $p(x) = 1$;

$$\lambda_n = \alpha_n^2, \qquad \phi_n(x) = \sqrt{\frac{2}{c}}\cos\alpha_n x \qquad (n = 1, 2, \ldots),$$

where
$$\alpha_n = \frac{(2n-1)\pi}{2c}.$$

7. $X'' + \lambda X = 0, \qquad X(0) = 0, \qquad X'(c) = 0.$

Solutions: Weight function $p(x) = 1$;

$$\lambda_n = \alpha_n^2, \qquad \phi_n(x) = \sqrt{\frac{2}{c}}\sin\alpha_n x \qquad (n = 1, 2, \ldots),$$

where
$$\alpha_n = \frac{(2n-1)\pi}{2c}.$$

8. $X'' + \lambda X = 0, \qquad X(0) = 0, \qquad X(1) - X'(1) = 0.$

Solutions: Weight function $p(x) = 1$;

$$\lambda_0 = 0, \qquad \lambda_n = \alpha_n^2 \qquad (n = 1, 2, \ldots),$$

$$\phi_0(x) = \sqrt{3}\,x, \qquad \phi_n(x) = \frac{\sqrt{2\left(\alpha_n^2 + 1\right)}}{\alpha_n}\sin\alpha_n x \qquad (n = 1, 2, \ldots),$$

where
$$\tan\alpha_n = \alpha_n \qquad (\alpha_n > 0).$$

9. $X'' + \lambda X = 0, \qquad hX(0) - X'(0) = 0 \qquad (h > 0), \qquad X(1) = 0.$

Solutions: Weight function $p(x) = 1$;

$$\lambda_n = \alpha_n^2, \qquad \phi_n(x) = \sqrt{\frac{2h}{h + \cos^2 \alpha_n}} \sin \alpha_n (1 - x) \qquad (n = 1, 2, \ldots),$$

where

$$\tan \alpha_n = -\frac{\alpha_n}{h} \qquad (\alpha_n > 0).$$

10. $(x^2 X')' + \lambda X = 0, \qquad X(1) = 0, \qquad X(b) = 0.$

Solutions: Weight function $p(x) = 1$;

$$\lambda_n = \frac{1}{4} + \alpha_n^2, \qquad \phi_n(x) = \sqrt{\frac{2}{x \ln b}} \sin(\alpha_n \ln x) \qquad (n = 1, 2, \ldots),$$

where

$$\alpha_n = \frac{n\pi}{\ln b}.$$

INDEX

Note: Page numbers followed by *n* refer to footnotes.

D

E

F

T